AF553720

INORGANIC BIOCHEMISTRY

INORGANIC BIOCHEMISTRY

By

Dr. Arvind N. Shukla

School of Studies of Zoology & Biotechnology

Vikram University

Ujjain

(India)

DISCOVERY PUBLISHING HOUSE PVT. LTD.

NEW DELHI-110 002

Published by:
Tilak Wasan
DISCOVERY PUBLISHING HOUSE PVT. LTD.
4383/4B, Ansari Road, Darya Ganj
New Delhi-110 002 (India)
Phone : +91-11-23279245; 23253475; 43596065
E-mail: discoverybooksindia@gmail.com
discoverypublishinghouse@gmail.com
namitwasan9@gmail.com
web : www.discoverypublishinggroup.com

Reprinted: **2019**
First Edition: **2010**

ISBN: 978-81-8356-640-7

Inorganic Biochemistry

Printed at:
Infinity Imaging Systems
Delhi

Preface

Biochemistry today has made spectacular progress in unraveling the mysteries of animate nature. This progress has allowed us to gain deeper insight into the principles of vital activity and has to a very significant extent stimulated the development of applied disciplines, especially medicine. The present title *"Inorganic Biochemistry"* is intended for those who wish to understand living organisms, especially man. Biochemistry is essential for this purpose, but it would be almost impossible for a student to survey on his own the massive body of existing knowledge, constantly augmented by a remarkable torrent of brilliat discoveries. The purpose of the book, then is to organize our knowledge into something that can be comprehended in a relatively short time and still convey a reasonable complete picture of the chemical structure and function of man. Readability without sacrifice of coverage has been a prime goal. An important device in gaining that goal is to keep attention constantly focused on function, with repeated use of rationalization to show that the chemical facts are not isolated, but part of a whole.

Biochemistry has two major goals as a fundamental science, it treats the vital functions from the standpoint of physical chemistry, and as an applied discipline, it points out practical applications for the wealth of scientific knowledge it has acquired. The dual purpose has been, as far as possible, take into account in the writing of this book. We have also tried to summarize our pedagogical experience in teaching biochemistry to students specializing in medicine and pharmacy. The material of this book has been organized according to the principle of functionally in order to trace the close relationship between the functions of the living organism and the structures of its constituents molecules as well as the chemical and physico-chemical process in which they are involved.

The aim of this book is to present a core of biochemical knowledge that is desirable for undergraduate and postgraduate students and also those involved in the field of medical, microbiology, biotechnology and pharmaceutical. Every attempt has been made to keep abrest of the advances in the subject and at the same time to include the fundamentals.

To make the work more comprehensive and informative, the author has consulted many authoritative books, research journals, abstracts, monographs etc. He is grateful to all those great scholars whose work are cited or substantially reproduced.

There can be no claim to originality except in the manner of treatment and much of the information has been obtained from the books and scientific journals available in the different libraries.

The author expresses his thanks to his friends and colleagues whose continue inspirations have initiated him to bring out this book.

The author expresses his gratitude to Mr. Wasan and Staff of M/s Discovery Publishing House Pvt. Ltd., for their whole hearted co-operation in the publication of this book.

In the mean time, the author will remain sincerely responsible for any shortcomings of the book and be grateful to the readers for their suggestions and constructive criticism for the continuous betterment of the book. He takes this opportunity to appeal to the readers to send their suggestions straightway to his publisher.

Author

CONTENTS

1. Introduction 1

Fundamental Unit—Cell Organelles and Inclusions—Complex Three-Dimensional Structures—Primary Factor—Biochemical Reactions—Catalytic Surfaces of Enzymes—Energy Utilization—Localized Reactions in the Cell—Pathways—Regulation—Biochemically Dependent Organism—Synthesis of Proteins—evaluation of Biochemical Systems—A Common Evolution—Concluding remarks.

2. Matter and Atomic Theory 28

Some General Types of Matter—Mixtures and Pure Substances—Homogeneous and Heterogeneous Mixtures—*Summary of the Classification of Matter*—Quantity of Matter: The Mole—The Mole and Avogadro's Number—Physical Properties of Matter—Density—Specific Gravity—Colour—States of Matter—Gases—The Gas Laws—Boyle's Law—Charles' Law—Avogadro's Law—The General Gas Law—Gas Law Calculations—Charles' Law Calculation—Boyle's Law Calculation—Calculations Using the Ideal Gas Law—Liquids and Solutions—Evaporation and Condensation of Liquids—Vapor Pressure—Solutions—Solubility—Solution Concentration—Solids—Thermal Properties—Melting Point—Boiling Point—Specific Heat—Heat of Vaporization—Heat of Fusion—Separation And Characterization of Matter—Distillation—Separation in Waste Treatment—Phase Transitions—Phase Transfer—Molecular Separation—Atomic Theory—Dalton's Atomic Theory—Size of Atoms—Atomic Mass—Subatomic Particles—Basic Structure of the Atom—Atomic Number, Isotopes, and Mass Number of Isotopes—Electrons in Atoms—Attraction Between Electrons and the Nucleus—Development of the Periodic Table—The Simplest Atom—Designation of Hydrogen in the Periodic Table—Showing Electrons in Hydrogen Atoms and Molecules—Properties of Elemental Hydrogen—Production and Uses of Elemental Hydrogen—Helium—Uses of Helium—Lithium—Uses of Lithium—The Second Period, Elements 4-10—Beryllium, Atomic Number 4—Boron, Atomic Number 5—Carbon, Atomic Number 6—Nitrogen, Atomic Number 7—Oxygen, Atomic Number 8—Fluorine, Atomic Number 9—Neon, Atomic Number 10—Stability of the Neon Noble Gas Electron Octet—Elements 11-20, and Beyond—The Elements Beyond Calcium—Detailed Look at Atomic Structure—Electromagnetic Radiation—Models of Electrons in Atoms—Wave Mechanical Model—Schrodinger Equation—Multielectron Atoms and Quantum Numbers—The Principal Quantum Number, n—The Azimuthal Quantum Number, l—The Magnetic Quantum Number, ml—Spin Quantum Number, ms—Energy Levels of Atomic Orbitals—Hund's Rule of Maximum Multiplicity—Shapes of Atomic Orbitals—Electron Configuration—Electrons in the First 20 Elements—Electron Configuration of Hydrogen—Electron Configuration of Helium—Electron Configurations of Elements 2-20—Lithium—Valence Electrons—Beryllium—Filling the 2p Orbitals—Filling the 3s, 3p, and 4s Orbitals—Electron Configurations and the Periodic Table.

3. Metals **81**

Factors Influencing Toxicity—Metal Chelation—Toxic Effects—Acquired Tolerance—Lead—Introduction—Metabolism—Absorption—Distribution—Excretion—Biologic Effects—Central Nervous System—Peripheral Nervous System—Kidney—Hematopoiesis and Heme Synthesis—Other Effects—Treatment of Lead Poisoning —Mercury—Introduction—Metabolism—Absorption—Distribution and Metabolism—Excretion—Biologic Effects—Dose-Effect and Dose-Response Relationships—Treatment of Mercury Poisoning—Introduction—Metabolism—Absorption—Distribution—Excretion—Biologic Effects—Dose-Effect and Dose-Response Relationships—Treatment of Cadmium Poisoning—Aluminum—Arsenic—Barium—Beryllium—Bismuth—Boron—Cesium—Chromium—Boron—Cesium—Chromium—Cobalt—Copper—Gallium—Germanium—Gold—Hafnium—Indium—Iron—Lanthanons (Rare Earths)—Lithium—Magnesium—Manganese—Molybdenum—Nickel—Niobium—Platinum-group Metals—Rhenium—Rubidium—Selenium—Silver—Strontium—Tantalum—Thallium—Tin—Titanium—Tungsten—Uranium—Vanadium—Zinc—Zirconium.

4. Metals in Catalysis and Electron Transport **162**

A. Iron—Uptake by Living Cells—Siderophores—Uptake of Iron by Eukaryotic Cells—Transferrins—Storage of Iron—Heme Proteins—Some Names to Remember —Hemes and Heme Proteins—The Cytochromes—The c-type Cytochromes—Cytochromes B, A, and O—Mechanisms of Biological Electron Transfer—Electron-transfer Pathways?—"Docking."—Coupling and Gating of Electron Transfer—Effects of Ionic Equilibria on Electron Transfer—Reactions of Heme Proteins with Oxygen or Hydrogen Peroxide—Oxygen-carrying Proteins—Catalases and Peroxidase—Mechanisms of Catalase and Peroxidase Catalysis —Haloperoxidases—The Iron-Sulfur Proteins—Ferredoxins, High-potential Iron Proteins, and Rubredoxins—Rubredoxins—Chloroplast-type Ferredoxins—The 3Fe-4S clusters—Properties of Iron-sulfur Clusters—Functions of Iron-sulfur Enzymes—The (m-oxo) Diiron Proteins—Hemerythrin—Purple Acid Phosphatases—Diiron Oxygenases ane Desaturases—Ribonucleotide Reductases—Superoxide Dismutases—B. Cobalt and Vitamin B12—Coenzyme Forms—Reduction of Cyanocobalamin and Synthesis of Alkyl Cobalamins—Three Nonenzymatic Cleavage Reactions of Vitamin B12 Coenzymes—Enzymatic Functions of BI2 Coenzymes—Cobalamin-dependent Ribonucleotide Reductase—The Isomerization Reactions—Stereochemistry of the Isomerization Reactions—Aminomutases—Transfer Reactions of Methyl Groups—Corrinoid-dependent Synthesis of Acetyl-AoA—C. Nickel—Urease—Hydrogenases—Cofactor F430 and Methyl-Coenzyme M Reductase—Tunichlorins—Carbon Monoxide Dehydrogenases and Carbon Monoxide

Dehydrogenase/ Acetyl-CoA Synthase—D. Copper—Electron-Transferring Copper Proteins—Copper, Zinc-Superoxide Dismutase—Nitrite and Nitrous Oxide Reductases—Hemocyanins—Copper Oxidases—Cytochrome c Oxidase—E. Manganese—F. Chromium—G. Vanadium—H. Molybdenum—Molybdenum Ions and Coenzyme Forms—Enzymatic Mechanisms—Nutritional Need for Mo—I. Tungsten—Magnetic Iron Oxide In Organisms—Cobalamin (Vitamin B12).

5. **Phosphorus Removal** 247

Normal Phosphorus Uptake Into Biomass—Precipitation By Metal-salts Addition To A—Biological Process—Enhanced Biological Phosphorus Removal—Problems.

6. **Inorganic Pollutants** 257

Minute Pollutants—Formation of Inorganic Particles—Composition of Particles—Fly Ash—Asbestos—Toxic Metals—Radioactive Particles—Effects of Particles—Emissions of Particles—SEDIMENTATION PROCESS—Inertial Mechanisms—Filtration Process—Scrubbers—Electrostatic Removal—Carbon Oxides—Carbon Monoxide—Regulation of Carbon Monoxide Emissions—Fate of Atmospheric CO—Global Warming—Cycle Sulfur—Sulfur Dioxide in the Atmosphere—Effects of Atmospheric Sulfur Dioxide—Removal of Sulfur Dioxide—Oxides of Nitrogen—Reactions of Nox in Atmosphere—Harmful Effects of Nitrogen Oxides—Regulation of Nitrogen Oxides—Ammonia in the Atmosphere—Acid Rain—Fluorine, Chlorine, and their Gaseous Inorganic Compounds—Chlorine and Hydrogen Chloride—Hydrogen Sulfide, Carbonyl Sulfide, and Carbon Disulfide.

1 Introduction

The most unique feature of the living cell is the way in which so many reactions are organized to serve a single purpose.

Aside from a number of striking geological features, the most prominent and unique aspect of the Earth's surface is that it is virtually covered with living organisms. In general biology we are introduced to the extraordinary diversity of living organisms, a diversity so great that it is generally acknowledged that there are more species than could ever be classified. Life seems incredibly complicated. If we limit our inspection of living things to the gross organismic level or what we can see with the naked eye, it is very easy to be overwhelmed by this complexity. Furthermore we will not find explanations for why organisms are constructed the way they are at this level of inquiry. If we probe more deeply into organismic structure, we find that all organisms are composed of much smaller units called cells. If we probe even further we find that cells are composed of a limited number of small molecules and macromolecules. We also find that the macromolecules, which make up the bulk of the solid matter of cells, are constructed from a small number of building blocks that are closely related in structure.

It is at the molecular level that we find the ultimate explanations for organismic structure and behaviour. We may revel in the discovery that life is not as complicated as we first thought. Thanks to the investigations of many biochemists over the past half century we are on the verge of a thorough understanding of life at the molecular level. This is a most satisfying and exciting time for biochemists; their investigations and accomplishments place biochemistry in the fore-front of the biological sciences. The object of this text is to acquaint the beginning student of biochemistry with the basic facts and principles of this subject.

FUNDAMENTAL UNIT

Microscopic examination of any organism reveals that it is composed of membrane-enclosed structures called cells. The enclosing membrane is called the cell membrane, or the plasma membrane. Cells vary enormously in size and shape, but even the largest cells would have to be much larger to be visible to the naked eye. Within this tiny object thousands of chemical reactions are taking place, all regulated, all designed to serve a specific function. Collectively these reactions serve the function of maintaining the cell and permitting it to replicate when the time is right. Perhaps the most amazing thing about the living cell is that so much organized activity takes place in such a small space. Fig. 1.1 shows prototypical animal and plant cells, along with some common shapes and sizes of bacteria. Bacteria are single celled organisms, but sometimes the cells are connected into long chains.

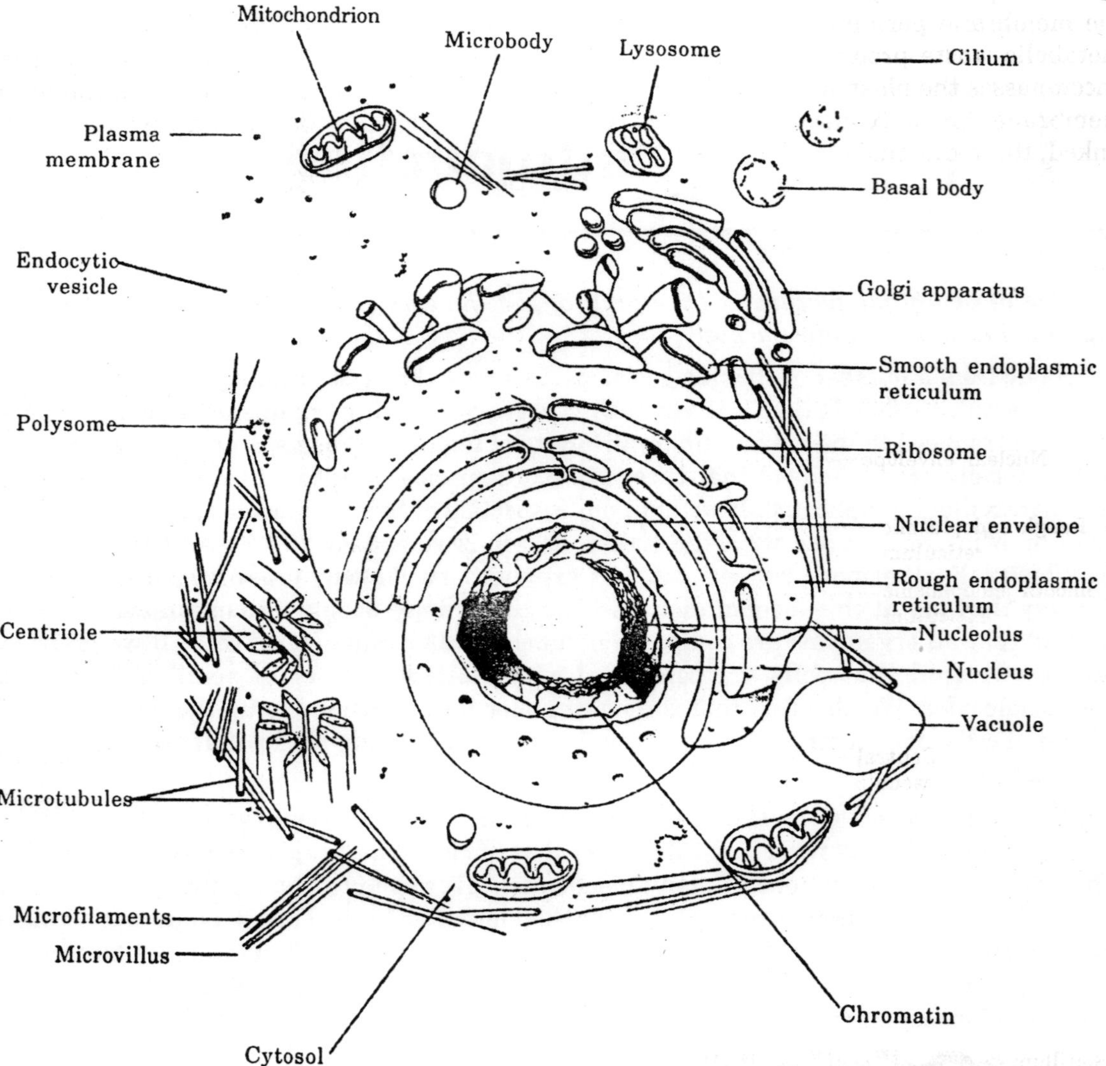

Fig. 1.1. Generalized representations of the internal structures of animal and plant cells (eukaryotic cells). Cells are the fundamental units in all living systems, and they vary tremendously in size and shape. All cells are functionally separated from their environment by the plasma membrane that encloses the cytoplasm. Plant cells have two structures not found in animal cells: a cellulose cell wall, exterior to the plasma membrane, and chloroplasts. The many different types of bacteria (prokaryotes) are all smaller than most plant and animal cells. Bacteria, like plant cells, have an exterior cell wall, but it differs greatly in chemical composition and structure from the cell wall in plants. Like all other cells, bacteria have a plasma membrane that functionally separates them from their environment. Some bacteria also have a second membrane, the outer membrane, which is exterior to the cell wall.

In multicellular organisms the cells associate to form specialized tissues. The plasma membrane is a delicate, semipermeable, sheetlike covering for the entire cell. Forming an enclosure prevents gross loss of the intracellular contents; the semi-permeable character of the membrane permits the selective absorption of nutrients and the selective removal of metabolic waste products. In many plant and bacterial (but not animal) cells, a cell wall encompasses the plasma membrane. The cell wall is a more porous structure than the plasma membrane, but it is mechanically stronger because it is constructed of a covalently cross-linked, three-dimensional network.

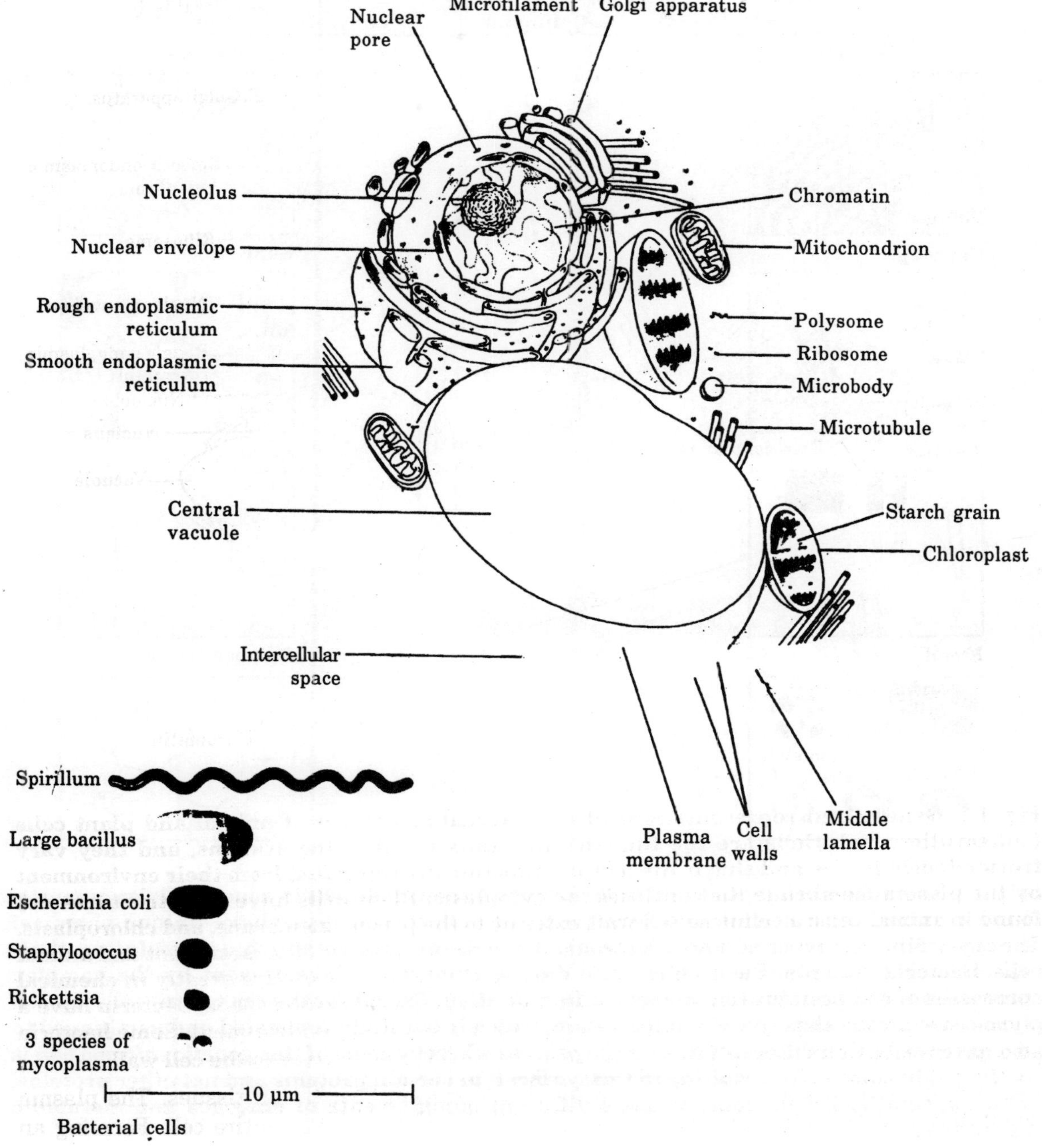

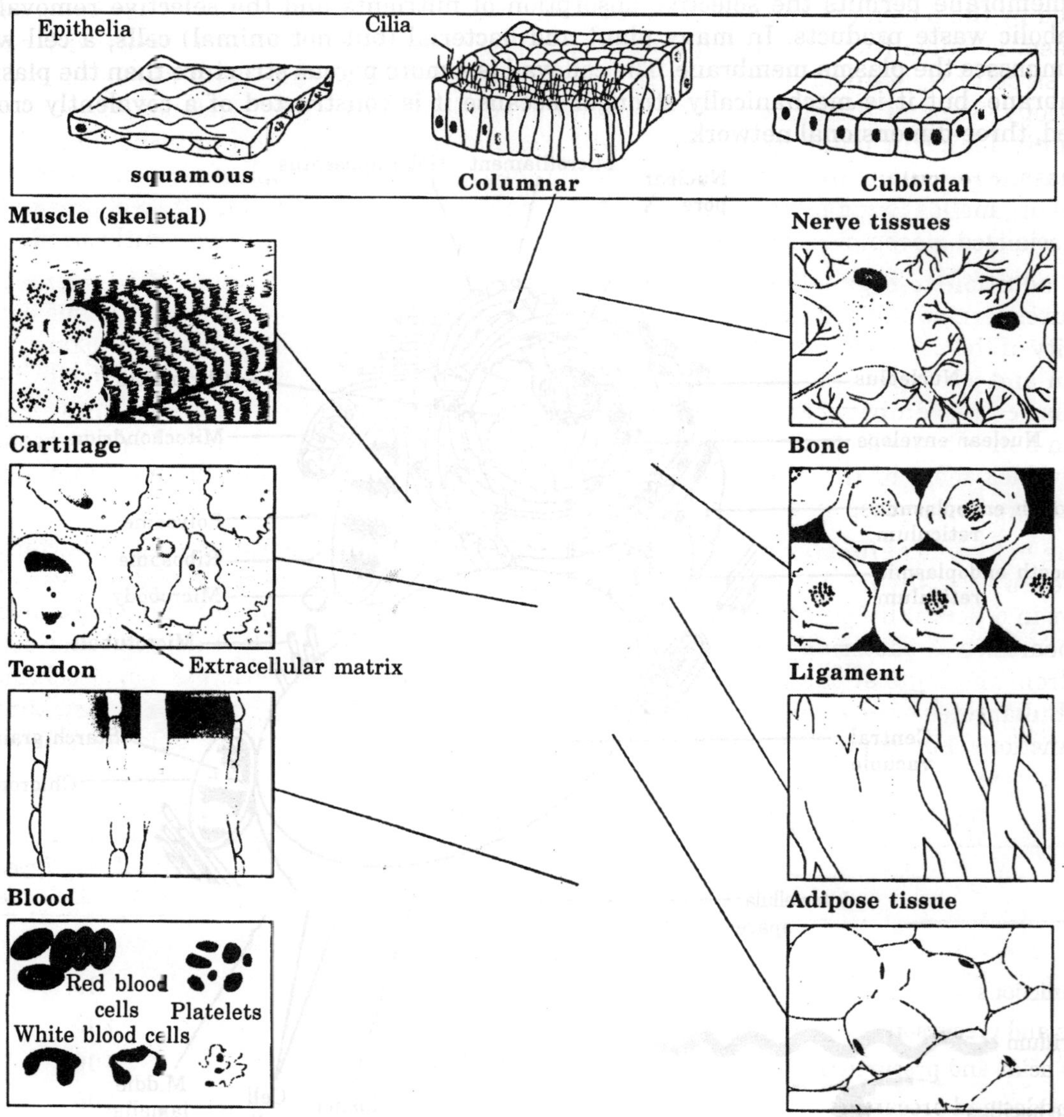

Fig. 1.2. Specialized cell types found in the human. Although all cells in a multicellular organism have common constituents and functions, specialized cell types have unique chemical compositions, structures, and biochemical reactions that establish and maintain their specialized functions. Such cells arise during embryonic development by the complex processes of cell proliferation and cell differentiation. Except for the sex (germ) cells, all cell types contain the same genetic information, which is faithfully replicated and partitioned to daughter cells. Cell differentiation is the process whereby some of this genetic information is activated in some cells, resulting in the synthesis of certain proteins and not other proteins. Thus, specialized cells come to have different complements of enzymes and metabolic capacities.

The cell wall maintains a cell's three-dimensional form when it is under stress. The contents enclosed by the plasma membrane constitute the cytoplasm. The purely liquid portion of the cytoplasm is called the cytosol. Within the cytoplasm are a number of macromolecules and larger structures, many of which can be seen by high-power light microscopy or by electron microscopy. Some of the structures are membranous and are called organelles. Organelles commonly found in plant and animal cells include the nucleus, the mitochondria, the endoplasmic reticulum, the Golgi apparatus, and the lysosomes. Chloroplasts are an important class of organelles found in many plant cells but never in animal cells. Each type of organelle is a specialized biochemical factory in which certain biochemical products are synthesized.

In addition to organelles, animal and plant cells contain a collection of filamentous structures termed the cytoskeleton, which is important in maintaining the three-dimensional integrity of the cell. As we will see, the evolutionary tree is bisected into a lower prokaryotic domain and an upper eukaryotic domain. The terms prokaryote and eukaryote refer to the most basic division between cell types. The fundamental difference is that eukaryotic cells contain a membrane-bounded nucleus, whereas prokaryotes do not. The cells of prokaryotes usually lack most of the other membrane-bounded organelles as well. Plants, fungi, and animals are eukaryotes, and bacteria are prokaryotes.

The biochemical functions associated with organelles are frequently present in bacteria, but they are usually located on the inner plasma membrane. Cells are organized in a variety of ways in different living forms. Prokaryotes of a given type produce cells that are very similar in appearance. A bacterial cell replicates by a process in which two identical daughter cells arise from an identical parent cell. Simple eukaryotes can also exist as single nonassociating cells. Eukaryotes of increasing complexity can contain many cells with specialized structures and functions. For example, humans contain about 10^{14} cells of more than a hundred different types.

Table 1.1. The Approximate Chemical Composition of a Bacterial Cell

	Percent to Total Cell Weight	*Number of Types of Each Molecule*
Water	70	1
Inorganic ions	1	20
Sugars and precursors	3	200
Amino acids and precursors	0.4	100
Nucleotides and precursors	0.4	200
Lipids and precursors	2	50
Other small molecules	0.2	≈ 200
Macromolecules	22	≈ 5,000
(proteins, nucleic acids, and polysaccharides)		

Specialized cells make up the skin, connective tissue, nerve tissue, muscles, blood, sensory functions, and reproductive organs. In such a complex organism, the capacity of different cells for replication is limited. When a skin cell or a muscle cell precursor replicates, it makes more cells of the same type. The only cells in a complex eukaryote capable of reproducing an entire organism are the germ cells, that is, the sperm and the egg.

CELL ORGANELLES AND INCLUSIONS

Of the many different types of molecules in the various organelles and the cytosol that constitute the living cell, water is by far the most abundant, constituting about 70% by weight of most living matter. As a result, most other components exist essentially in an aqueous environment. Except for water, most of the molecules found in the cell are lipids or macromolecules, which can be classified into four different categories: lipids, carbohydrates, proteins, and nucleic acids.

Each type of macromolecule possesses distinct chemical properties that suit it for the functions it serves in the cell. Lipids are primarily hydrocarbon structures. They tend to be poorly soluble in water and are therefore particularly well suited to serve as a major component of the various membrane structures found in cells. Lipids also serve as a compact means of storing chemical energy to drive the metabolism of the cell. Carbohydrates, like lipids, contain a carbon backbone, but they also contain many polar hydroxyl (—OH) groups and are therefore very soluble in water.

Large carbohydrate molecules called polysaccharides consist of many small, ringlike sugar molecules; these sugar monomers are attached to one another by glycosidic bonds in a linear or branched array to form the sugar polymer. In the cell, such polysaccharides often form storage granules that may be readily broken down into their component sugars. With further chemical breakdown these sugars release chemical energy and may also provide the carbon skeletons for the synthesis of a variety of other molecules. Important structural functions are also served by polysaccharides. Linear polysaccharides form a major component of plant cell walls, and bacterial cell walls are composed of linear polysaccharides that are cross-linked by short polypeptide chains. Proteins are the most complex macromolecules found in the cell. They are composed of linear polymers called polypeptides, which contain amino acids connected by peptide bonds.

Each amino acid contains a central carbon atom attached to four substituents: (1) a carboxyl group, (2) an amino group, (3) a hydrogen atom, and (4) an R group. The R group gives each amino acid its unique characteristics. Twenty different amino acids occur in proteins. Some R groups are charged, some are neutral but still polar, and some are apolar. The linear polypeptide chains of a protein fold in a highly specific way that is determined by the sequence of amino acids in the chains. Many proteins are composed of or more polypeptides.

Certain proteins function in structural roles. Some structural proteins interact with lipids in membrane structures. Others aggregate to form part of the cytoskeleton that helps to give the cell its shape. Still others are the chief components of muscle or connective tissue. Enzymes constitute yet another major class of proteins, which function as catalysts that accelerate and direct biochemical reactions. Nucleic acids are the largest macromolecules in the cell. They are very long, linear polymers, called polynucleotides, composed of many nucleotides.

A nucleotide contains (1) a five-carbon sugar molecule, (2) one or more phosphate groups, and (3) a nitrogenous base. It is the nitrogenous base that gives each nucleotide a distinct character. Five different types of nitrogenous bases are found in the two main types of nucleic acids, deoxyribonucleic acid (DNA) and ribonucleic acid (RNA). DNA contains the genetic information that is inherited when cells divide and organisms reproduce. This genetic information is used in the cell to make ribonucleic acids and proteins. In addition to water

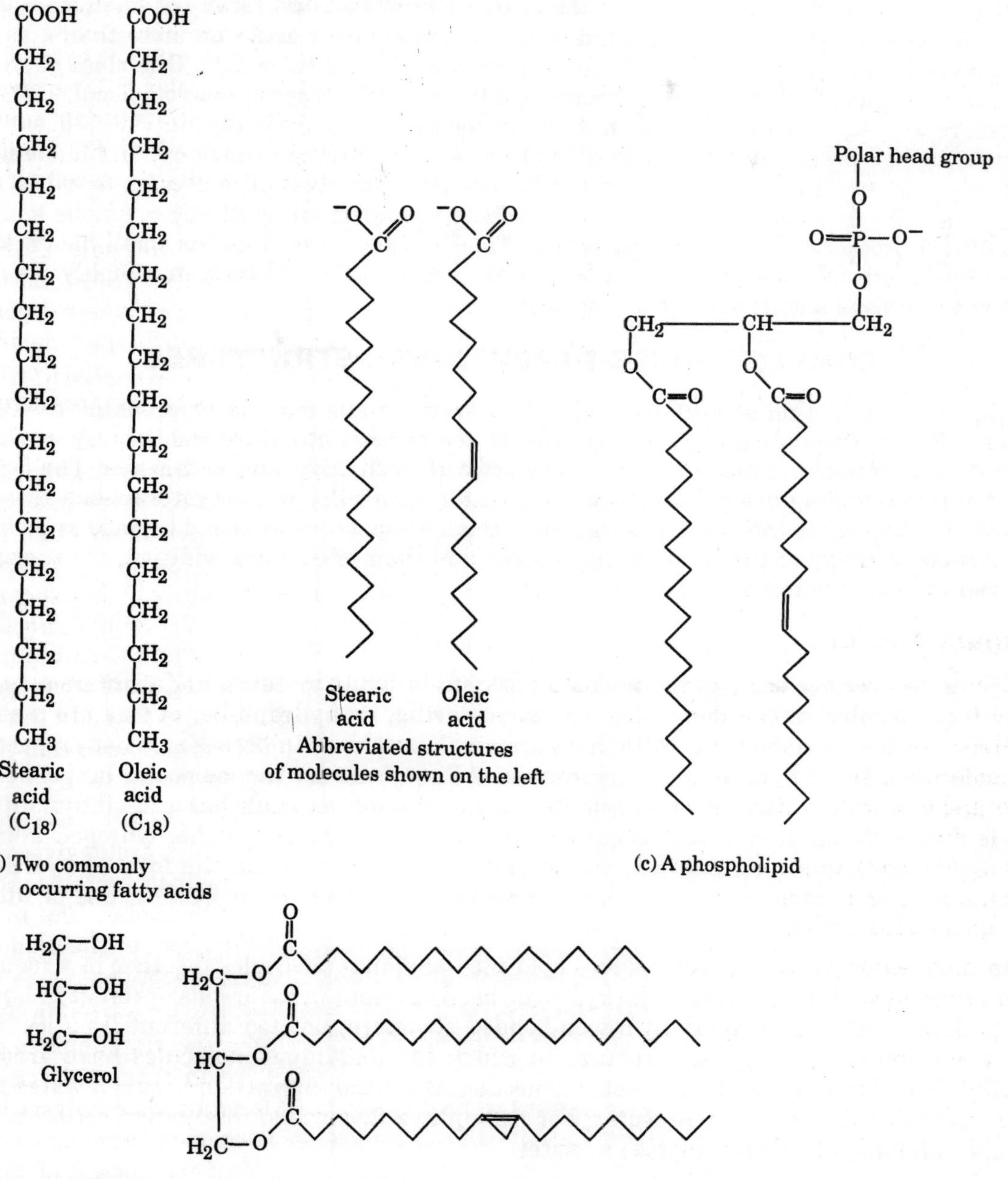

Fig. 1.3. The structures of common lipids. (*a*) The structures of saturated and unsaturated fatty acids, represented here by stearic acid and oleic acid. (*b*) Three fatty acids covalently linked to glycerol by ester bonds form a triacylglycerol. (*c*) The general structure for a phospholipid consists of two fatty acids esterified to glycerol, which is linked through phosphate to a polar head group. The polar head group may be any one of several different compounds—for example, choline, serine, or ethanolamine.

and the macromolecules and organelles, the cytosol contains a large variety of small molecules that differ greatly in both structure and function. These never make up more than a small fraction of the total cell mass despite their great variety (see table 1.1). One class of small molecules consists of the monomer precursors of the different types of macromolecules. These monomers are derived by a series of chemical modifications from the nutrients absorbed through the cell membrane. The intermediate molecules between nutrients and monomers are present in small concentrations in the cytosol. Another class of molecules found in the cytosol includes molecules formed as side products in important synthetic reactions and as degradation products of the macromolecules. Finally, the cytosol contains small bio-organic molecules known as coenzymes, which act in concert with the enzymes in a highly specific manner to catalyze a wide variety of reactions.

COMPLEX THREE-DIMENSIONAL STRUCTURES

The complex folding of biomacromolecules rarely entails making or breaking covalent linkages. Rather the folding process is dictated by the primary structure and the way in which different elements of the macromolecule interact with each other and with water. The forces that determine folding are noncovalent in character. As a rule, specific interactions amount to only a fraction of the interaction energy that occurs when a covalent bond is made or broken, but because so many of these interactions occur and their effects are additive, the energies involved can be quite large.

Primary Factor

Water, as we have seen, is the major component of living systems, and it interacts with many biomolecules. Some molecules are water-loving, or hydrophilic, others are water-abhorring, or hydrophobic, and still others are amphipathic, or in between. What properties of a molecule make it hydrophilic or hydrophobic? First, consider the molecular properties of water and how water interacts with itself. An individual water molecule has a significant dipole that is due to the greater electronegativity of the oxygen atom over the hydrogen atoms. This dipole leads to strong interactions between water molecules, in the form of hydrogen bonds. A hydrogen bond is a noncovalent interaction between polar molecules, one of which is an unshielded proton.

In solid water, or ice, the polar forces hold the individual molecules together in a regular three-dimensional lattice. Most of the hydrogen bonds present in ice are also-present in liquid water. Hence water is a highly hydrogen-bonded structure, not too different from ice, but with a somewhat less regular structure in which the individual molecules have greater mobility. The dipolar properties of water molecules affect the interaction between water and other molecules that dissolve in water. For example, a favourable interaction accounts for the high solubility of sodium chloride in water.

The kinds of ion—dipole interactions that take place between water and simple ions such as Na^+ and C^- are also important in the interactions between the charged, or polar, groups on biomolecules and water. Thus biomolecules that contain charged residues, hydrogen-bond-forming substituents, or other kinds of polar groups are hydrophilic. In the form of small molecules such groups tend to be very soluble in water. When attached to biopolymers they determine which parts of the molecule will be oriented on the exposed surface, where they can make contact with water. Apolar groups such as neutral hydrocarbon side chains do not contain significant dipoles or the capacity for forming hydrogen bonds.

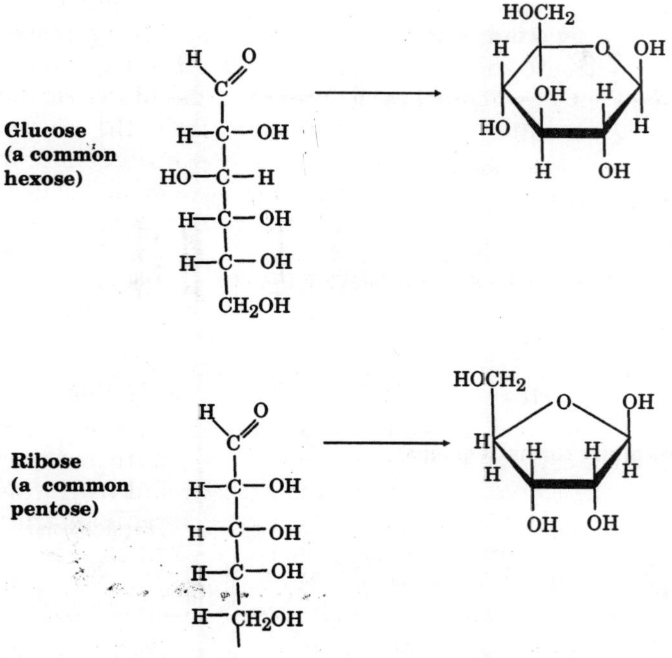

Glucose (a common hexose)

Ribose (a common pentose)

(a) Two common monosaccharides that circularize in aqueous solution

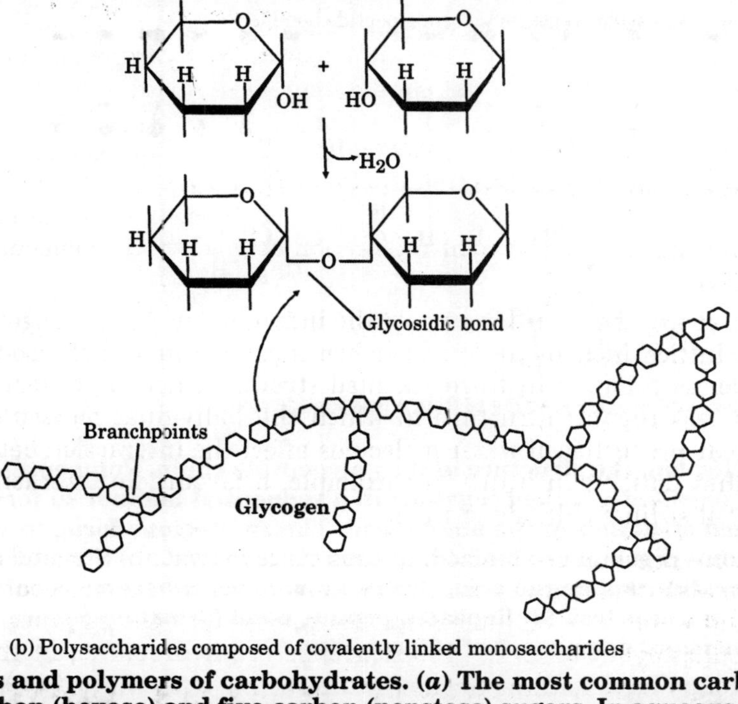

H_2O

Glycosidic bond

Branchpoints

Glycogen

(b) Polysaccharides composed of covalently linked monosaccharides

Fig. 1.4. Monomers and polymers of carbohydrates. (*a*) The most common carbohydrates are the simple six -carbon (hexose) and five-carbon (penstose) sugars. In aqueous solution, these sugar monomers form ring structures. (*b*) Polysacharides are usually composed of hexose monosaccharides covalently linked together by glycosidic bonds to form long straight-chain or branched-chain structures.

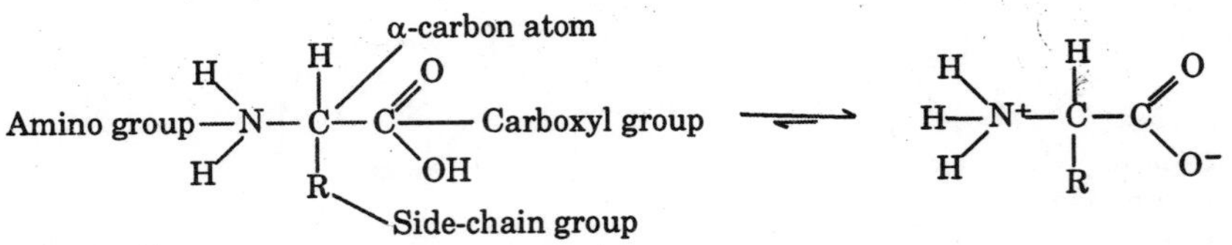

α-carbon atom
Amino group
Carboxyl group
Side-chain group

(a) Generalized structure ofamino acid

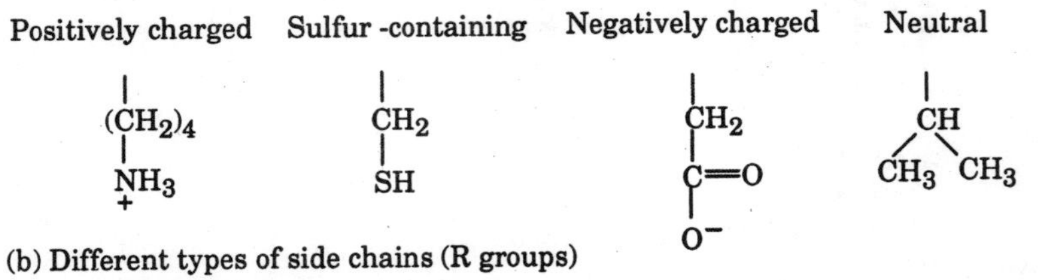

Positively charged Sulfur -containing Negatively charged Neutral

(b) Different types of side chains (R groups)

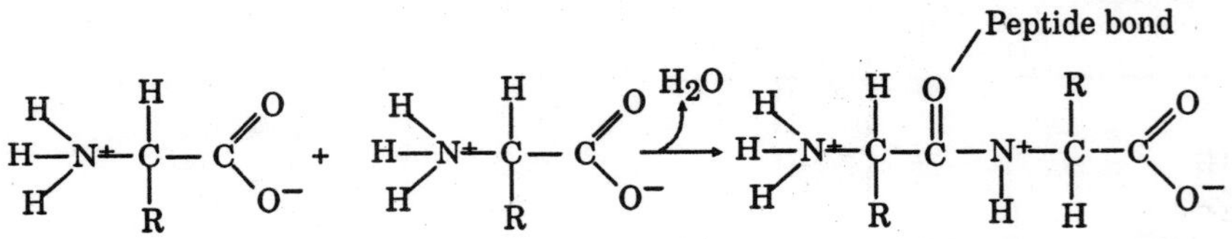

Peptide bond

(c) Two amino acides reactign to form a peptide bond

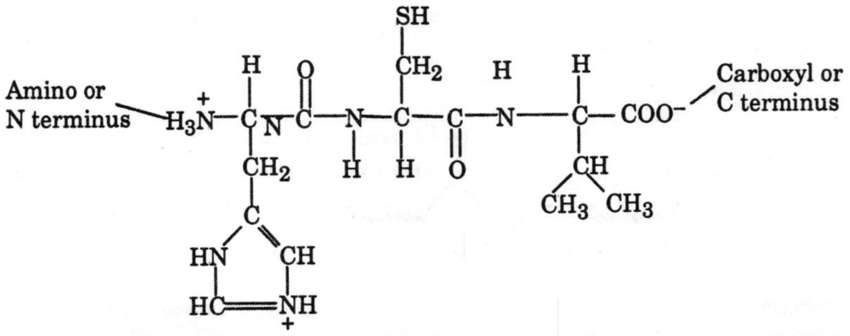

Amino or N terminus
Carboxyl or C terminus

(d) Many amino acids reacting to from a polypeptide chain

Fig. 1.5. Amino acids and the structure of the polypeptide chain. Polypeptides are composed of L-amino acids covalently linked together in a sequential manner to form linear chains. (*a*) The generalized structure of the amino acid. The zwitterions form, in which the amino group and the carboxyl group are ionized, is strongly favoured. (*b*) Structures of some of the R groups found for different amino acids. (*c*) Two amino acids become covalently linked by a peptide bond, and water is lost. (*d*) Repeated peptide bond formation generates a polypeptide chain, which is the major component of all proteins.

Consequently, they have nothing to gain by interacting with water, as evidenced by their poor solubility in water. When such hydrophobic molecules are present in water, the water forms a rigid clathrate (cagelike) structure around them. A polar groups in biopolymers tend to bury themselves within the structure of the biopolymer, where they are in the proximity

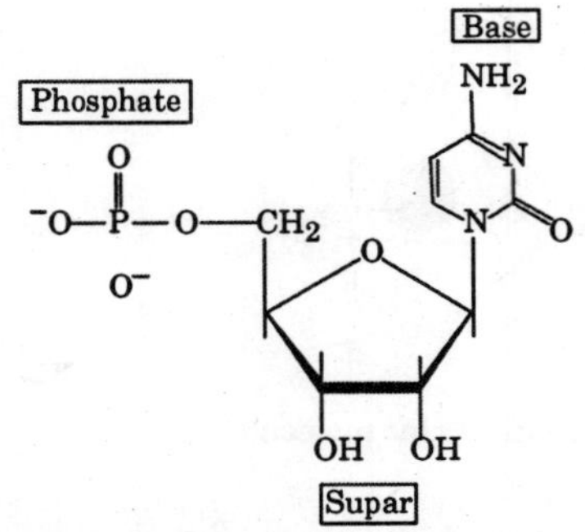
Base
NH2
Phosphate
O
-O—P—O—CH2
O−
N
O
OH OH
Supar

(a) Generalized structurew of a nucleotide

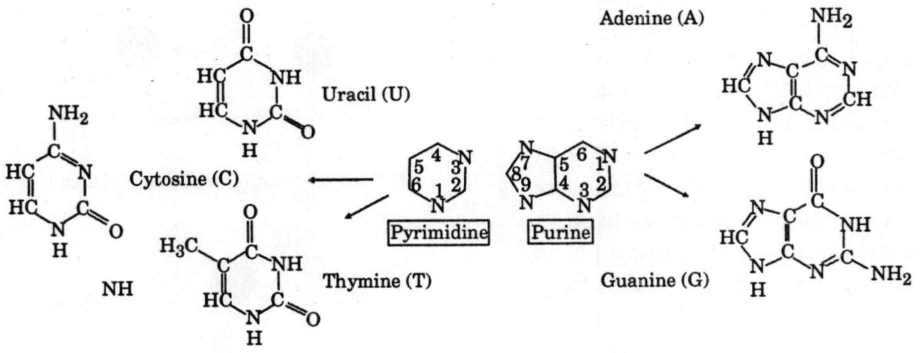
Uracil (U)
Cytosine (C)
Thymine (T)
Pyrimidine
Purine
Adenine (A)
Guanine (G)
NH2
NH
H3C
HC
N
H
O

(b) Different bases found in nucleotides

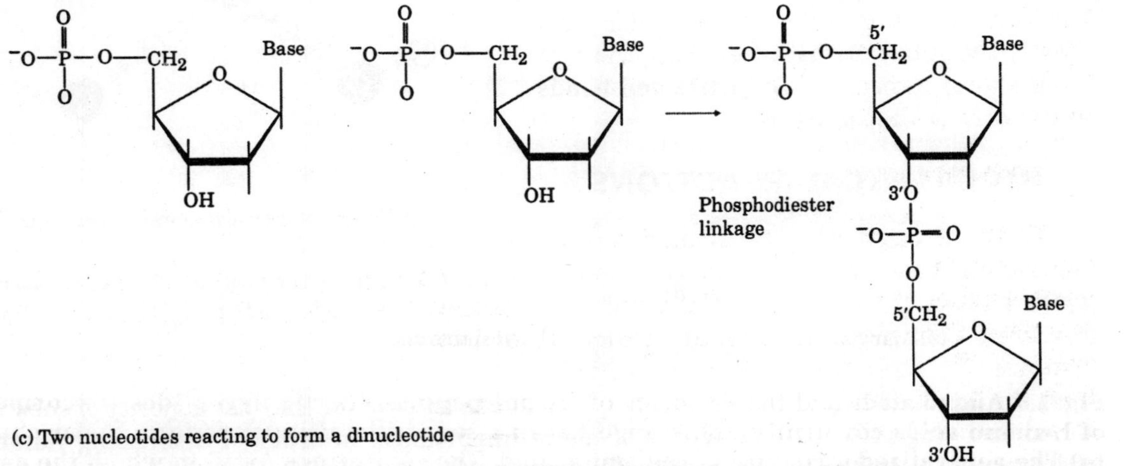
-O—P—O—CH2
Base
OH
5′
3′O
Phosphodiester
linkage
-O—P=O
5′CH2
3′OH

(c) Two nucleotides reacting to form a dinucleotide

Fig. 1.6. The structural components of nucleic acids. Nucleic acids are long linear polymers of nucleotides, called polynucleotides. (*a*) The nucleotide consists of a five-carbon sugar (ribose in RNA or deoxyribose in DNA) covalently linked at the 5′ carbon to a phosphate, and at the 1′ carbon to a nitrogenous base. (*b*) Nucleotides are distinguished by the types of bases they contain. These are either of the two-ring purine type or of the one-ring pyrimidine type. (*c*) When two nucleotides become linked they form a dinucleotide, which contains one phosphodiester bond. Repetition of this process produces a polynucleotide.

of other apolar groups and avoid contact with water. Phospholipids which have a hydrophilic polar group on one end and long hydrophobic side chains attached to it, produce multimolecular aggregates in an aqueous environment. These phospholipid aggregates form monomolecular layers at the air—water interface or bilayer vesicles within the water.

In all of these structures, the polar head groups of the lipid are in contact with water, whereas the apolar side chains are excluded from the solvent structure. As another example of polarity effects on macromolecular structure, consider polypeptide chains, which usually contain a mixture of amino acids with hydrophilic and hydrophobic side chains. Enzymes fold into complex three-dimensional globular structures with hydrophobic residues located on the inside of the structure and hydrophilic residues located on the surface, where they can interact with water. DNA forms a complementary structure of two helically oriented polynucleotide chains. The polar sugar and phosphate groups are situated on the surface, where they can interact with water; the nitrogenous bases from the two chains form intermolecular hydrogen bonds in the core of the structure.

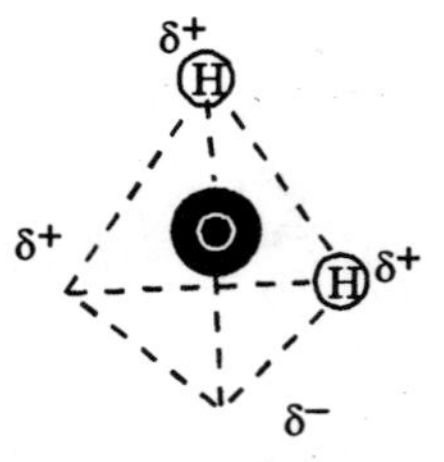

(a) Single water molecule

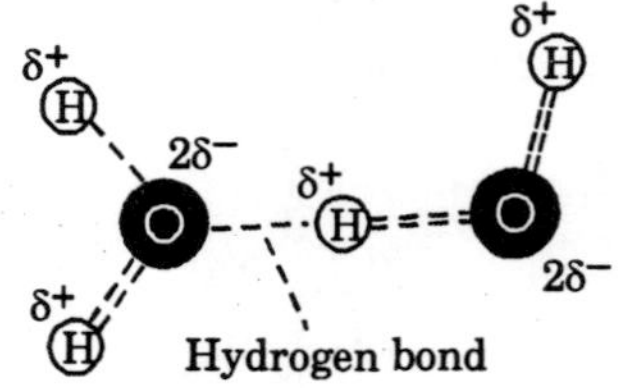

(b) Two interacting water molecules

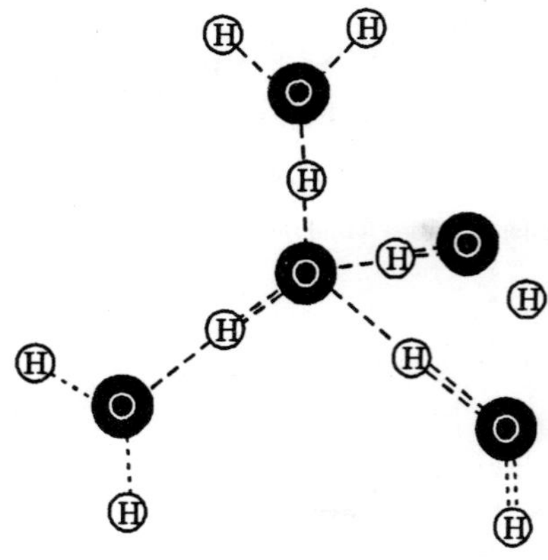

(c) Cluster of interacting water molecules

Fig. 1.7. The structure of water and the interaction of water with other water molecules.

BIOCHEMICAL REACTIONS

Even though the total number of biochemical reactions is very large, it is still much smaller than the potential number of reactions that occur in ordinary chemical systems. This simplification results partly from the fact that only a limited number of elements account for the vast majority of substances found in living cells. The elements of major importance, in order of decreasing numerical abundance, are hydrogen (H), carbon (C), oxygen (O), nitrogen (N), phosphorus (P), and sulfur (S). Certain metal ions are also important; these include Na^+, K^+, Mg^{2+}, Ca^{2+}, Zn^{2+}, and Fe^{2+} or Fe^{3+}. Other metals and elements that are needed in very small amounts are iodine, cobalt, molybdenum, selenium, vanadium, nickel, chromium, tin, fluorine, silicon, and arsenic.

In some cases we don't know the biological roles of these "trace elements" but only that they are needed by some organisms for normal growth or development. The types of covalent linkages most commonly found in biomolecules are also quite limited. Only 16 different types

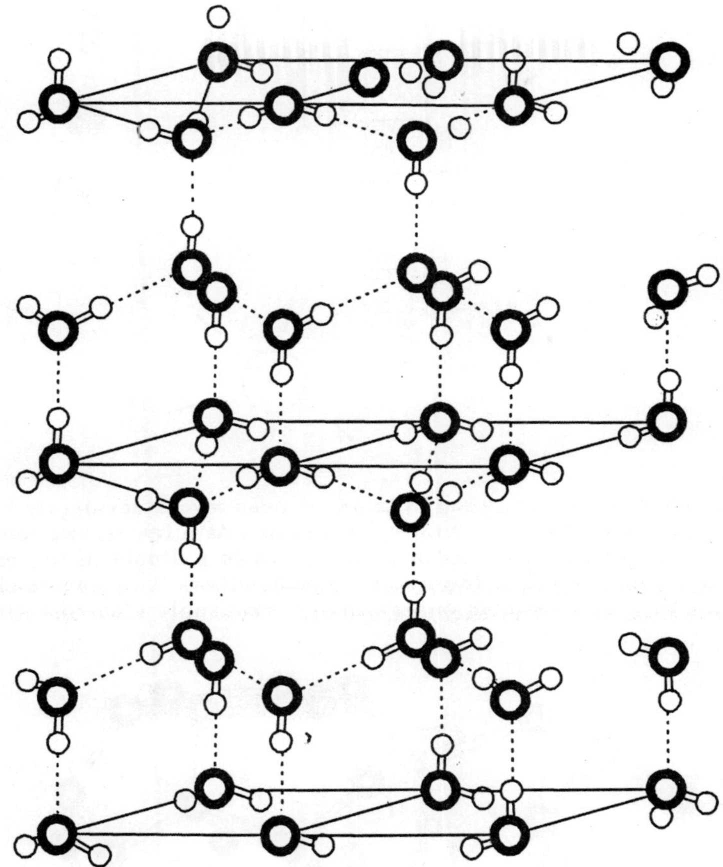

Fig. 1.8. The arrangement of molecules in an ice crystal. Water molecules are oriented so that one proton along each oxygen—oxygen axis is closer to one or the other of the two oxygen atoms.

of linkages account for more than 95% of the linkages found in biomolecules. All the elements can form single or double bonds, except for hydrogen, which only makes single bonds; all the elements exist primarily in one valence state, except for carbon and sulfur, which are frequently found in more than one valence state.

Despite this overall simplicity, many other valence states can be found in unusual cases, and some of these are very important. For example, the biochemistry of nitrogen involves consideration of all the valence states of nitrogen from +5 to 0 to –3. A major source of nitrogen available to biosystems is gaseous nitrogen found in the atmosphere (valence state 0). Biochemical reactions convert gaseous nitrogen into other forms of nitrogen by reactions which occur uniquely in a select group of microorganisms. Biochemical reactions involving the different classes of substances use a limited number of functional groups. All of the functional groups depicted are electrostatically neutral in organic solvents. In water or the cell cytosol, however, many of these functional groups either lose or gain protons to become charged species.

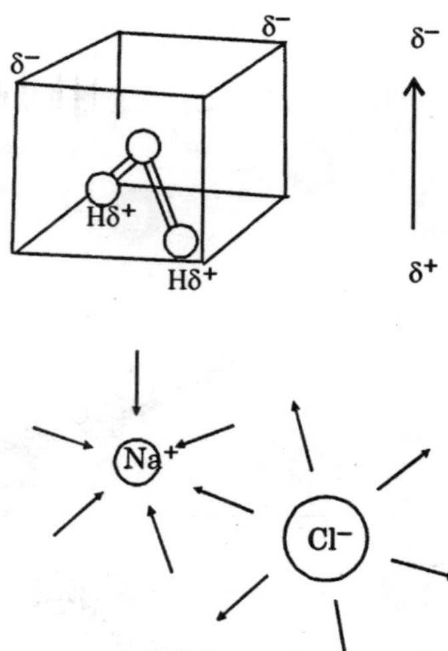

Fig. 1.9. The water molecule is composed of two hydrogen atoms covalently bonded to an oxygen atom with tetrahedral (sp^3) electron orbital hybridization. As a result, two lobes of the oxygen *sp^3* orbital contain pairs of unshared electrons, giving rise to a dipole in the molecule as a whole. The presence of an electric dipole in the water molecule allows it to solvate charged ions because the water dipoles can orient to form energetically favourable electrostatic interactions with charged ions.

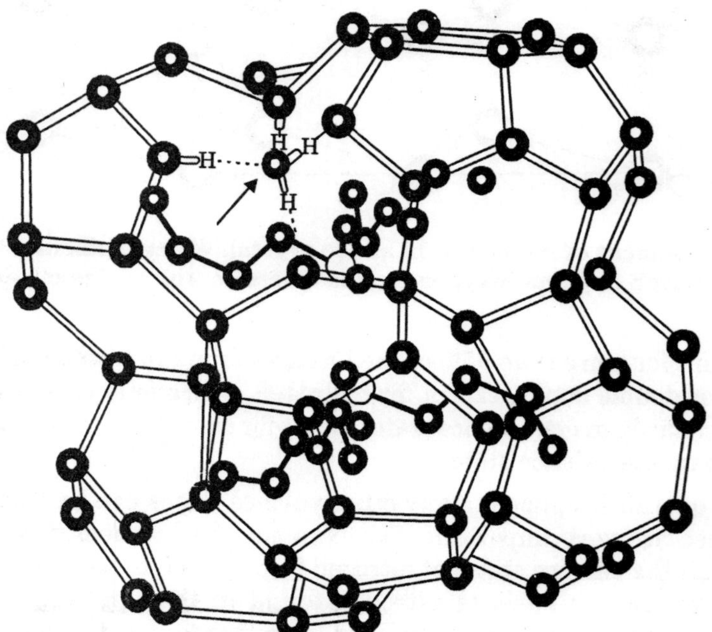

Fig. 1.10. Clathrate structures are ordered cages of water molecules around hydrocarbon chains. A portion of the cage structure of $(nC_4H_9)_3S^+F^- \cdot 23\ H_2O$ is shown. The trialkyl sulfur ion nests within the hydrogen-bonded framework of water molecules. In the intact framework, each oxygen is tetrahedrally coordinated to four others. One such oxygen atom and its associated hydrogens are shown by the arrow. (Illustration copyright by Irving Geis. Reprinted by permission.)

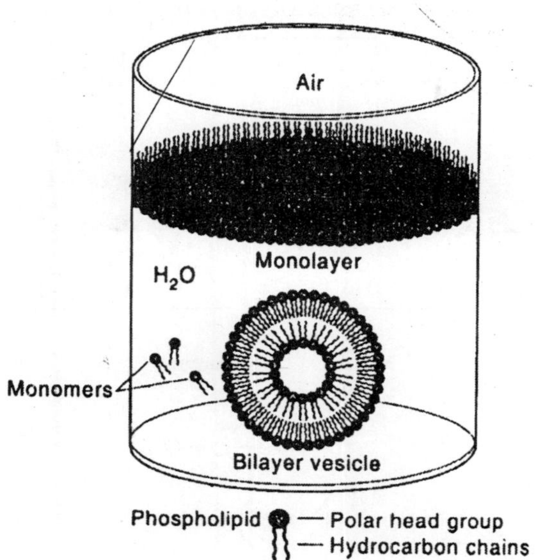

Fig. 1.11. Structures formed by phospholipids in aqueous solution. Phospholipids may form a monomolecular layer at the air-water interface, or they may form spherical aggregations surrounded by water. A vesicle consists of a double molecular layer of phospholipids surrounding an internal compartment of water.

Table 1.2. Types of Covalent Linkages Most Commonly Found in Biomolecules

	H	*C*	*O*	*N*	*P*	*S*
H						
C	—C—H	—C—C—				
		C=C				
O	—O—H	—C—O—				
		C=O				
N	N—H	—C—N=				
		C=N—	—			
P	—	—	P—O	—		
			P=O	—		
S	—S—H	—C—S—	S—P—	—	—	—S—S—
			S=O			

Most of the reactive groups in biomolecules contain one or more of these functional groups or ones closely related to these groups. Many cellular reactions involving these functional groups closely resemble reactions that take place in nonliving systems under different conditions. These extracellular reactions are studied in organic chemistry. For example, peptide bond formation can occur between two amino acids by a dehydration resulting from simple heating.

Table 1.3. Most Common Valences Displayed by Atoms in Covalent Linkages

Element	*Valence*
H	+1
C	–4 to +4
O	–2
P	+5
N	–3
S	+6, –2, –1

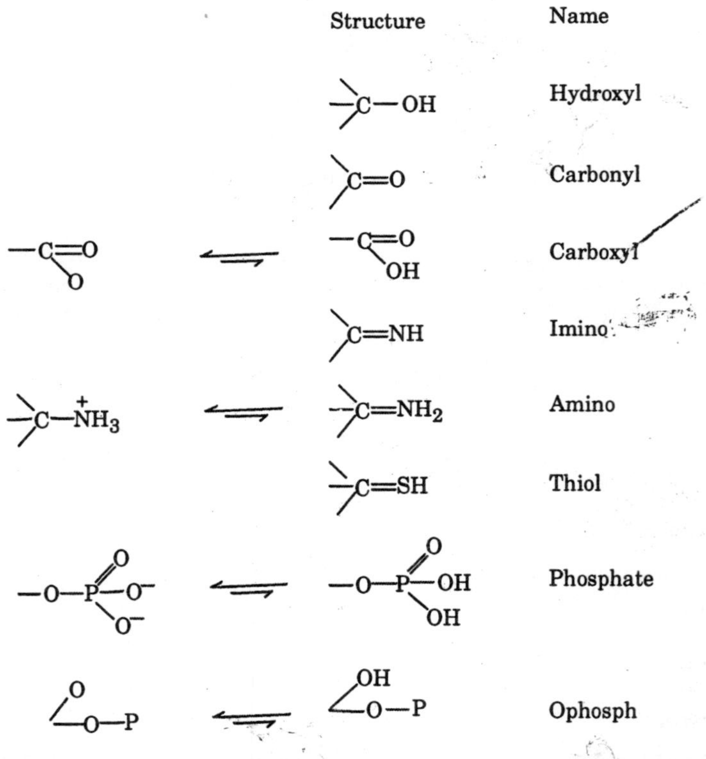

	Structure	Name
	>C—OH	Hydroxyl
	>C=O	Carbonyl
—C(=O)O ⇌	—C(=O)OH	Carboxyl
	>C=NH	Imino
>C—$\overset{+}{N}H_3$ ⇌	>C=NH_2	Amino
	>C=SH	Thiol
—O—P(=O)(O^-)O^- ⇌	—O—P(=O)(OH)OH	Phosphate
O—O—P ⇌	OH—O—P	Ophosph

Fig. 1.12. Different functional groups found in biomolecules. This figure includes the major functional groups. Other functional groups are found in minor amounts.

In the cell, peptide bond formation also takes place, but several intermediate steps are involved and the reaction takes place not by dehydrating but in the wet environment of the cytosol. Similarly the phosphodiester bond depicted in figure 1.6*c* is not formed by a simple dehydration reaction in the cell but rather when one nucleotide in the triphosphate form loses a pyrophosphate group as it becomes linked to the hydroxyl group of another nucleotide.

CATALYTIC SURFACES OF ENZYMES

Although biochemical reactions resemble ordinary chemical reactions, they differ in some important ways. Chemical reactions are frequently carried out in nonaqueous solvents, using elevated temperatures and pressures, acids or bases, or other harsh reagents—conditions that would destroy the functional organization of a living cell. Biochemical reactions usually take place under very mild conditions in aqueous solution. However, many chemical reactions do not proceed at reasonable rates under such conditions. Biochemical reactions proceed at substantially faster rates because of the very special nature of the enzyme catalysts that accelerate them. Enzymes are structurally complex, highly specific catalysts; each enzyme usually catalyzes only one type of reaction.

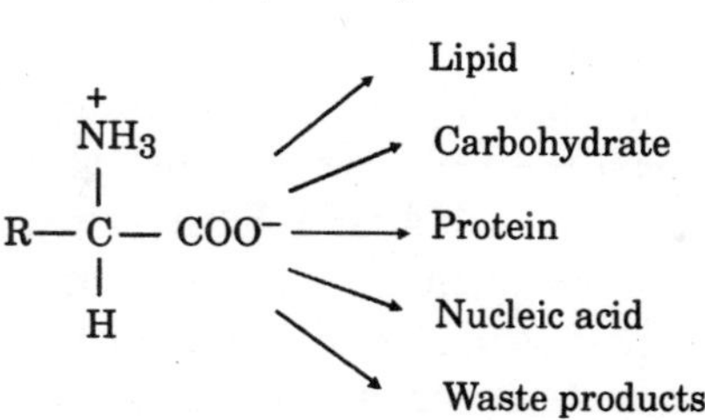

Fig. 1.13. The different fates of an amino acid. Depending on which enzymes are present and active and on the needs of the organism, an amino acid can be metabolized in different ways. Each of these conversions involves one or more steps, and usually each step requires a specific enzyme.

The enzyme surface binds the interacting molecules, or substrates, so that they are favourably disposed to react with one another. The specificity of enzyme catalysis also has a selective effect, so that only one of several potential reactions takes place. For example, a simple amino acid can be used in the synthesis of any of the four major classes of macromolecules or can simply be secreted as waste product. The fate of the amino acid is determined as much by the presence of specific enzymes as by its reactive functional groups.

ENERGY UTILIZATION

An appreciable amount of energy is needed to build a cell. Even maintaining a cell in a steady nongrowing state requires energy input. Chemical energy is needed to drive many biochemical reactions, to do mechanical work, and for transport of substances across the plasma membrane. The ultimate source of energy that drives a cell's reactions is sunlight. Light energy is converted into chemical energy in the chloroplasts of plant cells or in the photosynthetic structures of certain micro-organisms. The main form of chemical energy produced in the chloroplast is a nucleotide containing three phosphoric acid groups attached in sequence, adenosine triphosphate, or ATP.

Organisms that cannot harness the light rays of the sun themselves to make ATP are able to make ATP from the breakdown of organic nutrients originating from plants or other organisms. Most biochemical reactions fall into one of two classes: degradative or synthetic. Degradative, or catabolic, reactions result in the breakdown of organic compounds to simpler substances. Synthetic, or anabolic, reactions lead to the assembly of biomolecules from simpler molecules. Anabolic processes require energy to drive them. This energy is usually supplied by coupling the energy-requiring biosynthetic reactions to energy-releasing catabolic reactions.

LOCALIZED REACTIONS IN THE CELL

Biochemical reactions are organized so that different reactions occur in different parts of the cell. This organization is most apparent in eukaryotes, where membrane-bounded structures are visible proof for the localization of different biochemical processes. For example, the synthesis of DNA and RNA takes place in the nucleus of a eukaryotic cell. The RNA is subsequently transported across the nuclear membrane to the cytoplasm, where it takes part in protein synthesis. Proteins made in the cytoplasm are used in all parts of the cell.

A limited amount of protein synthesis also occurs in chloroplasts and mitochondria. Proteins made in these organelles are used exclusively in organelle-related functions. Most ATP synthesis occurs in chloroplasts and mitochondria. A host of reactions that transport nutrients and metabolites occur in the plasma membrane and the membranes of various organelles. The localization of functionally related reactions in different parts of the cell concentrates reactants and products at sites where they can be most efficiently utilized.

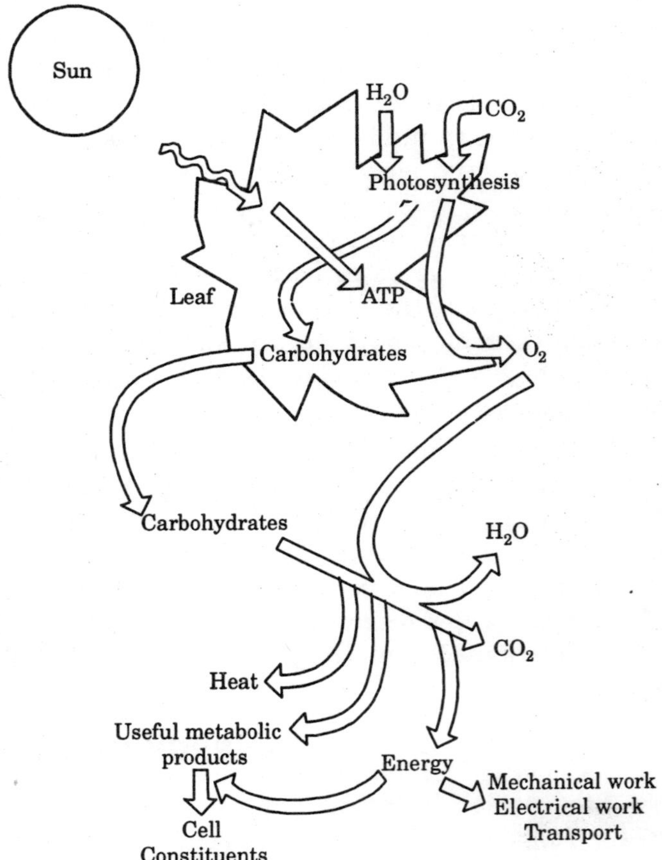

Fig. 1.14. Flow of energy in the biosphere. The sun's rays are the ultimate source of energy. These rays are absorbed and converted into chemical energy (ATP) in the chloroplasts. The chemical energy is used to make carbohydrates from carbon dioxide and water. The energy stored in the carbohydrates is then used, directly or indirectly, to drive all the energy-requiring processes in the biosphere.

PATHWAYS

Most biochemical reactions are integrated into multistep pathways using several enzymes. For example, the break-down of glucose into CO_2 and H_2O involves a series of reactions that begins in the cytosol and continues to completion in the mitochondrion. A complex series of reactions like this is referred to as a biochemical pathway. Synthetic reactions, such as the biosynthesis of amino acids in the bacterium *Escherichia coli,* are similarly organized into pathways. Frequently pathways have branch-points. For example, the synthesis of the amino acids threonine and lysine starts with oxaloacetate.

After three steps, a branchpoint is reached with the formation of the organic compound aspartic-β-semialdehyde. One branch of this pathway leads to the synthesis of the amino acid lysine, and another branch leads to the synthesis of the amino acids methionine, threonine, and isoleucine. To understand the role of each biochemical reaction we must identify its position in a pathway and also consider how that pathway interacts with others.

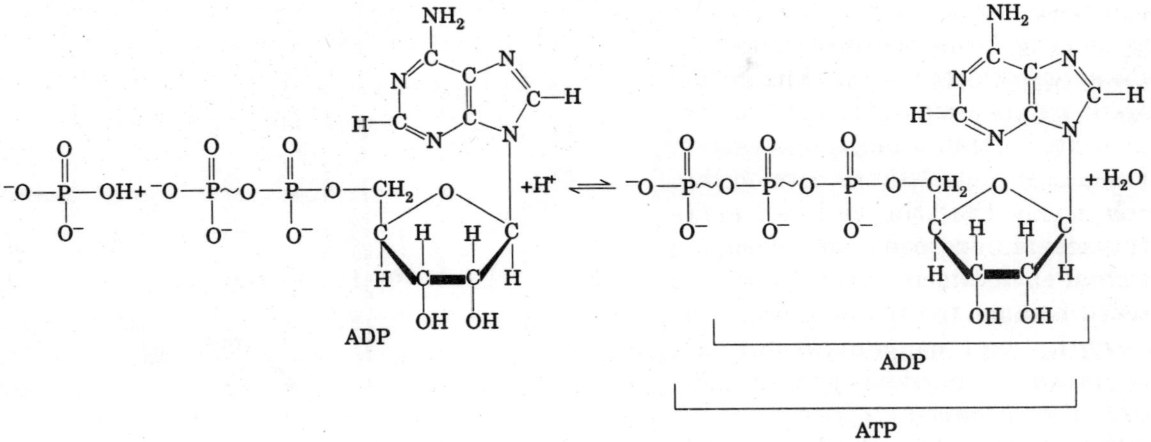

Fig. 1.15. The structures of ATP and ADP and their interconversion. The two compounds differ by a single phosphate group.

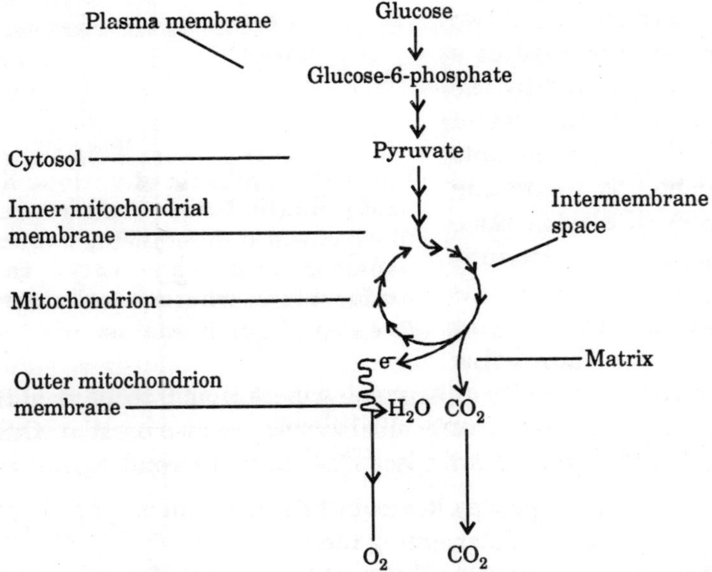

Fig. 1.16. Summary diagram of the breakdown of glucose to carbon dioxide and water in a eukaryotic cell. As depicted here, the process starts with the absorption of glucose at the plasma membrane and its conversion into glucose-6-phosphate. In the cytosol, this six-carbon compound is then broken down by a sequence of enzyme-catalyzed reactions into two molecules of the three-carbon compound pyruvate. After absorption by the mitochondrion, pyruvate is broken down to carbon dioxide and water by a sequence of reactions that requires molecular oxygen.

REGULATION

Hundreds of biochemical reactions take place even in the cells of relatively simple microorganisms. Living systems have evolved a sophisticated hierarchy of controls that permits them to maintain a stable intracellular environment. These controls ensure that substances required for maintenance and growth are produced in amounts that are adequate without being excessive. Biochemical controls have developed in such a way that the cell can make adjustments in response to a changing external environment. Adjustments are needed because the temperature, ionic strength, acid concentration, and concentration of nutrients present in the external environment vary over much could be tolerated inside the cell. The rate of intracellular reactions availability of substrates and enzymes.

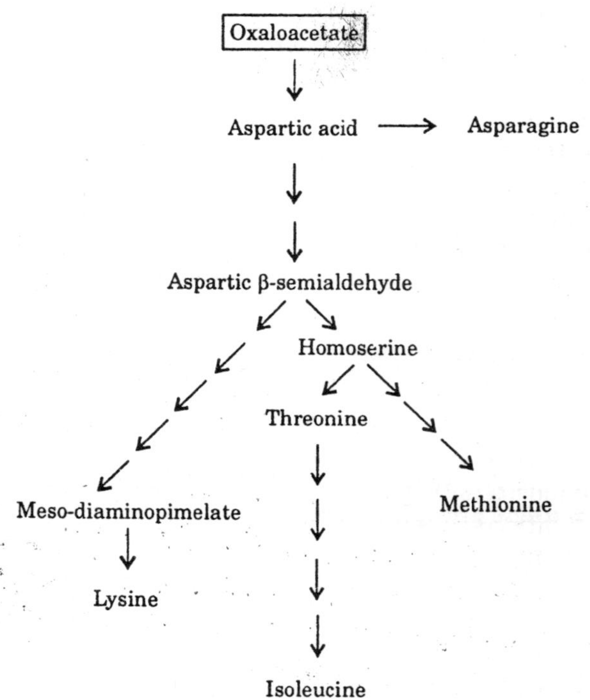

Fig. 1.17. Synthesis of various amino acids from oxaloacetate. Each arrow represents a discrete biochemical step requiring a unique enzyme. Thus aspartic acid is produced in one step from oxaloacetate, whereas isoleucine is produced in five steps from threonine.

Enzyme activity is controlled at two different levels. First, the rate of a catalyzed reaction is regulated by the amount of the catalyzing enzyme present in the cell. Control of enzyme amounts is usually accomplished by regulating the rate of enzyme synthesis; in some cases the rate of enzyme degradation is also regulated. We can think of controls that regulate the total amount of enzyme present as coarse controls. They define the limits of possible enzyme activity as being anywhere from 0 to 100% of the full activity of the enzyme. Fine controls that act directly on enzymes are also present. Only certain special enzymes, called regulatory enzymes, are susceptible to this second type of regulation.

Regulatory enzymes usually occupy key points in biochemical pathways, and their state of activity frequently is decisive in determining the utilization of the pathway. The underlying principle in regulation is maintaining a favourable intracellular environment in the most economical manner. The cell makes products in the amounts that are needed. Each pathway is regulated in a somewhat different way, ensuring that biochemical energy and substrates are efficiently utilized.

BIOCHEMICALLY DEPENDENT ORGANISM

Between 3 and 4 billion years ago the first self-replicating molecules appeared on earth. These entities had to have the capacity for extracting nutrients from the chemical compounds that existed in prebiotic times. We have some general notions about what types of substances were present at that time. One of the most important substances that was not present at that time in significant amounts was molecular oxygen, O_2. Currently this form of oxygen is required by all forms of life visible to the naked eye. The O_2 used by most organisms is ultimately converted by them into CO_2.

Oxygen is utilized at a rapid rate 'and it would soon disappear if it were not for special classes of photosynthetic organisms that are constantly producing more O_2 by the oxidation of water. The oxygen story is an example of the dependence of one class of organisms on another for certain chemicals. A similar situation exists with the elements carbon and nitrogen, which must be converted from gaseous forms, CO_2 and N_2, to organic forms usable by most organisms. Reduced carbon compounds are constantly being lost by oxidation to gaseous CO_2.

The supply of organic carbon compounds required by all forms of life is replenished by photosynthetic organisms; these include most plants and certain microorganisms. Similarly, nitrogen in organic molecules is constantly being lost to the atmosphere in the form of gaseous nitrogen.

The reactions required for the conversion of nitrogen to a reduced form more usable to the majority of organisms occurs in only a limited number of microorganisms; yet without these nitrogen-fixing organisms life as we know it would soon vanish. As we ascend the evolutionary tree, we find increasingly complex multicellular forms. Such organisms generally require more complex nutrients, which must ultimately be supplied to them by simpler living forms. Bacteria like *E. coli* can make all of their own amino acids from a reduced form of nitrogen, such as NH_3, and a reduced form of carbon, such as glucose. Humans, on the other hand, must receive most of their amino acids as nutrients.

Humans and other complex organisms have gained new biochemical Pacities, which permit them to synthesize the components associated with highly specialized differentiated tissues. At the same time, they have lost many of the biochemical systems required to survive on simpler nutrients. Many biochemical reactions of great importance take place in only a limited number of organisms. This fact increases the complexity of the study of biochemistry. We must learn many reactions; we must also be aware of the biochemical potentials of different organisms. This is the only way we can understand the biochemical interdependency of organisms.

SYNTHESIS OF PROTEINS

DNA contains the genetic information transmitted to each daughter cell when cells divide. The DNA usually exists in the form of nucleoprotein (DNA-protein) complexes called chromosomes. A prokaryotic cell contains a single chromosome. Prior to cell division this chromosome duplicates and segregates so that an identical complement of DNA goes to each of two newly formed daughter cells. Eukaryotic cells are more complex than prokaryotic cells and usually contain more DNA, which is partitioned between several chromosomes. In both prokaryotes and eukaryotes, almost all cells of the same organism contain the same number of chromosomes.

In eukaryotes most of the chromosomes are localized in the nucleus. Thus the DNA is isolated from the main body of the cytoplasm—a unique feature of eukaryotes and the primary distinction between prokaryotes and eukaryotes. Some organelles, notably the mitochondria and the chloroplasts, contain a single circular chromosome. Eukaryotic chromosomes are detectable by light microscopy at the stage just prior to cell duplication. At this stage, called mitosis, chromosomes appear as elongated refractile structures that can be seen to segregate in equal numbers and types to each of the daughter cells before cell division.

Each chromosome carries hereditary (genetic) information necessary for the synthesis of specific compounds essential for cell maintenance, growth, and replication. Each chromosome contains a single very long DNA molecule composed of 10^6 or more nucleotides in a specific sequence. The sequence of nucleotides in the chromosomal DNA determines the sequence of amino acids in the protein polypeptide chains of the organism. The relationship between base sequences and resultant amino acids is known as the genetic code. Each grouping of three bases, called a triplet, represents a specific amino acid and is called a codon.

The genetic code ensures that the organism's characteristics are reflected by the sequence of nucleotides in its DNA. When chromosomes replicate, the DNA replicates precisely, so that the same nucleotide sequence is passed along to each of the daughter cells resulting from mitosis and cell division. The DNA does not transfer its genetic information directly to protein. Rather, this information passes through an intermediary, the messenger RNA (mRNA). The mRNA is made on a DNA template in the nucleus of a eukaryotic cell and then passes into the cytoplasm, where it serves in turn as a template for the synthesis of the polypeptide chain. The overall process of information transfer from DNA to mRNA (transcription) and from mRNA to protein (translation) is depicted in figure 1.22.

EVOLUATION OF BIOCHEMICAL SYSTEMS

Biochemical systems are conservative and opportunistic. They tend to evolve one step at a time, using a readily accessible route that leads to an advantage. To appreciate biochemical systems today it is useful to have some understanding of how they came to be. The earth was formed by a process of accretion about 4.6 billion years ago. Initially it was a molten mass lacking the gravitational pull to retain its gases at the prevalent elevated temperatures. And yet, within a mere 700 million years of the planet's birth, as calculated from the isotopic record of sediments, cellular life almost certainly existed.

What raw materials were available to bring about this amazing turn of events? What were the sources of energy used to drive the necessary reactions? Where did the important reactions take place? Was it in the atmosphere, in the oceans, on dryland, or all three? Table 1.4 shows a distribution of the major elements found in the earth's crust, the ocean water, and the human body. The composition of the human body, which is reasonably representative of living organisms, differs appreciably from that of the earth's crust. The four most abundant elements in the human body are hydrogen, carbon, nitrogen, and oxygen. Of these, only oxygen belongs to the class of elements in highest abundance that make up the earth's crust. Nevertheless, the elements needed to make living things were present in sufficient quantities in the earth's crust and its primitive oceans and the atmospherere. The original water and air associated with the newly formed planet were lost because of the high temperatures.

As the earth cooled, water and various gases on the surface and in the atmosphere were produced by an outgassing process. Today the earth is only about 0.5% water by weight, but because of water's low density, most of it is present on the earth's surface, where it has a major impact on the environment. Water cycles through its gaseous form in the atmosphere and its liquid form in the oceans and bodies of fresh water. Geological evidence indicates that appreciable amounts of the total water mass have always been present as liquid water. Thus the temperature of the planet has for the most part been between 0° and 100°C, a range that is conducive to the formation of biomolecules and the origin and propagation of life?

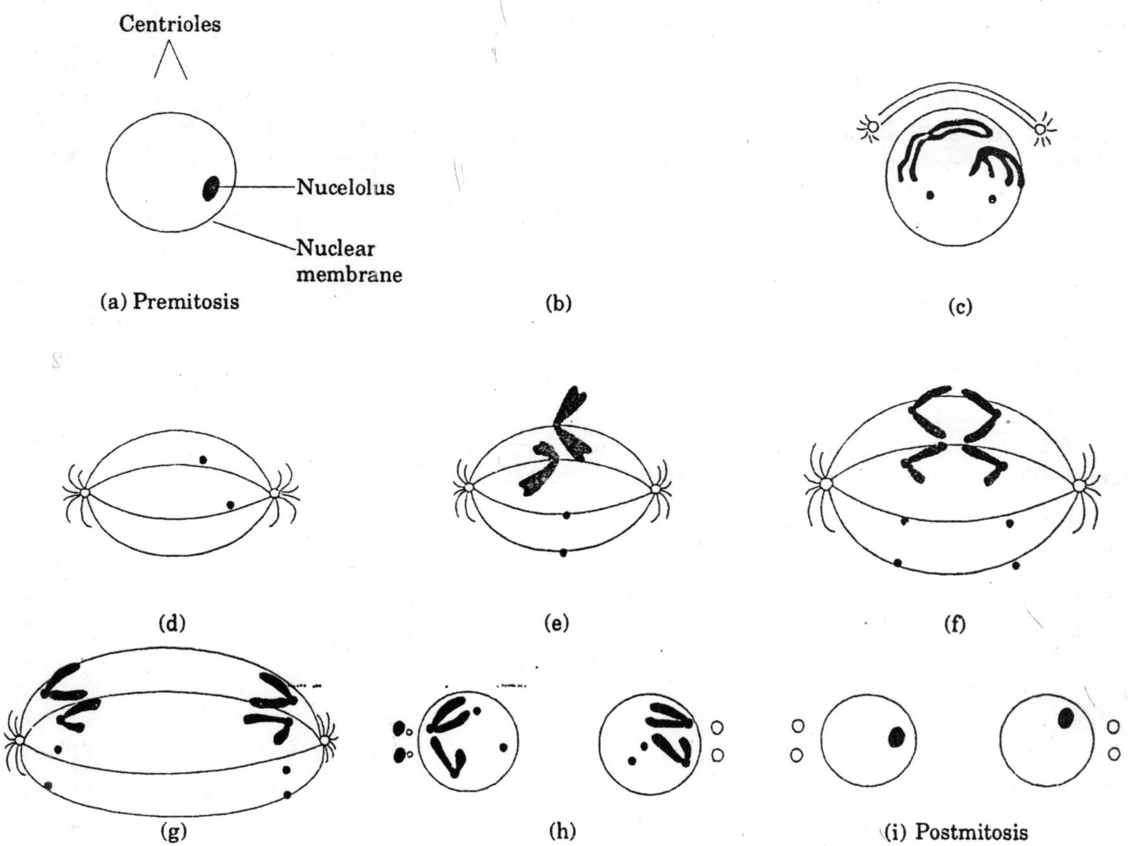

Fig. 1.18. Mitosis and cell division in eukaryotes. After DNA duplication has occurred, mitosis is the process by which quantitatively and qualitatively identical DNA is delivered to daughter cells formed by cell division. Mitosis is traditionally divided into a series of stages characterized by the appearance and movement of the DNA-bearing structures, the chromosomes, (*a*) Premitosis. (*b*) through (*h*) Successive stages of mitosis, (*i*) Postmitosis.

The earth has also been kind to living things in other ways. The buffering action of various clays and minerals is believed to have maintained the pH level of the oceans between 8.0 and 8.5, which is close to the pH inside living cells. Unlike the temperature and pH of the surface water, the composition of the atmosphere has changed drastically since the origin of life. In fact, the processes taking place in living things are primarily responsible for these changes. Today's atmosphere, which is about one millionth of the mass of the earth itself, is mainly composed of nitrogen (78%) and oxygen (21%).

Most of the remaining atmosphere is argon (0.9%), water (variable up to 4%), and carbon dioxide (0.034%). The gases that made up the primitive atmosphere were quite different; especially conspicuous was the absence of gaseous oxygen. Most of the oxygen in the present atmosphere is due to the oxidation of water by photosynthetic organisms. The main forms of carbon and nitrogen in the primitive atmosphere were probably CO_2 and N_2 as they are today. In addition, and probably of great significance to the origin of life, small amounts of the more reduced forms of carbon and H_2 gas were present. Thus the primitive earth probably had a weakly reducing atmosphere as contrasted with today's highly oxidizing atmosphere.

This situation was most favourable to the origin of life, because organic compounds that enter the biomass tend to be in a reduced state and they are readily oxidized in the presence of gaseous oxygen. It is generally believed that the first organics formed in this primitive atmosphere and then rained down to form larger bioorganic molecules in the liquid phase.

Table 1.4. Distribution of the 24 Elements Used in Biological Systems[a]

Elements	*Atomic No*	*Earth's Crust*	*Ocean*	*Human Body*
Hydrogen (H)	1	2,882	66,200	60,562
Carbon (C)	6	56	1.4	10,680
Nitrogen (N)	7	7	<1	2,440
Oxygen (O)	8	60,425	33,100	25,670
Fluorine (F)	9	77	<1	<1
Sodium (Na)	11	2,554	290	75
Magnesium (Mg)	12	1,784	34	11
Silicon (Si)	14	20,475	<1	<1
Phosphorus (P)	15	79	<1	130
Sulfur (S)	16	33	17	130
Chlorine (Cl)	17	11	340	33
Potassium (K)	19	1,374	6	37
Calcium (Ca)	20	1,878	6	230
Vanadium (V)	23	4	<1	<1
Chromium (Cr)	24	8	<1	<1
Manganese (Mn)	25	37	<1	<1
Iron (Fe)	26	1,858	<1	<1
Cobalt (Co)	27	1	<1	<1
Nickel (Ni)	28	3	<1	<1
Copper (Cu)	29	1	<1	<1
Zinc (Zu)	30	2	<1	<1
Selenium (Se)	34	<1	<1	<1
Molybdenum (Mo)	42	<1	<1	<1
Iodine (I)	53	<1	<1	<1

[a]Amounts are given in atoms per 100,000.

The origin of life probably occurred in three phases, (1) The earliest phase was a period of chemical evolution during which the compounds needed for the nucleation of life must have been formed. These compounds include the most important class of biological macromolecules, the nucleic acids. In this phase of evolution, the synthesis of nucleic acids was "noninstructed." (2) As soon as some nucleic acids were present, physical forces between them must have led to an "instructed" synthesis, in which the already formed molecules served as templates for the synthesis of new polymers. It seems likely that "feedback loops" selected out certain nucleic acids for preferential synthesis. At some point during this period of instructed synthesis more nucleic acids and possibly protein macromolecules were formed.

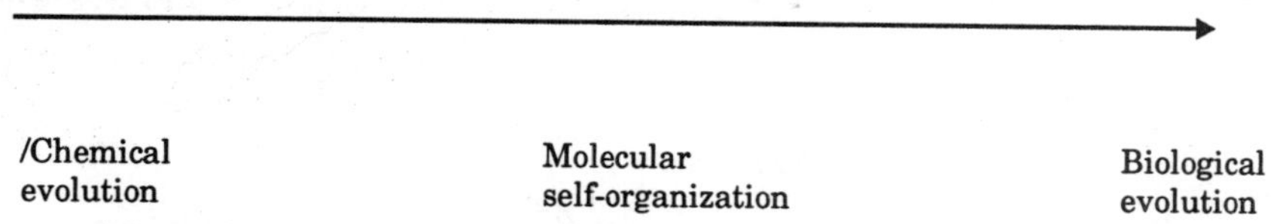

/Chemical evolution   Molecular self-organization   Biological evolution

Fig. 1.19. The origin of life probably occurred in three overlapping phases: Phase I, chemical evolution, involved the noninstructed synthesis of biological macromolecules. In phase II, biological macromolecules self-organized into systems that could reproduce. In phase III, organisms evolved from simple genetic systems to complex multicellular organisms. The arrow pointing from left to right emphasizes the unidirectional nature of the overall process.

The products of this phase of molecular self-organization must sooner or later have begun to resemble the complex organized units that we observe in the self-reproducing biosynthetic cycles of living cells. (3) In the final phase of the origin of life we find the beginnings of the divergent process of biological evolution, the development of the simplest single-celled organisms, and their differentiation into complex multicellular beings.

A Common Evolution

The classical view of evolution, based on morphological differences among organisms, can be diagrammed as a branching tree in which all existing organisms are shown at the tips of the branches. An evolutionary tree starts from the simple ancestral prokaryotic cell, which branches off in three main directions into the archaebacteria, the eubacteria, and the eukaryotes. Each of these kingdoms has continued to branch in elaborate ways; we have shown only some of the main branchpoints in figure 1.20. Prokaryotes for the most part have remained as relatively undifferentiated single-celled organisms containing a single chromosome. By contrast, eukaryotes have changed dramatically.

Although the organisms on many branches of the eukaryotic part of the evolutionary tree have remained as relatively undifferentiated single-celled forms, significant numbers of eukaryotic organisms have evolved into multicellular forms in which the individual cells of the total organism have differentiated to serve different functions. This process has given rise to plants, animals, and fungi. A somewhat different view of biological evolution has arisen from a comparison of nucleic acid sequences in different organisms.

The best-known sequences for such studies have come from the 16S ribosomal RNAs in different organisms. This comparative study has provided us with the evolutionary pattern shown in figure 1.21 in which distances are proportional to sequence differences. A most striking characteristic of this unrooted tree is the distinctness of the three primary kingdoms as evidenced by the large sequence distances that separate each of the kingdoms.

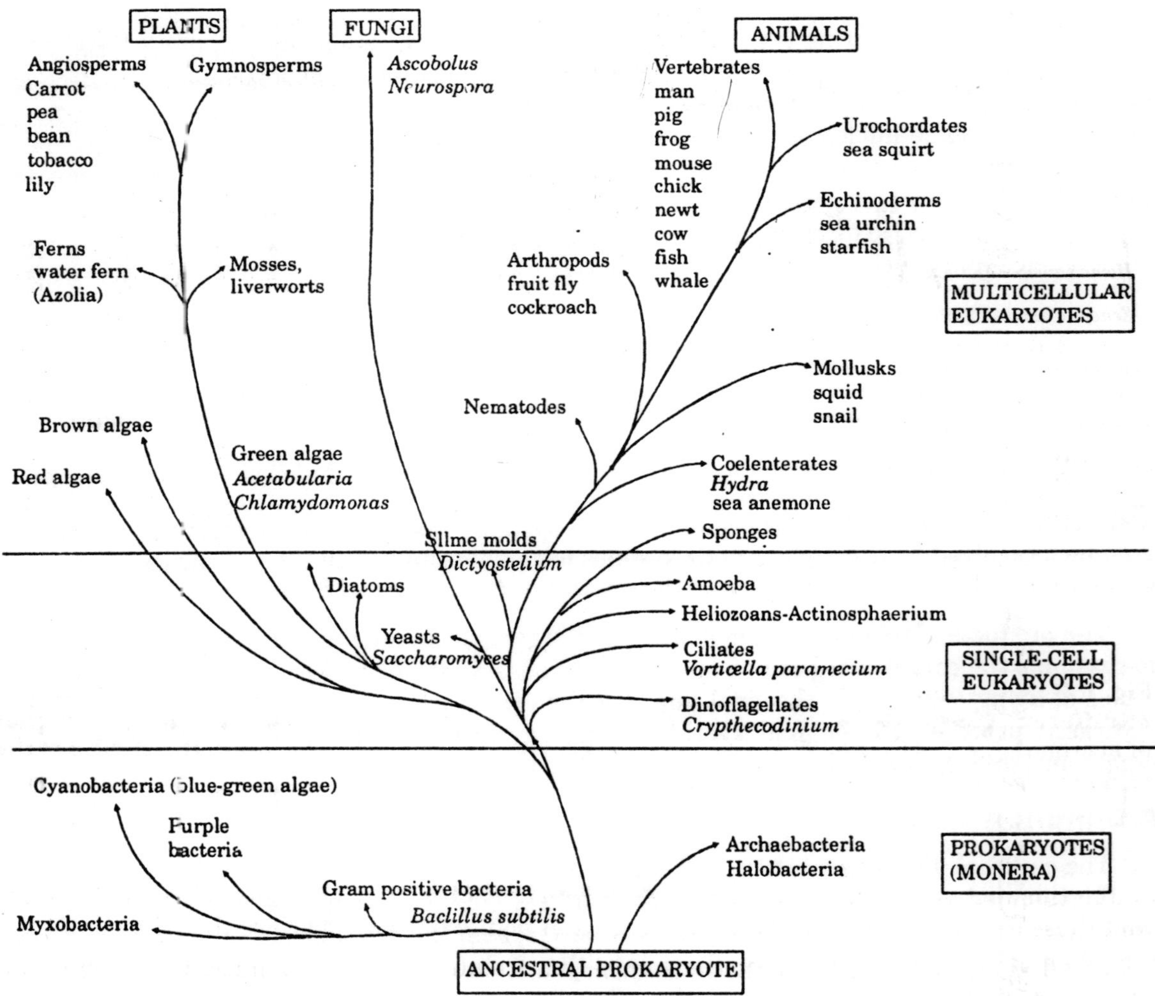

Fig. 1.20. Classical evolutionary tree. All living forms have a common origin, believed to be the ancestral prokaryote. Through a process of evolution some of these prokaryotes changed into other organisms with different characteristics. The evolutionary tree indicates the main pathways of evolution.

Although we do not know the position of the root of the tree, it seems likely that the archaebacteria are closer to the common ancestor of all three kingdoms than are the eubacteria or the prokaryotes. The organisms most familiar to us, the multicellular plants and animals, occupy a shallow domain within the eukaryotic line of descent. True, the developmental programs of the multicellular forms have generated an incredible diversity in form and function. Nevertheless, both bacteria and unicellular eukaryotes span far greater evolutionary histories. This fact is reflected in the greater biochemical diversity that we find in the unicellular microorganisms.

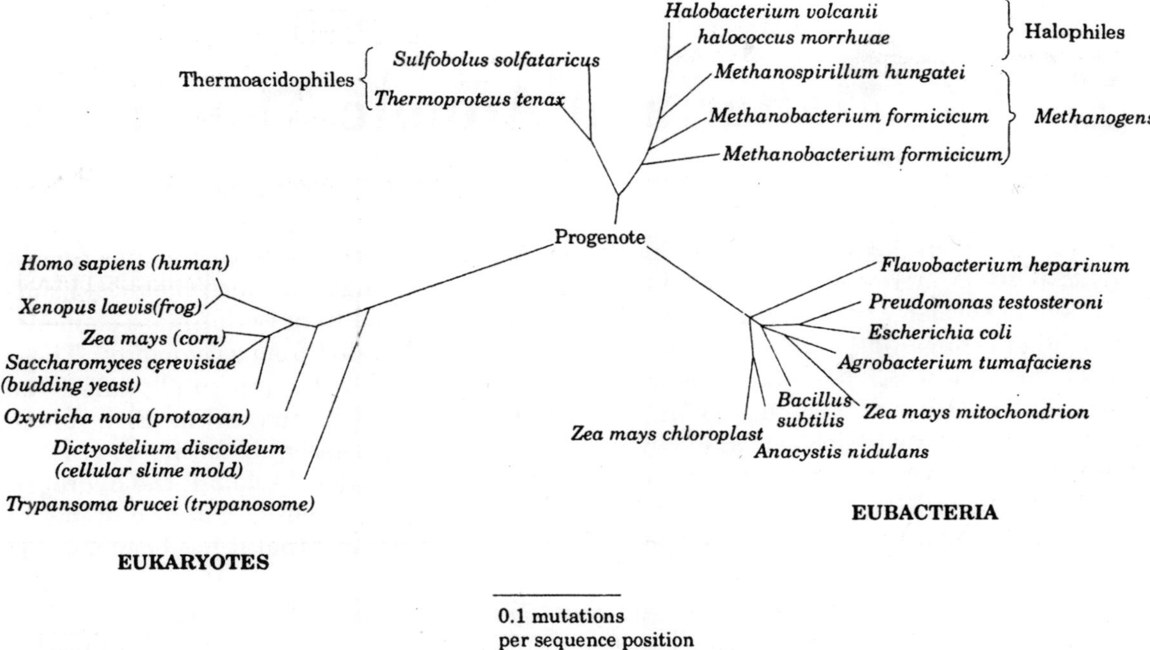

Fig. 1.21. An evolutionary tree can be constructed by comparing the complete sequences of 21 different 16S and l6S-like ribosomal RNAs (rRNAs). The scale bar represents the number of accumulated nucleotide differences (mutations) per sequence position in the rRNAs of the various organisms.

CONCLUDING REMARKS

In this chapter we discussed the ways in which biochemistry parallels ordinary chemistry and those in which it is quite different. The chief points to remember are the following.

1. The basic unit of life is the cell, which is a membrane-enclosed, microscopically visible object.
2. Cells are composed of small molecules, macromolecules, and organelles. The most prominent small molecule is water, which constitutes 70% of the cell by weight. Other small molecules are present only in quite small amounts; they are precursors or break-down products of macromolecules or coenzymes. There are four types of macromolecules: lipids, carbohydrates, proteins, and nucleic acids.
3. Noncovalent intermolecular forces largely determine the. folded structure adopted by a macromolecule, particularly the relative affinity of different groupings on the macromolecule for water. In general, hydrophobic groupings are buried within the folded macromolecular structure, whereas hydrophilic groupings are located on the surface, where they can interact with water.
4. Biochemical reactions utilize a limited number of elements, most prominently carbon, hydrogen, oxygen, nitrogen, sulfur, and phosphorus. Many biochemical reactions are simple organic reactions.

2 Matter and Atomic Theory

Matter exists in the form of either *elements* or *compounds*. Recall that compounds consist of atoms of two or more elements bonded together. Chemical changes in which chemical bonds are broken or formed are involved when a compound is produced from two or more elements or from other compounds, or when one or more elements are isolated from a compound. These kinds of changes, which are illustrated by the examples in Table 2.1, profoundly affect the properties of matter. For example, the first reaction in the table shows that sulfur (S), a yellow, crumbly solid, reacts with oxygen, a colourless, odorless gas that is essential for life, to produce sulfur dioxide, a toxic gas with a choking odor. In the second reaction Table 2.1, the hydrogen atoms in methane, CH_4, a highly flammable, light gas, are replaced by chlorine atoms to produce carbon tetrachloride, CCl_4, a dense liquid so non-flammable that it has been used as a fire extinguisher.

Table 2.1. Chemical Processes Involving Elements and Compounds

Type of Process	*Chemical Reaction*
Two elements combining to form a chemical compound, sulfur plus oxygen yield sulfur dioxide	$S + O_2 \rightarrow SO_2$
A compound reacting with an element to form two different compounds, methane plus cholrine yield carbon tetrachloride plus hydrogen chloride	$CH_4 + 4Cl_2 \rightarrow CCl_4 + 4HCl$
Two compounds reacting to form a different compound, calcium oxide plus water yield calcium hydroxide	$CaO + H_2O \rightarrow Ca(OH)_2$
A compound plus an element reacting to produce another compound and another pure element, iron oxide plus carbon yield carbon monoxide plus elemental iron (Fe)	$Fe_2O_3 + 3C \rightarrow 3CO + 2Fe$
A compound breaking down to its constituent elements, passing an electrical current through water yields hydrogen and oxygen	$2H_2O \rightarrow 2H_2 + O_2$

Some General Types of Matter

In discussing matter, some general terms are employed which are encountered frequently enough that they should be mentioned here. More exact definitions are given later in the text.

Elements are divided between metals and nonmetals. *Metals* are elements that are generally solid, shiny in appearance, electrically conducting, and malleable—that is, they can be pounded into flat sheets without disintegrating. Examples of metals are iron, copper, and silver. Most metallic objects that are commonly encountered are not composed of just one kind of elemental metal, but are alloys consisting of homogeneous mixtures of two or more metals. *Nonmetals* often have a dull appearance, are not at all malleable, and are frequently present as gases or liquids. Colourless oxygen gas, green chlorine gas (transported and stored as a liquid under pressure), and brown bromine liquid are common nonmetals. In a very general sense, we tend to regard as nonmetals substances that actually contain metal in a chemically combined form. One would classify table salt, sodium chloride, as a nonmetal even though it contains chemically-bound sodium metal.

Another general classification of matter is in the categories of inorganic and organic substances. *Organic substances* consist of virtually all compounds that contain carbon, including substances made by life processes (wood, flesh, cotton, wool), petroleum, natural gas (methane), solvents (dry cleaning fluids), synthetic fibers, and plastics. All of the rest of the chemical kingdom is composed of *inorganic substances* made up of virtually all substances that do not contain carbon. These include metals, rocks, table salt, water, sand, and concrete.

Mixtures and Pure Substances

Matter consisting of only one compound or of only one form of an element is a *pure substance*; all other matter is a *mixture* of two or more substances. The compound water with nothing dissolved in it is a pure substance. Highly purified helium gas is an elemental pure substance. The composition of a pure substance is defined and constant and its properties are always the same under specified conditions. Therefore, pure water at –1°C is always a solid (ice), whereas a mixture of water and salt might be either a liquid or a solid at that temperature, depending upon the amount of salt dissolved in the water. Air is a mixture of elemental gases and compounds, predominantly nitrogen, oxygen, argon, carbon dioxide, and water vapour. Drinking water is a mixture containing calcium ion (Ca^{2+}), hydrogen carbonate ion, (bicarbonate, $HCO3^{-}$), nitrogen gas, carbon dioxide gas, and other substances dissolved in the water. Mixtures can be separated into their constituent pure substances by *physical processes*. Liquefied air is distilled to isolate pure oxygen (used in welding, industrial processes, for some specialized wastewater treatment processes, and for breathing by people with emphysema), liquid nitrogen (used for quick-freezing frozen foods), and argon (used in specialized types of welding because of its chemically nonreactive nature).

Homogeneous and Heterogeneous Mixtures

A *heterogeneous mixture* is one that is not uniform throughout and possesses readily distinguishable constituents that can be isolated by means as simple as mechanical separation. Concrete is such a mixture; individual grains of sand and pieces of gravel may be separated from concrete with a pick or other tool. *Homogeneous mixtures* are uniform throughout; to observe different constituents would require going down to molecular levels. Homogeneous mixtures may have varying compositions and consist of two or more chemically distinct constituents, so they are not pure substances. Air, well filtered to remove particles of dust, pollen, or smoke is a homogeneous mixture. Homogeneous mixtures are also called *solutions*, a term that is usually applied to mixtures composed of gases, solids, and other liquids dissolved in a liquid. A hazardous waste leachate is a solution containing—in addition to water—contaminants such as dissolved acids, iron, heavy metals, and toxic organic compounds.

Whereas mechanical means including centrifugation and filtration can be used to separate the solids from the liquids in a heterogeneous mixture of solids and liquids, the isolation of components of homogeneous mixtures requires processes such as distillation and freezing.

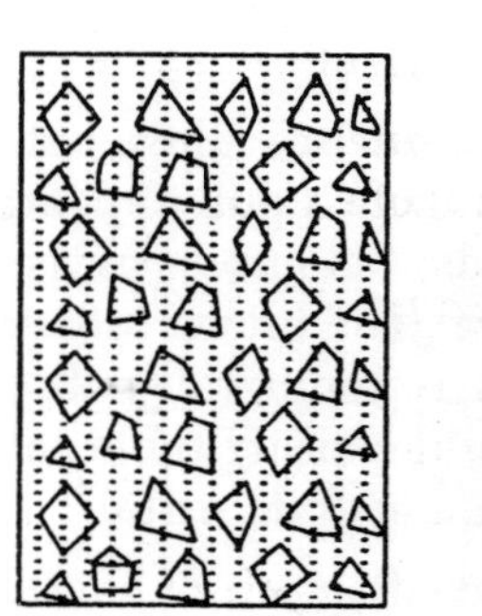

Fig. 2.1. Concrete (left) is a heterogeneous mixture with readily visible grains of sand, pieces of gravel, and cement dust. Gasoline is a homogeneous mixture consisting of a solution of petroleum hydrocarbon liquids, dye, and additives such as ethyl alcohol to improve its fuel qualities.

Summary of the Classification of Matter

As discussed above, all matter consists of compounds or elements. These may exist as pure substances or mixtures; the latter may be either homogeneous or heterogeneous. These relationships are summarized in Fig. 2.2.

QUANTITY OF MATTER: THE MOLE

One of the most fundamental characteristics of a specific mass of matter is the quantity of it. In discussing the quantitative chemical characteristics of matter it is essential to have a way of expressing quantity in a way that is proportional to the number of individual entities of the substance—that is, atoms, molecules, or ions—in numbers that are readily related to the properties of the atoms or molecules of the substance. The simplest way to do this would be as individual atoms, molecules, or ions, but for laboratory quantities these number in the order of 10^{23}, far too large to be used routinely in expressing quantities of matter. Instead, such quantities are readily expressed as moles of substance. A mole is defined in terms of specific entities, such as atoms of Ar, molecules of H_2O, or Na^+ and Cl^- ions, each pair of which composes a "molecule" of NaCl. A *mole* is defined as *the quantity of substance that contains the same number of specified entities as there are atoms of C in exactly* 0.012 *kg* (12 g) *of carbon*-12. It is easier to specify the quantity of a substance equivalent to its number of moles than it is to define the mole. To do so, simply state the atomic mass (of an element) or the molecular mass (of a compound) and affix "mole" to it as shown by the examples below:

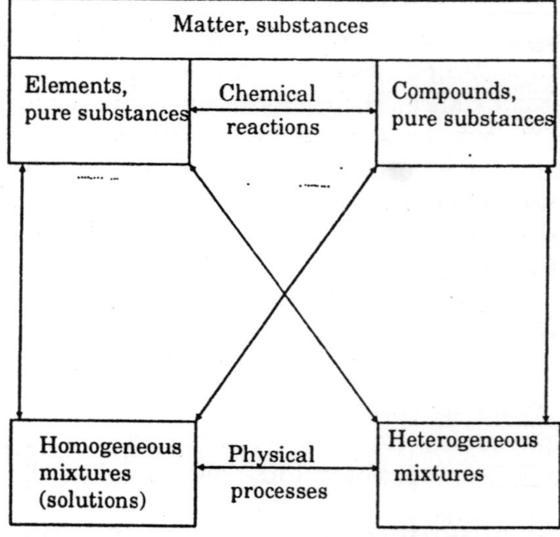

Fig. 2.2. Classification of matter.

- *A mole of argon, which always exists as individual Ar atoms*: The atomic mass of Ar is 40.0 Therefore exactly one mole of Ar is 40.0 grams of argon.

- *A mole of molecular elemental hydrogen, H_2:* The atomic mass of H is 1.0, the molecular mass of H_2 is, therefore, 2.0, and a mole of H_2 is 2.0 g of H_2.
- *A mole of methane, CH_4:* The atomic mass of H is 1.0 and that of C is 12.0, so the molecular mass of CH_4 is 16.0. Therefore a mole of methane has a mass of 16.0 g.

The Mole and Avogadro's Number

In Avogadro's number was mentioned as an example of a huge number, 6.02×10^{23}. *Avogadro's number is the number of specified entities in a mole of substance.* The "specified entities" may consist of atoms or molecules or they may be groups of ions making up the smallest possible unit of an ionic compound, such as 2 Na^+ ions and one S^{2-} ion in Na_2S. (It is not really correct to refer to Na_2S as a molecule because the compound consists of ions arranged in a crystalline structure so that there are 2 Na^+ ions for each S^{2-} ion.) A general term that covers all these possibilities is the formula unit. The average mass of a formula unit is called the formula mass. Examples of the terms defined are given in table 2.2.

Table 2.2. Relationships Involving Moles of Substance

Substance	*Formula Unit*	*Formula Mass*	*Mass of 1 mole*	*Numbers and Kinds of Individual Entities in 1 mole*
Helium	He atom	4.003*	4.003 g	6.02×10^{23} (Avogadro's number) of He atoms
Fluorine gas	F2 molecule	38.00**	38.00 g	6.02×10^{23} F_2 molecules $2 \times 6.02 \times 10^{23}$ F atoms
Methane	CH_4 molecules	16.04**	16.04 g	6.02×10^{23} CH_4 molecules 6.02×10^{23} C atoms $4 \times 6.02 \times 10^{23}$ H atoms
Sodium oxide	Na_2O	62.0 †	62.00 g	6.02×10^{23} Na_2O formula units $2 \times 6.02 \times 10^{23}$ Na^+ ions 6.02×10^{23} O^{2-}atoms

* Specifically, atomic mass.
** Specifically, molecular mass.
† Reference should be made to formula mass for this compound because it consists of Na^+ and O^{2-} ions in a ratio of 2/1.

PHYSICAL PROPERTIES OF MATTER

Physical properties of matter are those that can be measured without altering the chemical composition of the matter. A typical physical property is colour, which can be observed without changing matter at all. Malleability of metals, the degree to which they can be pounded into thin sheets, certainly alters the shape of an object but does not change it chemically. On the other hand, observation of a *chemical property,* such as whether or not sugar burns when ignited in air, potentially involves a complete change in the chemical composition of the substance tested.

Physical properties are important in describing and identifying particular kinds of matter. For example, if a substance is liquid at room temperature, has a lustrous metallic colour, conducts electricity well, and has a very high density (mass per unit volume) of 13.6 g/cm^3, it is doubtless elemental mercury metal. Physical properties are very useful in assessing the hazards and predicting the fates of environmental pollutants. An organic substance that readily forms a vapour (volatile organic compound) will tend to enter the atmosphere or to pose, an inhalation hazard. A brightly-coloured water soluble pollutant may cause deterioration of water quality by adding water "colour." Much of the health hazard of asbestos is its tendency to form extremely small diameter fibers that are readily carried far into the lungs and puncture individual cells.

Several important physical properties of matter are discussed in this section. The most commonly considered of these are density, colour, and solubility. Thermal properties are addressed separately.

Density

Density (*d*) is defined as mass per unit volume and is expressed by the formula

$$d = \frac{\text{mass}}{\text{volume}} \qquad ...(2.1)$$

Density is useful for identifying and characterizing pure substances and mixtures. For example, the density of a mixture of automobile system antifreeze and water, used to prevent the engine coolant from freezing in winter or boiling in summer, varies with the composition of the mixture. Its composition and, therefore, the degree of protection that it offers against freezing may be estimated by measuring its density and relating that through a table to the freezing temperature of the mixture.

Density may be expressed in any units of mass or volume. The densities of liquids and solids are normally given in units of grams per cubic centimeter (g/cm^3, the same as grams per milliliter, g/mL). These values are convenient because they are of the order of 1 g/cm^3 for common liquids and solids.

The volume of a given mass of substance varies with temperature, so the density is a function of temperature. This variation is relatively small for solids, greater for liquids, and very high for gases. Densities of gases vary a great deal with pressure, as well. The temperature-dependent variation of density is an important property of water and results in stratification of bodies of water, which greatly affects the environmental chemistry that occurs in lakes and reservoirs. The density of liquid water has a maximum value of 1.0000 g/mL at 4°C, is 0.9998 g/mL at 0°C, and is 0.9970 g/mL at 25°C. The combined temperature/pressure relationship for the density of air causes air to become stratified into layers, particularly the troposphere near the surface and the stratosphere from about 13 to 50 kilometers altitude.

Whereas liquids and solids have densities of the order of 1 to several g/cm^3, gases at atmospheric temperature and pressure have densities of only about one one-thousandth as much.

Table 2.3. Densities of Some Solids, Liquids, and Gases (at 1 atm Pressure and 0°C)

Solids	*d, g/cm^3*	*Liquids*	*d, g/cm^3*	*Gases*	*d· g/L*
Sugar	1.59	Benzene	0.879	Helium	0.178
Sodium chloride	2.16	Water (4°C)	1.0000	Methane	0.714
Iron	7.86	Carbon		Air	1.293
Lead	11.34	Tetrachloride	1.595	Chlorine	3.17
		Sulfuric acid	1.84		

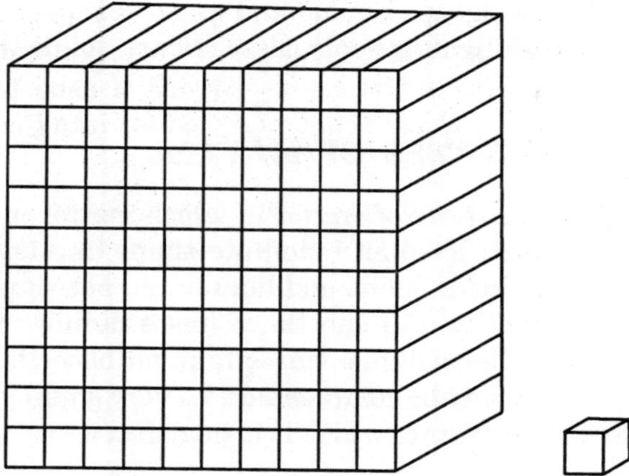

Fig. 2.3. A quantity of air that occupies a volume of 1000 cm³ of air at room temperature and atmospheric pressure (left) has about the same mass as 1 cm³ of liquid water (right).

Specific Gravity

Often densities are expressed by means of *specific gravity*, defined as the ratio of the density of a substance to that of a standard substance. For solids and liquids, the standard substance is usually water; for gases it is usually air. For example, the density of ethanol (ethyl alcohol) at 20°C is 0.7895 g/mL. The specific gravity of ethanol at 20°C referred to water at 4°C is given by

$$\text{Specific gravity} = \frac{\text{density of ethanol}}{\text{density of water}} = \frac{0.7895 \text{ g/mL}}{1.0000 \text{ g/ML}} = 0.7895 \qquad ...(2.2)$$

For an exact value of specific gravity, the temperatures of the substances should be specified. In this case the notation of "specific gravity of ethanol at 20°/4°C" shows that the specific gravity is the ratio of the density of ethanol at 20° C to that of water at 4°C.

Colour

Colour is one of the more useful properties for identifying substances without doing any chemical or physical tests. A violet vapor, for example, is characteristic of iodine. A red/brown gas is likely to be bromine or nitrogen dioxide (NO_2); a practiced eye can distinguish the two. Just a small amount of potassium permanganate, $KMnO_4$, in solution provides an intense purple colour. A characteristic yellow/brown colour in water may be indicative of organically-bound iron.

The human eye responds to colours of electromagnetic radiation ranging in wavelength from about 400 nanometers (nm) to somewhat over 700 nm. Within this wavelength range, humans see *light*; immediately below 400 nm is ultraviolet *radiation*, and somewhat above 700 nm is *infrared radiation*. Light with a mixture of wavelengths throughout the visible region, such as sunlight, appears white to the eye. Light over narrower wavelength regions has the following colours: 400-450 nm, blue; 490-550 nm, green; 550-580 nm, yellow; 580-650 nm, orange; 650 nm-upper limit of visible region, red. Solutions are coloured because of the light they absorb. Red, orange, and yellow solutions absorb violet and blue light; purple solutions absorb green and yellow light; and blue and green solutions absorb orange and red light. Solutions that do not absorb light are colourless (clear); solids that do not absorb light are white.

STATES OF MATTER

Figure 2.5 illustrates the three *states of matter* in which matter may exist. *Solids* have a definite shape and volume. *Liquids* have an indefinite shape and take on the shape of the container in which they are contained. Solids and liquids are not significantly compressible, which means that a specific quantity of a substance has a definite volume and cannot be squeezed into a significantly smaller volume. *Gases* take on both the shape and volume of their containers. A quantity of gas may be compressed to a very small volume and will expand to occupy the volume of any container into which it is introduced.

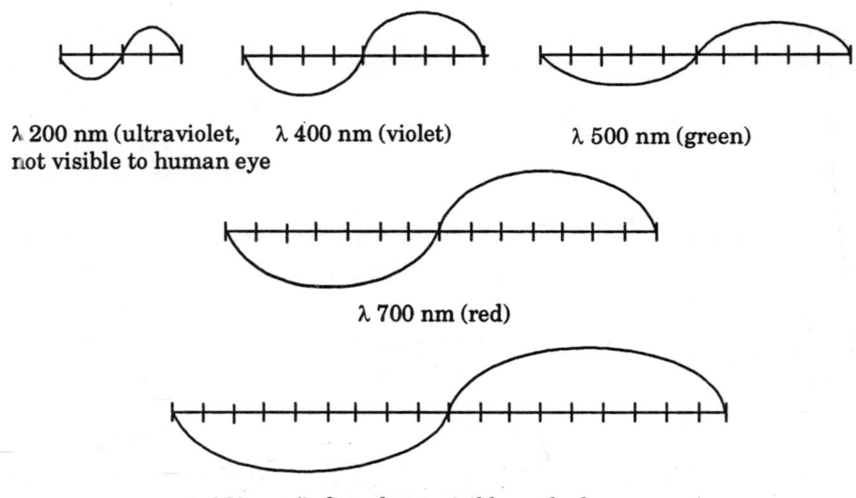

Fig. 2.4. Electromagnetic radiation of different wavelengths. Each division represents 50 nm.

Each of these separate phases is discussed in separate sections in this chapter. Everyone is familiar with the three states of matter for water. These are the following:

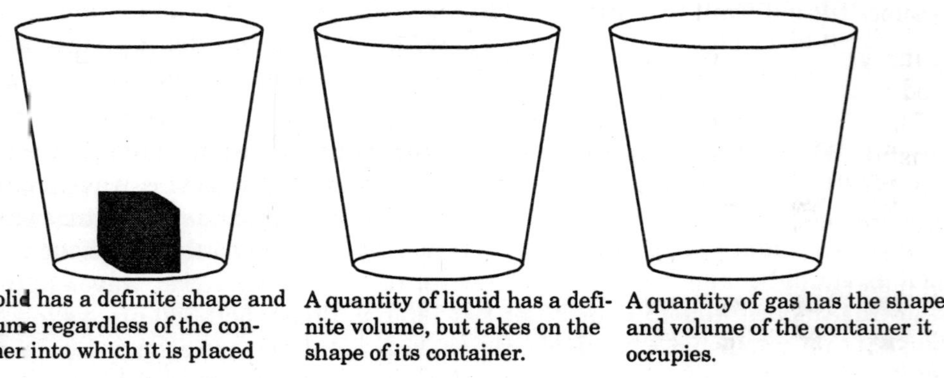

Fig. 2.5. Representations of the three states of matter.

- Gas: Water vapor in a humid atmosphere, steam
- Liquid: Water in a lake, groundwater
- Solid: Ice in polar ice caps, snow in snowpack

Changes in matter from one phase to another are very important in the environment. For example, water vapor changing from the gas phase to liquid results in cloud formation or precipitation. Water is desalinated by producing water vapor from sea water, leaving the solid salt behind, and recondensing the pure water vapor as a salt-free liquid. Some organic pollutants are extracted from water for chemical analysis by transferring them from the water to another organic phase that is immiscible with water. The condensation of organic pollutants from gaseous materials to solid products in the atmosphere is responsible for the visibility-reducing particulate matter in photochemical smog. Additional examples of phase changes are discussed some of the energy relationships involved are covered.

GASES

We live at the bottom of a "sea" of gas—the earth's atmosphere. It is composed of a mixture of gases, the most abundant of which are nitrogen, oxygen, argon, carbon dioxide, and water vapor. Although these gases have neither colour nor odor, so that we are not aware of their presence, they are crucial to the well-being of life on Earth. Deprived of oxygen, an animal loses consciousness and dies within a short time. Nitrogen extracted from the air is converted to chemically bound forms that are crucial to plant growth. Plants require carbon dioxide, and water vapor condenses to produce rain.

Gases and the atmosphere are addressed in more detail. At this point, however, it is important to have a basic understanding of the nature and behaviour of gases. Physically, gases are the "loosest" form of matter. A quantity of gas has neither a definite shape nor a definite volume so that it takes on the shape and volume of the container in which it is held. The reason for this behaviour is that gas molecules move independently and at random, bouncing off each other as they do so. They move very rapidly; at 0°C the average molecule of hydrogen gas moves at 3600 miles per hour, 6 to 7 times as fast as a passenger jet airplane. Gas molecules colliding with container walls exert *pressure*. An under-inflated automobile tire has a low pressure because there are relatively few gas molecules in the tire to collide with the container walls. As more air is added the pressure is increased and may become so high that it causes the tire wall to burst.

A quantity of a gas is mostly empty space, which is reflected in the high volume of gas compared to the same amount of material in a liquid or solid. For example, at 100°C a mole (18 g) of liquid water occupies a little more than 18 mL of volume, equivalent to just a few teaspoonsful. When enough heat energy is added to the water to convert it all to gaseous steam at 100°C, the volume becomes 30,600 mL, which is 1,700 times the original volume. This huge increase in volume occurs because of the large distances separating gas molecules compared to their own diameters. It is this great distance that allows gas to be compressed (pushed together).

The rapid, constant motion of gas molecules explains the phenomenon of gas *diffusion* in which gases move large distances from their sources. Diffusion is responsible for much of the hazard of volatile, flammable liquids, such as gasoline. Molecules of gasoline evaporated from an open container of this liquid can spread from their source. If the gaseous gasoline reaches an ignition source, such as an open flame, it may ignite and cause a fire or explosion.

The Gas Laws

To describe the physical and chemical behaviour of gases, it is essential to have a rudimentary understanding of the relationships among the following: Quantity of gas in numbers of moles, volume (V), temperature (T), and pressure (P). These relationships are expressed by the *gas laws*. The gas laws as discussed here apply to *ideal gases*, for which it is assumed that the molecules of gas have negligible volume, that there are no forces of attraction between them, and that they collide with each other in a perfectly elastic manner. Real gases such as H_2, O_2, He, and Ar behave in a nearly ideal fashion at moderate pressures and all but very low temperatures. The gas laws were derived from observations of the effects on gas volume of changing pressure, temperature, and quantity of gas, as stated by Boyle's law, Charles' law, and Avogadro's law, respectively. These three laws are defined below:

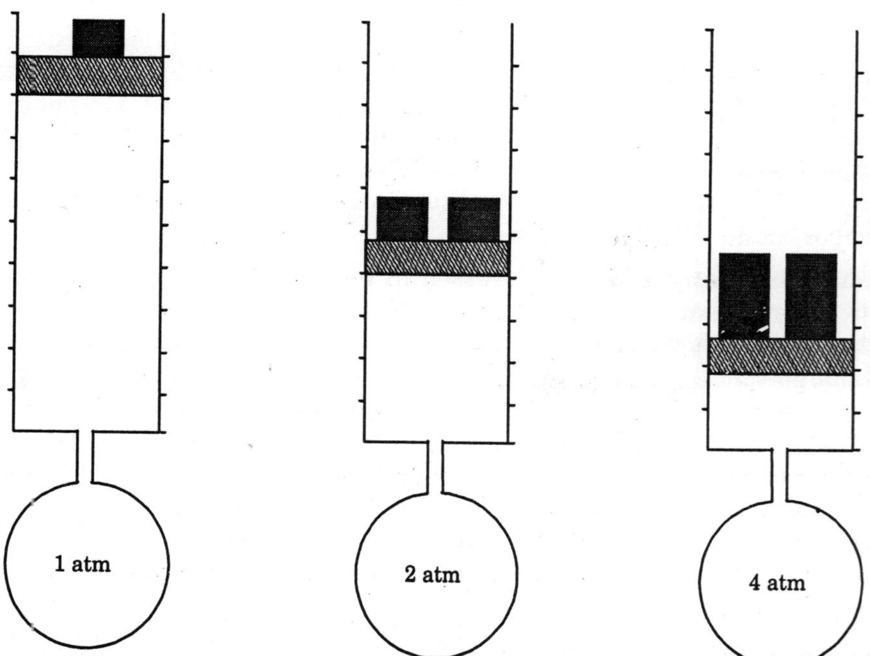

Fig. 2.6. In accordance with Boyle's law, doubling the pressure of a specific quantity of gas at a constant temperature makes its volume half as great.

Boyle's Law

As illustrated, in figure 2.6 doubling the pressure on a quantity of a gas halves its volume. This illustrates *Boyle's law*, according to which *at constant temperature, the volume of a fixed quantity of gas is inversely proportional to the pressure of the gas.* Boyle's law may be stated mathematically as

$$V = (\text{a constant}) \times \frac{1}{P} \qquad \text{...(2.3)}$$

This inverse relationship between the volume of a gas and its pressure is a consequence of the compressibility of matter in the gas phase (recall that solids and liquids are not significantly compressible).

Charles' Law

Charles' law gives the relationship between gas volume and temperature and states that *the volume of a fixed quantity of gas is directly proportional to the absolute temperature* (°C + 273) *at constant pressure.* This law may be stated mathematically as

$$V = (\text{a constant}) \times T \qquad ...(2.4)$$

where T is the temperature in K. According to this relationship, doubling the absolute temperature of a fixed quantity of gas at constant pressure doubles the volume.

Avogadro's Law

The third of three fundamentally important gas laws is *Avogadro's law* relating the volume of gas to its quantity in moles. This law states that *at constant temperature and pressure the volume of a gas is directly proportional to the number of molecules of gas, commonly expressed as moles.* Mathematically, this relationship is

$$V = (\text{a constant}) \times n \qquad ...(2.5)$$

where n is the number of moles of gas.

The General Gas Law

The three gas laws just defined may be combined into a ***general gas law*** stating that *the volume of a quantity of ideal gas is proportional to the number of moles of gas and its absolute temperature, and inversely proportional to 'its pressure.* Mathematically, this law is

$$V = (\text{a constant}) \times \frac{nT}{P} \qquad ...(2.6)$$

Designating the proportionality constant as the *ideal gas constant, R,* yields the *ideal gas equation*:

$$V = \frac{RnT}{P} \text{ or } PV = nRT \qquad ...(2.7)$$

The units of R depend upon the way in which the ideal gas equation is used. For calculations involving volume in liters and pressures in atmospheres, the value of R is 0.0821 L-atm/deg-mol.

The ideal gas equation shots that at a chosen temperature and pressure a mole of any gas should occupy the same volume. A temperature of 0°C (273.15 K) and 1 atm pressure have been chosen as *standard temperature and pressure (STP). At STP the volume of 1 mole of ideal gas is 22.4 L.* This volume is called the molar volume of a gas. It should be remembered that at a temperature of 273 K and a pressure of 1 atm the volume of 1 mole of ideal gas is 22.4 L. Knowing these values, it is always possible to calculate the value of R as follows:

$$R = \frac{PV}{nT} = \frac{1 \text{ atm} \times 22.4 \text{ L}}{1 \text{ mole} \times 273 \text{K}} = 0.0821 \frac{\text{L} \times \text{atm}}{\text{deg} \times \text{mol}} \qquad ...(2.8)$$

Gas Law Calculations

A common calculation is that of the volume of a gas, starting with a particular volume and changing temperature and/or pressure. This kind of calculation follows logically from the ideal gas equation, but can also be reasoned out knowing that an increase in temperature causes an increase in volume, whereas an increase in pressure causes a decrease in volume. Therefore, a second volume, V_2, is calculated from an initial volume, V_1, by the relationship

$$V_2 = V_1 \times \frac{\square}{\square} \times \frac{\square}{\square} \qquad \text{...(2.9)}$$

Ratio of temperatures   Ratio of pressures

This relationship requires only that one remember the following:

- If the temperature increases (if T_2 is greater than T_1), the volume increases. Therefore, T_2 is always placed over T_1 in the ratio of temperatures.
- If the pressure increases (if P_2 is greater than P_1), the volume decreases. Therefore, P_1 is always placed over P_2 in the ratio of pressures.
- If there is no change in temperature or pressure, the corresponding ratio remains 1.

These relationships can be illustrated by several examples.

Charles' Law Calculation

When the temperature of a quantity of gas changes while the pressure stays the same, the resulting calculation is a *Charles' law calculation.* As an example of such a calculation, calculate the volume of a gas with an initial volume of 10.0 L when the temperature changes from –11.0°C to 95.0°C at constant pressure. The first step in solving any gas law problem involving a temperature change is to convert Celsius temperatures to kelvin:

$$T_1 = 273 + (-11) = 262 \text{ K} \qquad T_2 = 273 + 95 = 368 \text{ K}$$

Substitution into Equation 2.9 gives

$$V_2 = V_1 \times \frac{368 \text{ K}}{262 \text{ K}} \times 1 \text{ (factor for pressure = 1 because } P_2 = P_1) \qquad \text{...(2.10)}$$

$$V_2 = 10.0 \text{ L} \times \frac{368 \text{ K}}{262 \text{ K}} = 14.1 \text{ L}$$

Note that in this calculation it is seen that the temperature increases; this increases the volume, so the higher temperature is placed over the lower. Furthermore, since $P_2 = P_1$, the factor for the pressure ratio simply drops out.

Calculate next the volume of a gas with an initial volume of 10.0 L after the temperature decreases from 111.0°C to 2.0°C at constant pressure:

$$T_1 = 273 + (111) = 384 \text{ K} \qquad T_2 = 273 + 2 = 275 \text{ K}$$

Substitution into Equation 2.9 gives

$$V_2 = V_1 \times \frac{275 \text{ K}}{384 \text{ K}} = 10.0 \text{ L} \times \frac{275 \text{ K}}{384 \text{ K}} \qquad \text{...(2.11)}$$

$$= 7.16 \text{ L}$$

In this calculation the temperature decreases; this decreases the volume, so the lower temperature is placed over the higher (mathematically it is still T_2 over T_1 regardless of which way the temperature changes).

Boyle's Law Calculation

When the pressure of a quantity of gas changes while the temperature stays the same, the resulting calculation is a *Boyle's law calculation.* As an example, calculate the new volume that results when the pressure of a quantity of gas initially occupying 12.0 L is changed from 0.856 atm to 1.27 atm at constant temperature. In this case, the pressure has increased; this decreases the volume of the gas so that V_2 is given by the following:

$$V_2 = V_1 \times \frac{0.856 \text{ atm}}{1.27 \text{ atm}} \times 1(\text{because } T_2 = T_1) = 12.0 \text{ L} \times \frac{0.856 \text{ atm}}{1.27 \text{ atm}} \quad ...(2.12)$$

$$= 8.09 \text{ L}$$

Calculate next the volume of a gas with an initial volume of 15.0 L when the pressure decreases from 1.71 atm to 1.07 atm at constant temperature. In this case the pressure decreases; this increases the volume, so the appropriate factor to multiply by is 1.71/1.07:

$$V_2 = V_1 \times \frac{1.71 \text{ atm}}{1.07 \text{ atm}} = 15.0 \text{ L} \times \frac{1.71 \text{ atm}}{1.07 \text{ atm}} = 24.0 \text{ L} \quad ...(2.13)$$

Calculations Using the Ideal Gas Law

The ideal gas law equation

$$\text{PV} = \text{nRT} \quad ...(2.14)$$

may be used to calculate any of the variables in it if the others are known. As an example, calculate the volume of 0.333 moles of gas at 300 K under a pressure of 0.950 atm:

$$V = \frac{\text{nRT}}{\text{P}} = \frac{0.333 \text{ mol} \times 0.0821 \text{ L atm/K mol} \times 300 \text{ K}}{0.950 \text{ atm}} = 8.63 \text{ L} \quad ...(2.15)$$

As another example calculate the temperature of 2.50 mol of gas that occupies a volume of 52.6 L under a pressure of 1.15 atm:

$$T = \frac{\text{PV}}{\text{nR}} = \frac{1.15 \text{ atm} \times 52.6 \text{ L}}{2.50 \text{ mol} \times 0.0821 \text{ L atm/K mol}} = 295 \text{ K} \quad ...(2.16)$$

Other examples of gas law calculations are given in the problem section at the end of this chapter.

The changes in a gas resulting from variations in one or more of the parameters of P, V, n, or T can be calculated by the relationship,

$$\frac{P_2V_2}{n_2T_2} = \frac{P_1V_1}{n_1T_1} \quad ...(2.17)$$

derived from the fact that R is a constant, where the subscripts 1 and 2 denote values of the parameters designated before and after these are changed. By holding the appropriate parameters constant, this equation can be used to derive the mathematical relationships used in calculations involving Boyle's law, Charles' law, or Avogadro's law, as well as any combination of these laws.

Exercise: Derive an equation that will give the volume of a gas after all three of the parameters P, T, and n have been changed.

Answer: Rearrangement of Equation 2.17 gives the relationship needed:

$$V_2 = V_1 \frac{P_1 n_2 T_2}{P_2 n_1 T_1} \quad ...(2.18)$$

by holding the appropriate parameters constant, this equation can be used to give mathematical expressions for Boyle's law, Charles' law, or Avogadro's law.

LIQUIDS AND SOLUTIONS

Whereas the molecules in gases are relatively far apart and have minimal attractive forces between them (none in the case of ideal gases, which are discussed in the preceding section), molecules of liquids are close enough together that they may be regarded as "touching" and are strongly attracted to each other. Like those of gases, however, the molecules of liquids move freely relative to each other. These characteristics give rise to several significant properties of liquids. Whereas gases are mostly empty space, most of the volume of a liquid is occupied by molecules; therefore, the densities of liquids are much higher than those of gases. Because there is little unoccupied space between molecules of liquid, liquids are only slightly *compressible;* doubling the pressure on a liquid causes a barely perceptible change in volume, whereas it halves the volume of an ideal gas. Liquids do not expand to fill their containers as do gases. Essentially, therefore, a given quantity of liquid occupies a fixed volume. Because of the free movement of molecules relative to each other in a liquid, it takes on the shape of that portion of the container that it occupies.

Evaporation and Condensation of Liquids

Consider the surface of a body of water as illustrated in Fig. 2.7. The molecules of liquid water are in constant motion relative to each other. They have a distribution of energies such that there are more higher energy molecules at higher temperatures. Some molecules are energetic enough that they escape the attractive forces of the other molecules in the mass of liquid and enter the gas phase. This phenomenon is called *evaporation.*

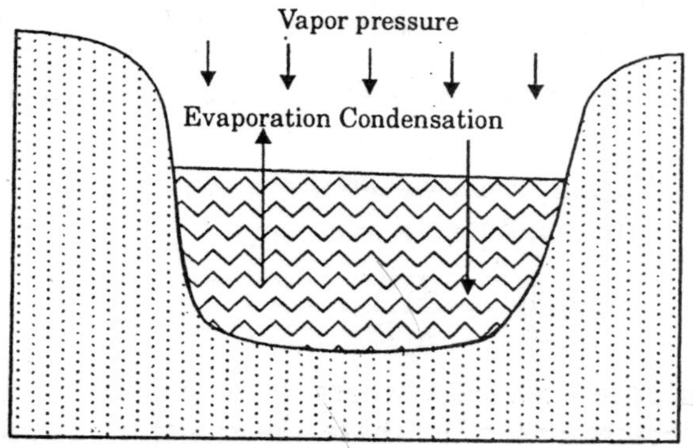

Fig. 2.7. Evaporation and condensation of liquid.

Molecules of liquid in the gas phase, such as water vapor molecules above a body of water, may come together or strike the surface of the liquid and re-enter the liquid phase. This process is called *condensation*. It is what happens, for example, when water vapor molecules in the atmosphere strike the surface of very small water droplets in clouds and enter the liquid water phase, causing the droplets to grow to sufficient size to fall as precipitation.

Vapor Pressure

In a confined area above a liquid, equilibrium is established between the evaporation and condensation of molecules from the liquid. For a given temperature, this results in a steady-state level of the vapor, which can be described as a pressure. Such a pressure is called the *vapor pressure* of the liquid.

Vapor pressure is very important in determining the fates and effects of substances, including hazardous substances, in the environment. Loss of a liquid by evaporation to the atmosphere increases with increasing vapor pressure and is an important mechanism by which volatile pollutants enter the atmosphere. High vapor pressure of a flammable liquid can result in the formation of explosive mixtures of vapor above a liquid, such as in a storage tank.

Solutions

Solutions were mentioned homogeneous mixtures. Here solutions are regarded as liquids in which quantities of gases, solids, or other liquids are dispersed as individual ions or molecules. Rainwater, for example, is a solution containing molecules of N_2, O_2 and CO_2 from air dispersed among the water molecules. Rainwater from polluted air may contain harmful substances such as sulfuric acid (H_2SO_4) in small quantities among the water molecules. The predominant liquid constituent of a solution is called the *solvent*. A substance dispersed in it is said to be *dissolved* in the solvent and is called the *solute*.

Solubility

Mixing and stirring some solids, such as sugar, with water results in a rapid and obvious change in which all or part of the substance seems to disappear as it dissolves. When this happens, the substance is said to *dissolve,* and one that does so to a significant extent is said to be *soluble* in the solvent. Substances that do not dissolve to a perceptible degree are called *insoluble*. *Solubility* is a significant physical property. It is a relative term. For example, Teflon is insoluble in water, limestone is slightly soluble, and ethanol (ethyl alcohol) is infinitely so in that it may be mixed with water in any proportion ranging from pure water to pure ethanol with only a single liquid phase consisting of an ethanol/water solution being observed.

Solution Concentration

Solutions are addressed specifically and in more detail. At this point, however, it is useful to have some understanding of the quantitative expression of the amount of solute dissolved in a specified amount of solution. In a relative sense, a solution that has comparatively little solute dissolved per unit volume of solution is called a *dilute solution*, whereas one with an amount of solute of the same order of magnitude as that of the solvent is a concentrated solution. Water saturated with air at 25°C contains only about 8 milligrams of oxygen dissolved in 1 liter of water; this composes a *dilute solution* of oxygen. A typical engine coolant solution contains about as much ethylene glycol (anti-freeze) as it does water, so it is a *concentrated solution.* A solution that is at equilibrium with excess solute so that it contains the maximum amount of solute that it can dissolve is called a *saturated solution.* One that can still dissolve more solute is called an *unsaturated solution.*

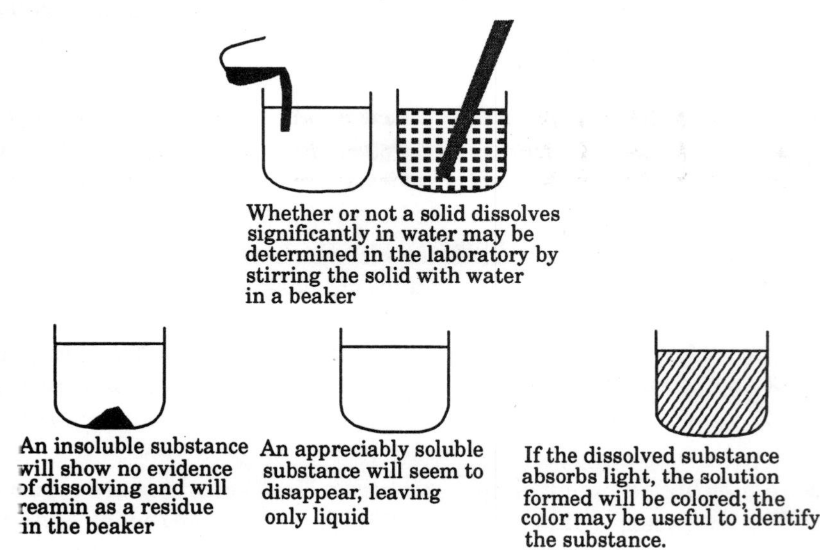

Fig. 2.8. Illustration of whether or not a substance dissolves in a liquid.

For the chemist, the most useful way to express the concentration of a solution is in terms of the number of moles of solute dissolved per liter of solution. The *molar concentration, M,* of a solution is *the number of moles of solute dissolved per liter of solution.* Mathematically,

$$M = \frac{\text{moles of solute}}{\text{number of liters of solution}} \quad \text{...(2.19)}$$

Example: Exactly 34.0 g of ammonia, NH_3, were dissolved in water and the solution was made up to a volume of exactly 0.500 L. What was the molar concentration, *M*, of ammonia in the resulting solution?

Answer: The molar mass of NH_3 is 17.0 g/mole. Therefore

$$\text{Number of moles of } NH_3 = \frac{34.0 \text{ g}}{17.0 \text{ g/mole}} = 2.00 \text{ mol} \quad \text{...(2.20)}$$

The molar concentration of the solution is calculated from Equation 2.19:

$$M = \frac{2.00 \text{ mol}}{0.500 \text{ L}} = 4.00 \text{ mol/L} \quad \text{...(2.21)}$$

Equations 2.19 and 2.20 are useful in doing calculations that relate mass of solute to solution volume and solution concentration in chemistry.

SOLIDS

Matter in the form of *solids* is said to be in the *solid state.* The *solid state* is the most organized form of matter in that the atoms, molecules, and ions in it are in essentially fixed relative positions and are highly attracted to each other. Therefore, solids have a definite shape, maintain a constant volume, are virtually non-compressible under pressure, and expand and contract only slightly with changes in temperature. Like liquids, solids have generally very high densities relative to gases. (Some solids such as those made from styrofoam appear to have very low densities, but that is because such materials are composed mostly of bubbles of air in a solid matrix.) Because of the strong attraction of the atoms, molecules, and ions of solids for each other, solids do not enter the vapor phase very readily at all; the phenomenon by which this happens to a limited extent is called *sublimation*.

Some solids (quartz, sodium chloride) have very well defined geometric shapes and form characteristic crystals. These are called *crystalline solids* and occur because the molecules or ions of which they are composed assume well defined, specific positions relative to each other. Other solids have indefinite shapes because their constituents are arranged at random; glass is such a solid. These are *amorphous solids*.

THERMAL PROPERTIES

The behaviour of a substance when heated or cooled defines several important physical properties of it. These include the temperature at which it melts or vaporizes. Also included are the amounts of heat required by a given mass of the substance to raise its temperature by a unit of temperature, to melt it, or to vaporize it.

Melting Point

The *melting point* of a pure substance is the temperature at which the substance changes from a solid to a liquid. At the melting temperature pure solid and pure liquid composed of the substance may be present together in a state of equilibrium. An impure substance does not have a single melting temperature. Instead, as the substance is melted by application of heat, it begins to melt at a temperature below the melting temperature of the pure substance; melting proceeds as the temperature increases until no solid remains. Therefore, melting temperature measurements serve two purposes in characterizing a substance. The melting point of a pure substance is indicative of the identity of the substance. The melting behaviour—whether melting occurs at a single temperature or over a temperature range—is a measure of substance purity.

Boiling Point

Boiling occurs when a liquid-is heated to a temperature so that bubbles of vapor of the substance are evolved. The boiling temperature depends upon pressure. When the surface of a pure liquid substance is in contact with the pure vapor of the substance at 1 atm pressure, boiling occurs at a temperature called the *normal boiling point*. Whereas a pure liquid remains at a constant temperature during boiling until it has all turned to vapor, the temperature of an impure liquid increases during boiling. As with melting point, the boiling point is useful for identifying liquids. The extent to which the boiling temperature is constant as the liquid is converted to vapor is a measure of the purity of the liquid.

Specific Heat

As the temperature of a substance is raised, energy must be put into it to enable the molecules of the substance to move more rapidly relative to each other and to overcome the attractive forces between them. The amount of heat energy required to raise the temperature of a unit mass of a solid or liquid substance by a degree of temperature varies with the substance. Of the common liquids, the most heat is required to raise the temperature of water. This is because of the molecular structure of water, the molecules of which are strongly attracted to each other by electrical charges between the molecules (each of which has a relatively positive and a relatively more negative end). Furthermore, water molecules tend to be held in molecular networks by special kinds of chemical bonds called hydrogen bonds, in which hydrogen atoms form bridges between O atoms in different water molecules. The very high amount of energy required to increase the temperature of water has some important environmental implications because it stabilizes the temperatures of bodies of water and of geographic areas close to bodies of water. In contrast to water, a hydrocarbon liquid such as those in gasoline requires relatively little heat for warming because the molecules interact much less with each other than do those of water.

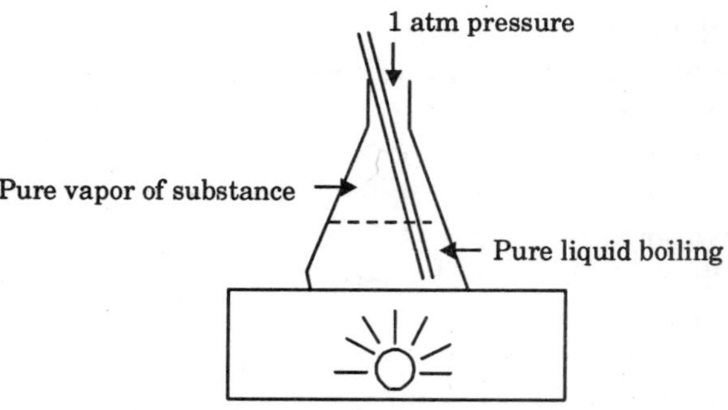

Fig. 2.9. Illustration of normal boiling liquid.

The heat required to raise the temperature of a substance is called the *specific heat* defined as the amount of heat energy required to raise the temperature of a gram of substance by 1 degree Celsius. It can be expressed by the equation

$$\text{Secific heat} = \frac{\text{heat energy absorbed, } J}{(\text{mass, g})\ (\text{increase in temperature, °C})} \quad ...(2.22)$$

where the heat energy is in units of joule, J. The most important value of specific heat is that of water, 4.18 J/g°C. From Equation 2.22, the amount of heat, q, required to raise the mass, m, of a particular substance over a temperature range of ΔT is

$$q = (\text{specific heat}) \times m \times \Delta T \quad ...(2.23)$$

This equation can be used to calculate quantities of heat used in increasing water temperature or released when water temperature decreases. As an example, consider the amount of heat required to raise the temperature of 11.6 g of liquid water from 14.3°C to 21.8°C:

$$q = 4.18\ \frac{\text{J}}{\text{g°C}} \times 11.6\ \text{g} \times (21.8\text{°C} - 14.3\ \text{°C}) = 364\ \text{J} \quad ...(2.24)$$

Heat of Vaporization

Large amounts of heat energy may be required to convert liquid to vapor compared to the specific heat of the liquid. This is because changing a liquid to a gas requires breaking the molecules of liquid away from each other, not just increasing the rates at which they move relative to one another. The vaporization of a pure liquid occurs at a constant temperature (normal boiling temperature, see above). For example, the temperature of a quantity of pure water is increased by heating to 100°C where it remains until all the water has evaporated. The *heat of vaporization* is the quantity of heat taken up in converting a unit mass of liquid entirely to vapor at a constant temperature. The heat of vaporization of water is 2,260 J/g

(2.26 kJ/g) for water boiling at 100°C at 1 atm pressure. This amount of heat energy is about 540 times that required to raise the temperature of 1 gram of liquid water by 1°C. The fact that water's heat of vaporization is the largest of any common liquid has significant environmental effects. It means that enormous amounts of heat are required to produce water vapor from liquid water in bodies of water. This means that the temperature of a body of water is relatively stable when it is exposed to high temperatures because so much heat is dissipated when a fraction of the water evaporates. When water vapor condenses, similar enormous amounts of heat energy called *heat of condensation* are released. This occurs when water vapor forms precipitation in storm clouds and is the driving force behind the tremendous releases that occur in thunderstorms and hurricanes.

The heat of vaporization of water can be used to calculate the heat required to evaporate a quantity of water. For example, the heat, q, required to evaporate 2.50 g of liquid water is

$$q = \text{(heat of vaporization) (mass of water)} = -2.26 \text{ kJ/g} \times 2.50 \text{ g} = 5.65 \text{ kJ} \quad (2.25)$$

and that released when 5.00 g of water vapor condenses is

$$q = -\text{(heat of vaporization) (mass of water)} = 2.26 \text{ kJ/g} \times 5.00 \text{ g} = -11.3 \text{ kJ} \quad (2.26)$$

the latter value is negative to express the fact that heat is released.

Heat of Fusion

Heat of fusion is the quantity of heat taken up in converting a unit mass of solid entirely to liquid at a constant temperature. The heat of fusion of water is 330 J/g for ice melting at 0°C. This amount of heat energy is 80 times that required to raise the temperature of 1 gram of liquid water by 1°C. When liquid water freezes, an exactly equal amount of energy per unit mass is released, but is denoted as a negative value.

The heat of fusion of water can be used to calculate the heat required to melt a quantity of ice. For example, the heat, q, required to melt 2.50 g of ice is

$$q = \text{(heat of fusion)(mass of water)} = 330 \text{ J/g} \times 2.50 \text{ g} = 825 \text{ J} \quad ...(2.27)$$

and that released when 5.00 g of liquid water freezes is

$$q = -\text{(heat of fusion) (mass of water)} = -330 \text{J/g} \times 5.00 \text{ g} = -1{,}650 \text{ J} \quad ...(2.28)$$

SEPARATION AND CHARACTERIZATION OF MATTER

A very important aspect of the understanding of matter is the separation of mixtures of matter into their constituent pure substances. Consider, for example, the treatment of an uncharacterized waste sludge found improperly disposed in barrels. Typically, the sludge will appear as a black goo with an obnoxious odor from several constituents. In order to do a proper analysis of this material, it often needs to be separated into its constituent components, such as by extracting water-soluble substances into water, extracting organic materials into an organic solvent, and distilling off volatile constituents. The materials thus isolated can be analyzed to determine what is present. With this knowledge and with information about the separation itself a strategy can be devised to treat the waste in an effective, economical manner.

Many kinds of processes are used for separations, and it is beyond the scope of this chapter to go into detail about them here. However, several of the more important separation operations are discussed briefly below.

Distillation

Distillation consists of evaporating a liquid by heating and cooling the vapor so that it condenses back to the liquid in a different container, as shown for water in figure 2.11. Less volatile constituents such as solids are left behind in the distillation flask; more volatile impurities such as volatile organic compound pollutants in water will distill off first and can be placed in different containers before the major liquid constituent is collected.

The apparatus shown in Fig. 2.10 could be used, for example, to prepare fresh water from seawater, which may be considered as a solution of sodium chloride in water. The seawater to be purified is placed in a round-bottom distillation flask heated with an electrically powered heating mantle. The seawater is boiled and pure water vapor in the gaseous state flows into the condenser where it is cooled and condensed back to a purified salt-free product in the receiving flask. When most of the seawater has been evaporated, the distillation flask and its contents are cooled and part of the sodium chloride separates out as crystals that can be removed. Sophisticated versions of this distillation process are used in some arid regions of the world to produce potable (drinking) water from seawater.

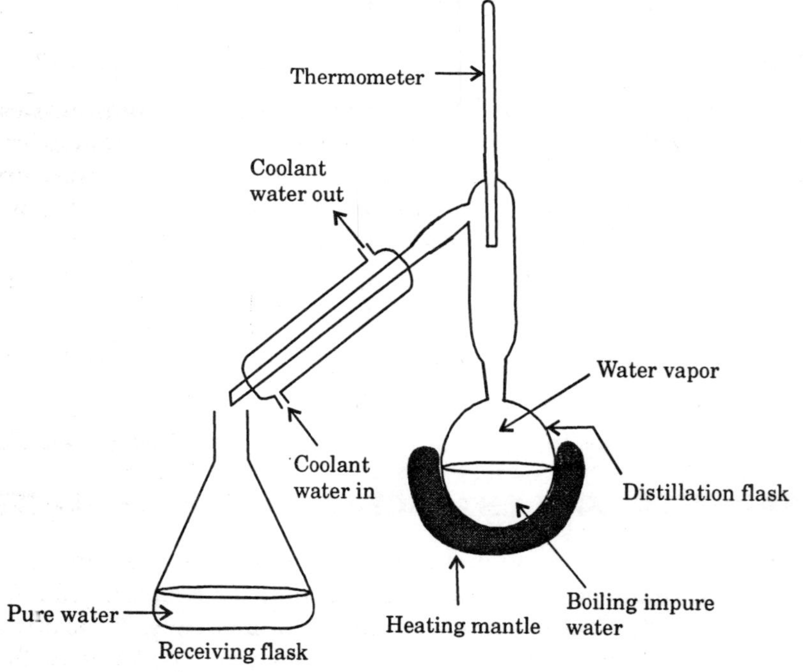

Fig. 2.10. Separation by distillation.

Distillation is widely used in the petroleum and chemical industries. A sophisticated distillation apparatus is used to separate the numerous organic components of crude oil, including petroleum ether, gasoline, kerosene, and jet fuel fractions. This requires *fractional distillation* in which part of the vapor recondenses in a *fractionating column* which would be mounted vertically on top of the distillation flask in the apparatus shown in Fig. 2.10. The most volatile components enter into the condenser and are condensed back to liquid products;

less volatile liquid constituents return to the distillation flask. The net effect is that of numerous distillations, which gives a very efficient separation.

In some cases the residues left from distillation, *distillation bottoms* ("still bottoms") are waste materials. Several important categories of hazardous wastes consist of distillation bottoms, which pose disposal problems.

Separation in Waste Treatment

In addition to distillation, several other kinds of separation procedures are used for waste treatment. These are addressed in more detail and are mentioned here as examples of separation processes.

Phase Transitions

Distillation is an example of a separation in which one of the constituents being separated undergoes a *phase transition* from one phase to another, then often back again to the same phase. In distillation, for example, a liquid is converted to the vapor state, then recondensed as a liquid. Several other kinds of separations by phase transition as applied to waste treatment are the following:

Evaporation is usually employed to remove water from an aqueous waste to concentrate it. A special case of this technique is *thin-film evaporation* in which volatile constituents are removed by heating a thin layer of liquid or sludge waste spread on a heated surface.

Drying—removal of solvent or water from a solid or semisolid (sludge) or the removal of solvent from a liquid or suspension—is a very important operation because water is often the major constituent of waste products, such as sludges. In *freeze drying*, the solvent, usually water, is sublimed from a frozen material. Hazardous waste solids and sludges are dried to reduce the quantity of waste, to remove solvent or water that might interfere with subsequent treatment processes, and to remove hazardous volatile constituents.

Stripping is a means of separating volatile components from less volatile ones in a liquid mixture by the partitioning of the more volatile materials to a gas phase of air or steam (steam stripping). The gas phase is introduced into the aqueous solution or suspension containing the waste in a stripping tower that is equipped with trays or packed to provide maximum turbulence and contact between the liquid and gas phases. The two major products are condensed vapor and a stripped bottoms residue. Examples of two volatile components that can be removed from water by air stripping are the organic solvents benzene and dichloromethane.

Physical precipitation is used here as a term to describe processes in which a solid forms from a solute in solution as a result of a physical change in the solution. The major changes that can cause physical precipitation are cooling the solution, evaporation of solvent, or alteration of solvent composition. The most common type of physical precipitation by alteration of solvent composition occurs when a water-miscible organic solvent is added to an aqueous (water) solution of a salt so that the solubility of a salt is lowered below its concentration in the solution.

Phase Transfer

Phase transfer consists of the transfer of a solute in a mixture from one phase to another. An important type of phase transfer process is *solvent extraction*, a process in which a substance is transferred from solution in one solvent (usually water) to another (usually an organic solvent) without any chemical change taking place. When solvents are used to leach substances from solids or sludges, the process is called *leaching*. One of the more promising approaches to solvent extraction and leaching of hazardous wastes is the use of *supercritical fluids*, most commonly CO_2, as extraction solvents. A supercritical fluid is one that has characteristics of both liquid and gas and consists of a substance above its supercritical temperature and pressure (31.1°C and 73.8 atm, respectively, for CO_2). After a substance has been extracted from a waste into a supercritical fluid at high pressure, the pressure can be released, resulting in separation of the substance extracted. The fluid can then be compressed again and recirculated through the extraction system. Some possibilities for treatment of hazardous wastes by extraction with supercritical CO_2 include removal of organic contaminants from wastewater, extraction of organohalide pesticides from soil, extraction of oil from emulsions used in aluminum and steel processing, and regeneration of spent activated carbon.

Transfer of a substance from a solution to a solid phase is called *sorption*. The most important sorbent used in waste treatment is *activated carbon*. It can also be applied to pretreatment of waste streams going into other processes to improve treatment efficiency and reduce fouling. Activated carbon sorption is most effective for removing from water organic materials that are poorly water soluble and that have high molecular masses.

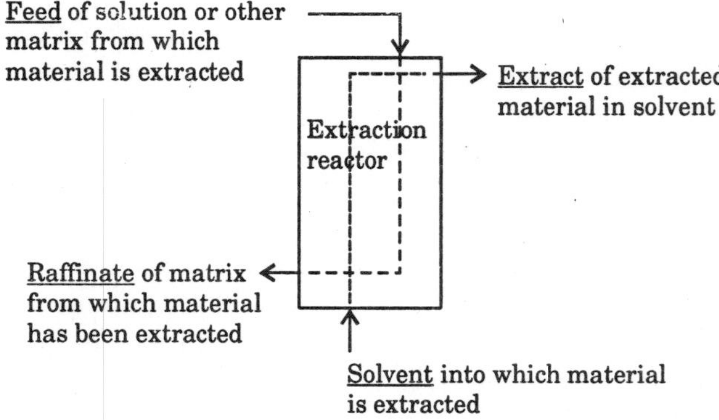

Fig. 2.11. Outline of solvent extraction/leaching process with important terms underlined.

Molecular Separation

A third major class of physical separation is *molecular separation*, often based upon *membrane processes* in which dissolved contaminants or solvent pass through a size-selective membrane under pressure. *Reverse osmosis* is the most widely used of the membrane techniques. It operates by virtue of a membrane that is selectively permeable to water and excludes ionic solutes. Reverse osmosis uses high pressures to force permeate through the membrane, producing a concentrate containing high levels of dissolved salts.

ATOMIC THEORY

The nature of atoms in relation to chemical behaviour is summarized in the *atomic theory*. This theory in essentially its modern form was advanced in 1808 by John Dalton, an English schoolteacher, taking advantage of a substantial body of chemical knowledge and the contributions of others. It has done more than any other concept to place chemistry on a sound, systematic theoretical foundation. Those parts of the atomic theory that are consistent with current understanding of atoms summarized in Fig. 2.12.

Dalton's Atomic Theory

The atomic theory outlined in Fig. 2.12 explains the following three laws that are of fundamental importance in chemistry :

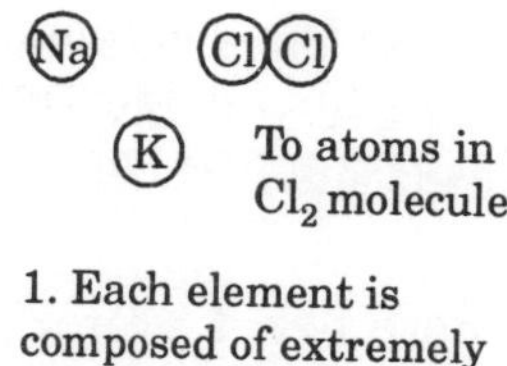

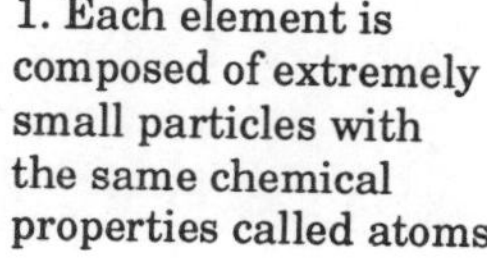

1. Each element is composed of extremely small particles with the same chemical properties called atoms

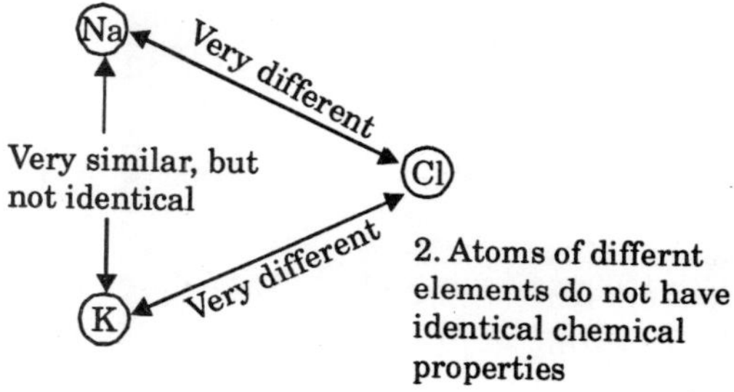

2. Atoms of differnt elements do not have identical chemical properties

3. Chemical compounds are formed by the combination of atoms of different elements in definite,constant ratios that usually can be expressed as integers or simple fractions.

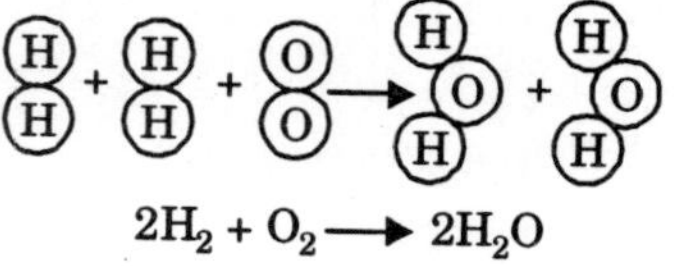

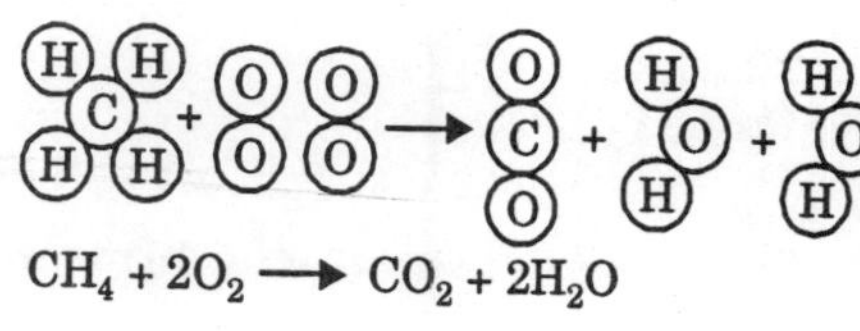

4. Chemical reactions involve the separation and combination of atoms as in this example where bonds are broken between C and H in CH_4 and between O and O in O_2, and bonds are formed between C and O in CO_2 and between H and O in H_2O.

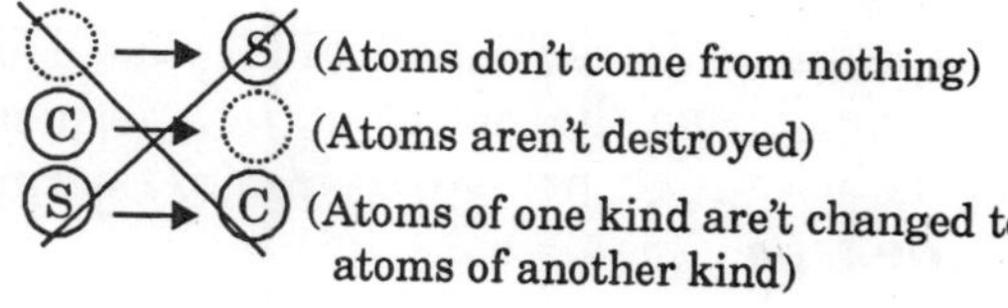

5. During the course of ordinary chemical reactions, the phenomena illustrated above do not occur: Atoms are not created, destroyed, or changed to atoms of other elemtns.

Fig. 2.12. Illustration of Dalton's atomic theory.

1. Law of Conservation of Mass : *There is no detectable change in mass in an ordinary chemical reaction* (This law was first stated in 1798 by "the father of chemistry," the Frenchman Antoine Lavoisier. Since, as shown in Item 5 of Fig. 2.12 no atoms are lost, gained, or changed in chemical reactions, mass is conserved.)

2. Law of Constant Composition: *A specific chemical compound always contains the same elements in the same proportions by mass.* (If atoms always combine in definite, constant ratios to form a particular chemical compound as implied in Item 3 of Fig. 2.12, the elemental composition of the compound by mass always remains the same.)

3. Law of Multiple Proportions: *When two elements combine to form two or more compounds, the masses of one combining with a fixed mass of the other are in ratios of small whole numbers.* (This law is illustrated for two compounds composed only of carbon and hydrogen below.)

For CH_4: Relative mass of hydrogen = $4 \times 1.0 = 4.0$ (because there are 4 atoms of H, atomic mass 1.0)

Relative mass of carbon = $1 \times 12.0 = 12.0$ (because there is 1 atom of C, atomic mass 12.0)

C/H ratio for CH_4 $= \dfrac{\text{Mass of C}}{\text{Mass of H}} = \dfrac{12.0}{4.0} = 3.0$

For C_2H_6: Relative mass of hydrogen = $6 \times 1.0 = 6.0$ (because there are 6 atoms of H, atomic mass 1.0)

Relative mass of carbon = $2 \times 12.0 = 24.0$ (because there are 2 atoms of C, atomic mass 12.0)

C/H ratio for C_2H_6 $= \dfrac{\text{Mass of C}}{\text{Mass of h}} = \dfrac{24.0}{6.0} = 4.0$

Comparing C_2H_6 and CH_4 : $\dfrac{\text{C/H ratio for } C_2H_6}{\text{C/H ratio for } CH_4} = \dfrac{4.0}{3.0}$ 4/3

Note that this is a ratio of small, whole numbers.

Size of Atoms

It is difficult to imagine just how small an individual atom is. An especially small unit of mass, the *atomic mass unit, u,* is used to express the masses of atoms. This unit was defined as a mass equal to exactly 1/12 that of the carbon-12 isotope. An atomic mass unit is only 1.66×10^{-24} g. An average atom of hydrogen, the lightest element, has a mass of only 1.0079 u. The average mass of an atom of uranium, the heaviest naturally-occurring element, is 238.03 u.

Fig. 2.13. The ballpoint pen ink in a typical signature might have a mass of 6×10^{19} u.

To place these values in perspective, consider that a signature written by ballpoint pen on a piece of paper typically has a mass of 0.1 mg (1×10^{-4} g). This mass is equal to 6×10^{9} u.

It would take almost 60 000 000 000 000 000 000 hydrogen atoms to equal the mass of ink in such a signature! Atoms visualized as spheres have diameters of around 1–3 × 10^{-10} meters (100-300 picometers, pm). By way of comparison, a small marble has a diameter of around 1 cm (1 × 10^{-2} m), which is about 100 000 000 times that of a typical atom.

Atomic Mass

The *atomic mass* of an element is the average mass of all atoms of the element relative to carbon-12 taken as exactly 12. Since atomic masses are *relative* quantities, they can be expressed without units. Or atomic masses may be given in atomic mass units. For example, an atomic mass of 14.0067 for nitrogen means that the average mass of all nitrogen atoms is 14.0067/12 as great as the mass of the carbon-12 isotope and is also 14.0067 u.

SUBATOMIC PARTICLES

Small as atoms are, they in turn consist of even smaller entities called ***sub-atomic particles.*** Although physicists have found several dozen of these, chemists need consider only three—*protons, neutrons,* and *electrons.* These subatomic particles differ in mass and charge. Like the atom, their masses are expressed in atomic mass units.

The *proton, p,* has a mass of 1.007277 u and a unit charge of + 1. This charge is equal to 1.6022 × 10^{-9} coulombs, where a coulomb is the amount of electrical charge involved in a flow of electrical current of 1 ampere for 1 second.

The *neutron, n,* has no electrical charge and a mass of 1.009665 u. The proton and neutron each have a mass of essentially 1 u and are said to have a *mass number* of 1. (Mass number is a useful concept expressing the total number of protons and neutrons, as well as the approximate mass, of a nucleus or sub-atomic particle.)

The *electron, e,* has an electrical charge of –1. It is very light, however, with a mass of only 0.00054859 u, about 1/1840 that of the proton or neutron. Its mass number is 0. The properties of protons, neutrons, and electrons are summarized in Table 2.4.

Table 2.4. Properties of Protons, Neutrons, and Electrons

Subatomic Particle	*Symbol*	*Unit Charge*	*Mass Number*[a]	*Mass in u*	*Mass in grams*
Proton	*p*	+1	1	1.007277	1.6726 × 10^{-24}
Neutron	*n*	0	1	1.008665	1.6749 × 10^{-24}
Electron	*e*	–1	0	0.000549	9.1096 × 10^{-28}

[a]The mass number and charge of each of these kinds of particles may be indicated by a superscript and subscript, respectively in the symbols ${}^{1}_{1}p$, ${}^{1}_{0}n$, and ${}^{0}_{-1}e$.

Although it is convenient to think of the proton and neutron as having the same mass, and each is assigned a mass number of 1, it is seen in Table 2.4 that their exact masses differ slightly from each other. Furthermore, the mass of an atom differs slightly from the sum of the masses of subatomic particles composing the atom. This is because of the energy relationships involved in holding the subatomic particles together in atoms so that the masses of the atom's constituent subatomic particles do not add up to exactly the mass of the atom.

BASIC STRUCTURE OF THE ATOM

Protons and neutrons are located in the *nucleus* of an atom; the remainder of the atom consists of a cloud of rapidly moving electrons. Since protons and neutrons have much higher masses than electrons, essentially all the mass of an atom is in its nucleus. However, the electron cloud makes up virtually all of the volume of the atom, and the nucleus is very small.

Atomic Number, Isotopes, and Mass Number of Isotopes

All atoms of the same element have the same number of protons and electrons equal to the *atomic number* of the element. (When reference is made to atoms here it is understood that they are electrically neutral atoms and not ions consisting of atoms that have lost or gained 1 or more electrons.) Thus all helium atoms have 2 protons in their nuclei, and all nitrogen atoms have 7 protons. *Isotopes* were defined as atoms of the same element that differ in the number of neutrons in their nuclei.

It was noted that both the proton and neutron have a *mass number* of exactly 1. Mass number is commonly used to denote isotopes. The *mass number of an isotope is the sum of the number of protons and neutrons in the nucleus of the isotope.* Since atoms with the same number of protons—that is, atoms of the same element—may have different numbers of neutrons, several isotopes may exist of a particular element. The naturally-occurring forms of some elements consist of only one isotope, but isotopes of these elements can be made artificially, usually by exposing the elements to neutrons produced by a nuclear reactor. It is convenient to have a symbol that clearly designates an isotope of an element.

Borrowing from nuclear science, a superscript in front of the symbol for the element is used to show the mass number and a subscript before the symbol designates the number of protons in the nucleus (atomic number). Using this notation denotes carbon-12 as ${}^{12}_{6}C$.

Exercise: Fill in the blanks designated with letters in the table below:

Element	*Atomic Number*	*Mass Number of Isotope*	*Number of Neutrons in Nucleus*	*Isotope Symbol*
Nitrogen	7	14	7	${}^{14}_{7}N$
Chlorine	17	35	(*a*)———	${}^{35}_{17}Cl$
Chlorine	(*b*)———	37	(*c*)———	(*d*)———
(*e*)———	6	(*f*)———	7	(*g*)———
(*h*)———	(*i*)———	(*j*)———	(*k*)———	${}^{11}_{5}B$

Answers: (*a*) 18, (*b*) 17, (*c*) 20, (*d*) ${}^{37}_{17}Cl$, (*e*) carbon, (*f*) 13, (*g*) ${}^{13}_{6}C$, (*h*) boron, (*i*) 5, (*j*) 11, (*k*) 6.

Electrons in Atoms

Electrons around the nucleus of an atom were depicted as forming a cloud of negative charge. The energy levels, orientations in space, and behaviour of electrons varies with the number of them contained in an atom. In a general sense, the arrangements of electrons in atoms are described by *electron configuration,* a term discussed in some detail later in this chapter.

Attraction Between Electrons and the Nucleus

The electrons in an atom are held around the nucleus by the attraction between their negative charges and the positive charges of the protons in the nucleus. Opposite electrical charges attract, and like charges repel. The forces of attraction and repulsion are expressed quantitatively by *Coulomb's law,*

$$F = \frac{Q_1 Q_2}{d^2} \qquad \text{...(2.29)}$$

where F is the force of interaction, Q_1 and Q_2 are the electrical charges on the bodies involved, and d is the distance between the bodies.

Coulomb's law explains the attraction between negatively charged electrons and the positively charged nucleus in an atom. However, the law does not explain why electrons move around the nucleus, rather than coming to rest on it. The behaviour of electrons in atoms—as well as the numbers, types, and strengths of chemical bonds formed between atoms—is explained by *quantum theory*.

DEVELOPMENT OF THE PERIODIC TABLE

The *periodic table,* which was mentioned briefly in connection with the elements, was first described by the Russian chemist Dmitry Mendeleyev in 1867. It was based upon Mendeleyev's observations of periodicity in chemical behaviour without any knowledge of atomic structure. The periodic table is discussed in more detail in this chapter along with the development of the concepts of atomic structure. Elements are listed in the periodic table in an ordered, systematic way that correlates with their electron structures. The elements are placed in rows or *periods* of the periodic table in order of increasing atomic number so that there is a periodic repetition of elemental properties across the periods.

The rows are arranged so that elements in vertical columns called *groups* have similar chemical properties reflecting similar arrangements of the outermost electrons in their atoms. With some knowledge of the ways that electrons behave in atoms it is much easier to develop the concept of the periodic table. This is done for the first 20 elements in the following sections. After these elements are discussed, they are placed in an abbreviated 20-element version of the periodic table as shown in figure 2.20.

THE SIMPLEST ATOM

The simplest atom is that of *hydrogen,* having only one positively charged proton in its nucleus, which is surrounded by a cloud of negative charge formed by only one electron. By far the most abundant kind of hydrogen atom has no neutrons in its nucleus. Having only one proton with a mass number 1, and 1 electron with a mass number of zero, the mass number of this form of hydrogen is 1 + 0 = 1. There are, however, two other forms of hydrogen atoms having, in addition to the proton, 1 and 2 neutrons, respectively, in their nuclei. The three different forms of elemental hydrogen are *isotopes* of hydrogen that all have the same number of protons, but different numbers of neutrons in their nuclei. Only about 1 of 7,000 hydrogen atoms is ${}^{2}_{1}H$, deuterium, mass number 2 (from 1 proton + 1 neutron). The mass number of *tritium,* which has 2 neutrons, is 3 (from 1 proton + 2 neutrons) = 3. These three forms of hydrogen atoms may be designated as ${}^{1}_{1}H$, ${}^{2}_{1}H$, and ${}^{3}_{1}H$. The meaning of this notation is reviewed below:

Designation of Hydrogen in the Periodic Table

The atomic mass of hydrogen is 1.0079. This means that the average atom of hydrogen has a mass of 1.0079 u (atomic mass units); hydrogen's atomic mass is simply 1.0079 relative to the carbon-12 isotope taken as exactly 12. With this information it is possible to place hydrogen in the periodic table with the designation shown in Fig. 2.13.

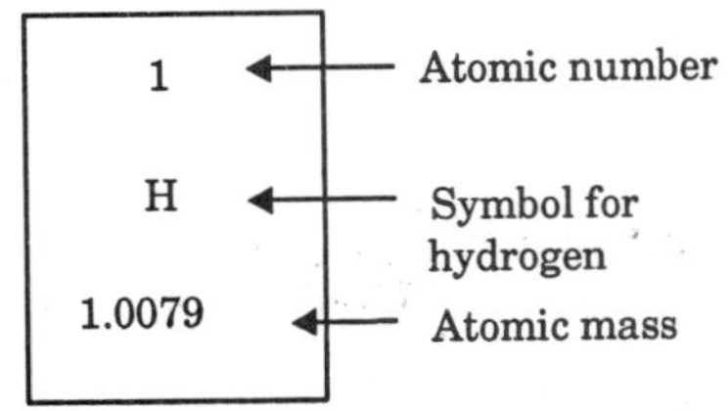

Fig. 2.13. Designation of hydrogen in the periodic table.

Showing Electrons in Hydrogen Atoms and Molecules

It is useful to have some simple way of showing the hydrogen atom's single electron in chemical symbols and formulas. This is accomplished with *electron-dot symbols* or *Lewis symbols* (after G. N. Lewis), which use dots around the symbol of an element to show outer electrons (those that may become involved in chemical bonds). The Lewis symbol for hydrogen is

Elemental hydrogen consists of molecules made up of 2 H atoms held together by a chemical bond consisting of two shared electrons. Just as it is useful to show electrons in atoms with a Lewis symbol, it is helpful to visualize molecules and the electrons in them with *electron-dot formulas* or *Lewis formulas* as shown for the H_2 molecule in Fig. 2.14.

H • → ← • G	H_2	H:H
Elemental hydrogen does not exist as individual H atoms.	Instead, it exists as molecules made up to 2 H atoms with the chemical formula H_2.	The covalent bond holding the two H atoms together consists of 2 shared electrons shown in the Lewis formula for H_2 above.

Fig. 2.14. Lewis formula for hydrogen molecules.

Properties of Elemental Hydrogen

Pure hydrogen is colourless and odorless. It is a gas at all but very low temperatures; liquid hydrogen boils at –253°C and solidifies at –259°C. Hydrogen gas has the lowest density of any pure substance. It reacts chemically with a large number of elements, and hydrogen-filled balloons that float readily in. the atmosphere also explode convincingly when touched with a flame because of the following very rapid reaction with oxygen in the air:

$$2H_2 + O_2 \rightarrow 2H_2O \qquad ...(2.30)$$

Production and Uses of Elemental Hydrogen

Elemental hydrogen is one of the more widely produced industrial chemicals. On an industrial scale, it is commonly made by *steam reforming* of methane (natural gas, CH_4) under high-temperature, high-pressure conditions:

$$CH_4 + H_2O \xrightarrow[30\text{ atm } P]{800°C\,T} CO + 3H_2 \qquad ...(2.31)$$

Hydrogen is used to manufacture a number of chemicals. One of the most important of these is ammonia, NH_3. Methanol (methyl alcohol, CH_3OH) is a widely used industrial chemical and solvent synthesized by the following reaction between carbon monoxide and hydrogen:

$$CO + 2H_2 \rightarrow CH_3OH \qquad ...(2.32)$$

Methanol made by this process can be blended with gasoline to yield a fuel that produces relatively less pollutant carbon monoxide; such "oxygenated gasoline additives" are now required for use in some cities during winter months. Gasoline is upgraded by the chemical addition of hydrogen to some petroleum fractions. Synthetic petroleum can be made by the addition of hydrogen to coal at high temperatures and pressures. A widely used process in the food industry is the addition of hydrogen to unsaturated vegetable oils to give margarine and other hydrogenated fats and oils.

HELIUM

The *electron configurations* of atoms determine their chemical behaviour. Electrons in atoms occupy distinct *energy levels.* At this point it is useful to introduce the concept of the *electron shell* to help explain electron energy levels and their influence on chemical behaviour. Each electron shell can hold a maximum number of electrons. An atom with a *filled electron shell* is especially content in a chemical sense, with little or no tendency to lose, gain, or share electrons. Elements with these characteristics exist as gas-phase atoms and are called *noble gases*. Examined in order of increasing atomic number, the first element consisting of atoms with filled electron shells is helium.

He, atomic number 2. All helium atoms contain 2 protons and 2 electrons. Virtually all helium atoms are ${}^{4}_{2}He$ that contain 2 neutrons in their nuclei; the ${}^{3}_{2}He$ isotope containing only 1 neutron in its nucleus occurs to a very small extent. The atomic mass of helium is 4.00260. The two electrons in the helium atom are shown by the Lewis symbol illustrated in Fig. 2.14. These electrons constitute a *filled electron shell,* so that helium is a *noble gas* composed of individual helium atoms that have no tendency to form chemical bonds with other atoms. Helium gas has a very low density of only 0.164 g/L at 25°C and 1 atm pressure.

Uses of Helium

Helium is extracted from some natural gas sources that contain up to 10°7o helium by volume. It has many uses that depend upon its unique properties. Because of its very low density compared to air, helium is used to fill weather balloons and airships. Helium is nontoxic, odorless, tasteless, and colourless. Because of these properties and its low solubility in blood, helium is mixed with oxygen for breathing by deep sea divers and persons with some respiratory ailments. Use of helium by divers avoids the very painful condition called "the bends" caused by bubbles of nitrogen forming from nitrogen gas dissolved in blood. Liquid helium, which boils at a temperature of only 4.2 K above absolute zero is especially useful in the growing science of *cryogenics*, which deals with very low temperatures.

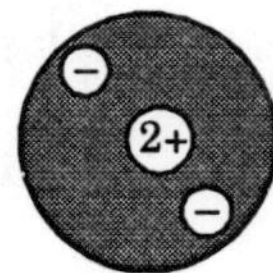

He:

A helium atom has a filled electron shell containing 2 electrons.

It can be represented by the Lewis symbol above.

Fig. 2.14. Two representations of the helium atom having a filled electron shell.

Some metals are superconductors at such temperatures so that helium is used to cool electromagnets that develop very powerful magnetic fields for a relatively small magnet. Such magnets are components of the very useful chemical tool known as nuclear magnetic resonance (NMR). The same kind of instrument modified for clinical applications and called MRI is used as a medical diagnostic tool.

LITHIUM

The third element in the periodic table is lithium (Li), atomic number 3, atomic mass 6.941. The most abundant lithium isotope has 4 neutrons in its nucleus, so it has a mass number of 7 and is designated $^{7}_{3}Li$. A less common isotope, $^{6}_{3}Li$, has only 3 neutrons. Lithium is the first element in the periodic table that is a *metal*. Mentioned metals tend to have the following properties:

- Characteristic *luster* (like freshly-polished silverware or a new penny)
- *Malleable* (can be pounded or pressed into various shapes without breaking)
- *Conduct electricity*
- Chemically, tend to *lose electrons* and form cations with charges of + 1 to +3.

Lithium is the lightest metal, with a density of only 0.531 g/cm^3.

Uses of Lithium

Lithium compounds have a number of important uses in industry and medicine. Lithium carbonate, Li_2CO_3, is one of the most important lithium compounds and is used as the starting material for the manufacture of many other lithium compounds. It is an ingredient of specialty glasses, enamels, and specialty ceramic ware having low thermal expansion coefficients (minimum expansion when heated). Lithium carbonate is widely prescribed as a drug to treat acute mania in manic-depressive and schizo-affective mental disorders.

Lithium hydroxide, LiOH, is an ingredient in the manufacture of some lubricant greases and in some long-life alkaline storage batteries. The lithium atom has three electrons. As lithium has both *inner electrons*—in this case 2 contained in an inner shell—as in the immediately preceding noble gas helium, and an outer electron that is farther from, and less strongly attracted to, the nucleus. The outer electron is said to be in the atom's outer shell. The inner electrons are, on the average, closer to the nucleus than is the outer electron, are very difficult to remove from the atom, and do not become involved in chemical bonds.

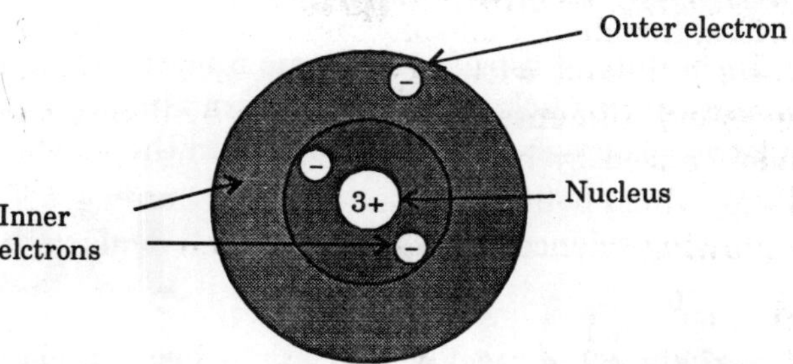

Fig. 2.15. An atom of lithium, Li, has 2 inner electrons and 1 outer electron. The latter can be lost to another-atom to produce the Li^+ ion, which is present in ionic compounds.

Lithium's outer electron is relatively easy to remove from the atom, which is what happens when ionic bonds involving Li^+ ion are formed. The distinction between inner and outer electrons is developed to a greater extent later in this chapter.

In atoms such as lithium that have both outer and inner electrons, the Lewis symbol shows only the outer electrons. Therefore, the Lewis symbol of lithium is

$$\mathbf{Li\cdot}$$

A lithium atom's loss of its single outer electron is shown by the *half-reaction* (one in which there is a net number of electrons on either the reactant or product side) in Fig. 2.16. The Li^+ product of this reaction has the very stable helium core of 2 electrons. The Li^+ ion is a constituent of ionic lithium compounds in which it is held by attraction for negatively charged anions (such as Cl^-) in the crystalline lattice of the ionic compound. The tendency to lose its outer electron and to be stabilized in ionic compounds determines lithium's chemical behaviour.

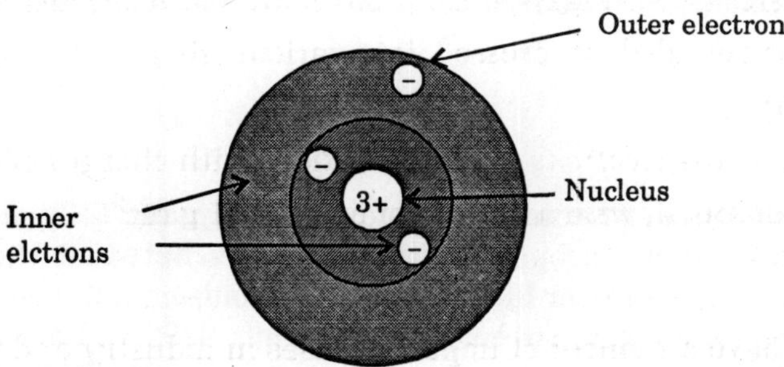

Fig. 2.16. Half-reaction showing the formation of Li^+ from an Li atom. The Li^+ ion has the especially stable helium core of just 2 electrons. The atom to which the electron is lost is not shown, so this is a half-reaction.

THE SECOND PERIOD, ELEMENTS 4-10

In this section elements 4-10 will be discussed and placed in the periodic table to complete a period in the table.

Beryllium, Atomic Number 4

Each atom of *beryllium*-atomic number 4, atomic mass 9.01218-contains 4 protons and 5 neutrons in its nucleus. The beryllium atom has two inner electrons and two outer electrons, the latter designated by the two dots in the Lewis symbol below:

$$\mathbf{Be:}$$

Beryllium can react chemically by losing 2 electrons from the beryllium atom. This occurs according to the half reaction

$$\text{Be:} \rightarrow \text{Be}^{2} + 2e^{-} \text{ (lost to another atom)} \quad ...(2.33)$$

in which the beryllium atom, Lewis symbol Be:, loses two e^- to form a beryllium ion with a charge of +2. The loss of these two outer electrons gives the beryllium atom the same stable helium core as that of the Li^+ ion discussed in the preceding. Beryllium is melted together with certain other metals to give homogeneous mixtures of metals called alloys. The most important beryllium alloys are hard, corrosion-resistant, nonsparking, and good conductors of electricity. They are used to make such things as springs, switches, and small electrical contacts.

A very high melting temperature of about 1,290°C combined with good heat absorption and conduction properties has led to the use of beryllium metal in aircraft brake components. Beryllium is an environmentally and toxicologically important element because it causes *berylliosis*, a disease marked by lung deterioration. Inhalation is particularly hazardous, and atmospheric standards have been set at very low levels.

Boron, Atomic Number 5

Boron, *B*, has an atomic number of 5 and an atomic mass of 10.81. Most boron atoms have 6 neutrons in addition to 5 protons in their nuclei; a less common isotope has 5 neutrons. Two of boron's 5 electrons are in a helium core and 3 are outer electrons, as shown by the Lewis symbol

$$\dot{\text{B}}\text{:}$$

Boron—along with silicon, germanium, arsenic, antimony, and tellurium—is one of a few elements called *metalloids* with properties intermediate between those of metals and nonmetals. Although they have a luster like metals, metalloids do not form positively-charged ions (cations). The melting temperature of boron is very high, 2190°C. Boron is added to copper, aluminum, and steel to improve their properties. It is used in control rods of nuclear reactors because of the good neutron-absorbing properties of the ${}^{10}_{5}B$ isotope. Some chemical compounds of boron, especially boron nitride, BN, are noted for their hardness. Boric acid, H_3BO_3, is used as a flame retardant in cellulose insulation in houses. The oxide of boron, B_2O_3, is an ingredient of fiberglass used in textiles and insulation.

Carbon, Atomic Number 6

Atoms of *carbon*, *C*, have 2 inner and 4 outer electrons, the latter shown by the Lewis symbol

$$\cdot\overset{\bullet}{C}:$$

The carbon-12 isotope with 6 protons and 6 neutrons in its nucleus, ${}^{12}_{6}C$, constitutes 98.9% of all naturally occurring carbon. The ${}^{13}_{6}C$ isotope makes up 1.1% of all carbon atoms. Radioactive carbon-14, ${}^{14}_{6}C$, is present in some carbon sources.

Carbon is an extremely important element with unique chemical properties, without which life could not exist. All of organic chemistry is based upon compounds of carbon, and it is an essential element in life molecules. Carbon atoms are able to bond to each other to form long straight chains, branched chains, rings, and three-dimensional structures. As a result of its self-bonding abilities, carbon exists in several elemental forms. These include powdery carbon black, very hard, clear diamonds, and graphite so soft that it is used as a lubricant. Activated carbon prepared by treating carbon with air, carbon dioxide, or steam at high temperatures is widely used to absorb undesirable pollutant substances from air and water. Carbon fiber has been developed as a structural material in the form of composites consisting of strong strands of carbon bonded together with special plastics and epoxy resins.

Nitrogen, Atomic Number 7

'*Nitrogen, N*, composes 78% by volume of air in the form of N_2 molecules. The atomic mass of nitrogen is 14.0067, and the nuclei of nitrogen atoms contain 7 protons and 7 neutrons. Nitrogen has 5 outer electrons, so its Lewis symbol is

$$\cdot\overset{\bullet}{\underset{\bullet}{N}}:$$

Like carbon, nitrogen is a nonmetal. Pure N_2 is prepared by distilling liquified air, and it has a number of uses. Since nitrogen gas is not very chemically reactive, it is used as an inert atmosphere in some industrial applications, particularly where fire or chemical reactivity may be a hazard. People have been killed by accidentally entering chambers filled with nitrogen gas, which acts as a simple asphyxiant with no warning odor of its presence. Liquid nitrogen boils at a very cold –190°C. It is widely used to maintain very low temperatures in the laboratory, for quick-freezing foods, and in freeze-drying processes.

Freeze-drying is used to isolate fragile biochemical compounds from water solution, for the concentration of environmental samples to be analyzed for pollutants, and for the preparation of instant coffee and other dehydrated foods. It has potential applications in the concentration and isolation of hazardous waste substances. Like carbon, nitrogen is an essential element for life processes.

Nitrogen is an ingredient of all of the amino acids found in proteins Nitrogen compounds are fertilizers essential for the growth of plants. The *nitrogen cycle*, which involves incorporation of N_2 from the atmosphere into living matter and chemically-bound nitrogen in soil and water, then back into the atmosphere again, is one of nature's fundamental cycles. Nitrogen compounds, particularly *ammonia* (NH_3), and nitric acid (HNO_3), are widely used industrial chemicals.

Oxygen, Atomic Number 8

Like carbon and nitrogen, oxygen, atomic number 8, is a major component of living organisms. Oxygen is *a* nonmetal existing as molecules of O_2 in the elemental gas state, and air is 21% oxygen by volume. Like all animals, humans require oxygen to breathe and to maintain their life processes. The nuclei of oxygen atoms contain 8 protons and 8 neutrons, and the atomic mass of oxygen is 15.9994. The oxygen atom has 6 outer electrons as shown by its Lewis symbol below:

In addition to O_2, there are two other important elemental oxygen species in the atmosphere. These are atomic oxygen, O, and ozone, O_3. These species are normal constituents of the stratosphere, a region of the atmosphere that extends from about 11 kilometers to about 50 km in altitude. Oxygen atoms are formed when high-energy ultraviolet radiation strikes oxygen molecules high in the stratosphere:

$$O_2 \xrightarrow[\text{radiation}]{\text{Ultraviolet}} O + O \quad ...(2.34)$$

The oxygen atoms formed by the above reaction combine with O_2 molecules,

$$O + O_2 \rightarrow O_3 \quad ...(2.35)$$

to form ozone molecules. These molecules make up the *ozone layer* in the stratosphere and effectively absorb additional high-energy ultraviolet radiation. If it weren't for this phenomenon, the ultraviolet radiation would reach the Earth's surface and cause painful sunburn and skin cancer in exposed people. However, ozone produced in photochemical smog at ground level is toxic to animals and plants. The most notable chemical characteristic of elemental oxygen is its tendency to combine with other elements in energy-yielding reactions.

Such reactions provide the energy that propels automobiles, heats buildings, and keeps body processes going. One of the most widely used chemical reactions of oxygen is that with hydrocarbons, particularly those from petroleum and natural gas. For example, butane (C_4H_{10}, a liquifiable gaseous hydrocarbon fuel) burns in oxygen from the atmosphere,

$$2C_4H_{10} + 13O_2 \rightarrow 8CO_2 + 10H_2O \quad ...(2.36)$$

a reaction that provides heat in home furnaces, water heaters, and other applications.

Fluorine, Atomic Number 9

Fluorine, F, has 7 outer electrons, so its Lewis symbol is

Under ordinary conditions elemental fluorine is a greenish-yellow gas consisting of F_2 molecules. Fluorine compounds have many uses. One of the most notable of these is the manufacture of chlorofluorocarbon compounds known by the trade name Freons.

These are chemical combinations of chlorine, fluorine, and carbon, an example of which is dichlorodifluoromethane, Cl_2CF_2. These compounds until recently were widely used as refrigerant fluids and blowing agents to make foam plastics; they were once widely employed as propellants in aerosol spray cans. Uses of chlorofluorocarbons are now being phased out because of their role in destroying stratospheric ozone (discussed with oxygen, above).

NEON, ATOMIC NUMBER 10

The last element in the period of the periodic table under discussion is *neon*. Air is about 2 parts per thousand neon by volume, and neon is obtained by the distillation of liquid air. Neon is especially noted for its use in illuminated signs that consist of glass tubes containing neon, through which an electrical current is passed causing the neon gas to emit a characteristic glow.

In addition to 10 protons, most neon atoms have 10 neutrons in their nuclei, although some have 12, and a very small percentage have 11. As shown by its Lewis symbol,

the neon atom has 8 outer electrons. These 8 electrons constitute a *filled electron shell,* just as the 2 electrons in helium give it a filled electron shell. Because of this "satisfied" outer shell, the neon atom has no tendency to acquire, give away, or share electrons. Therefore, neon is a *noble gas,* like helium, and consists of individual neon atoms.

Stability of the Neon Noble Gas Electron Octet

In going through the rest of the periodic table it can be seen that all other atoms with 8 outer electrons like neon are also noted for a high degree of chemical stability. In addition to neon, these noble gases are argon (atomic number 18), krypton (atomic number 36), xenon (atomic number 54), and radon (atomic number 86). Each of these may be represented by the Lewis symbol

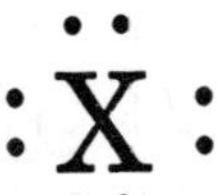

where *X* is the chemical symbol of the noble gas. It is seen that these atoms each have 8 outer electrons, a group known as an *octet* of electrons. In many cases atoms that do not have an octet of outer electrons acquire one by losing, gaining, or sharing electrons in chemical combination with other atoms; that is, they acquire a *noble gas outer electron configuration.* For all noble gases except helium, which has only 2 electrons, the noble gas outer electron configuration consists of eight electrons. The tendency of elements to acquire an 8-electron outer electron configuration, which is very useful in predicting the nature of chemical bonding and the formulas of compounds that result, is called the *octet, rule*. Although the use of the octet rule to explain and predict bonding is discussed, at this point it is useful to show how it explains bonding between hydrogen and carbon in methane:

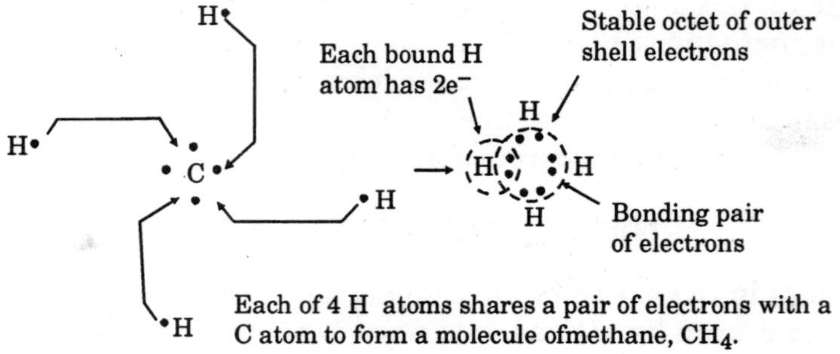

Stable octet of outer shell electrons

Each bound H atom has $2e^-$

Bonding pair of electrons

Each of 4 H atoms shares a pair of electrons with a C atom to form a molecule ofmethane, CH_4.

Fig. 2.17. Illustration of the octet rule in methane.

ELEMENTS 11-20, AND BEYOND

The abbreviated version of the periodic table will be finished with elements 11 through 20. The names, symbols, electron configurations, and other pertinent information about these elements in Table 2.6. An abbreviated periodic table with these elements in place is shown in Fig. 2.7. This table shows the Lewis symbols of all of the elements to emphasize their orderly variation across periods and similarity in groups of the periodic table. The first 20 elements in the periodic table are very important. They include the three most abundant elements on the earth's surface (oxygen, silicon, aluminum); all elements of any appreciable significance in the atmosphere (hydrogen in H_2O vapor, N_2, O_2, carbon in CO_2, argon, and neon); the elements making up most of living plant and animal matter (hydrogen, oxygen, carbon, nitrogen, phosphorus, and sulfur); and elements such as sodium, magnesium, potassium, calcium, and chlorine that are essential for life processes.

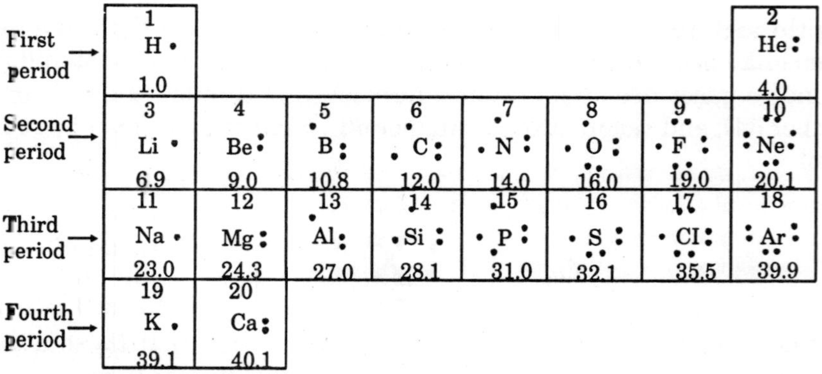

First period →	1 H • 1.0							2 He : 4.0
Second period →	3 Li • 6.9	4 Be : 9.0	5 B 10.8	6 C 12.0	7 N 14.0	8 O 16.0	9 F 19.0	10 Ne 20.1
Third period →	11 Na • 23.0	12 Mg : 24.3	13 Al 27.0	14 Si 28.1	15 P 31.0	16 S 32.1	17 CI 35.5	18 Ar 39.9
Fourth period →	19 K • 39.1	20 Ca : 40.1						

Fig. 2.18. Abbreviated 20-element version of the periodic table showing Lewis symbols of the elements.

The chemistry of these elements is relatively straightforward and reasonably easy to relate to their atomic structures. Therefore, emphasis is placed on them in the earlier chapters of this book. It is helpful to remember their names, symbols, atomic numbers, atomic masses, and Lewis symbols. As mentioned, the vertical columns of the table contain *groups* of elements that have similar chemical structures. One exception is hydrogen, H, which has unique chemical properties. All elements other than hydrogen in the first column of the abbreviated table are *alkali metals*—lithium, sodium, and potassium. These are generally soft silvery-white metals of low density that react violently with water to produce hydroxides (LiOH, NaOH, KOH) and with chlorine to produce chlorides (LiCl, NaCl, KC1). The *alkaline earth* metals—beryllium, magnesium, calcium—are in the second column of the table.

Table 2.5. Elements 11-20

Atomic Number	*Name and Lewis Symbol*	*Atomic Mass*	*Number of Outer e^-*	*Major Properties and Uses*
11	Sodium (Na·)	22.9898	1	Soft, chemically very reactive metal. Nuclei contain 11 *p* and 12 *n*.
12	Magnesium (Mg:)	24.312	2	Lightweight metal used in aircraft components, extension ladders, portable tools. Chemically very reactive. Three isotopes with 12, 13, 14 *n*.
13	Aluminum (Ȧl:)	26.9815	3	Lightweight metal used in aircraft, automobiles, electrical transmissio line. Chemically reactive, but forms self-protective coating.
14	Silicon (·Ṡi:)	28.086	4	Nonmetal, 2nd most abundant metal in Earth's crust. Rock constituent. Used in semiconductors.
15	Phosphorus (·Ṗ:)	30.9738	5	Chemically very reactive nonmetal. Highly toxic as elemental white phosphorus. Component of bones and teeth, genetic material (DNA), fertilizers, insecticides.
16	Sulfur (·Ṡ̤:)	32.064	6	Brittle, generally yellow nonmetal. Essential nutrient for plants and animals, occurring in amino acids. Used to manufacture sulfuric acid. Present in pollutant sulfar dioxide, SO_2.
17	Cholrine (·C̤̈:)	35.453	7	Greenish-yellow toxic gas composed of molecules of Cl_2. mnufactured in large quantities to disinfect water and to mnufacture plastics and solvents.
18	Argon (:Ar̤̈:)	39.948	8	Noble gas used to fill light bulbs and as a plasma medium in inductively coupled plasma atomic emission analysis of elemental pollutants.
19	Potassium (K·)	39.098	1	Chemically reactive alkali metal very similar to sodium in chemical and physical properties. Essential fertilizer for plant growth as K^+ ion.
20	Calcium (Ca:)	40.078	2	Chemically reactive alkaline earth metal with properties similar to those of magnesium.

When freshly cut, these metals have a grayish-white luster. They are chemically reactive and have a strong tendency to form doubly-charged cations (Be^{2+}, Mg^{2+}, Ca^{2+}) by losing two electrons from each atom. Another group notable for the very close similarities of the elements in it consists of the *noble gases* in the far right column of the table. Each of these—helium, neon, argon—is a monatomic gas that does not react chemically.

The Elements Beyond Calcium

The electron structures of elements beyond atomic number 20 are more complicated than those of the first 20 elements just discussed. As shown in the complete periodic table in Fig. 2.19, included among these heavier elements are transition metals, the lanthanides, and the actinides. The transition metals include many metals that are important in industry and in life processes. Among the transition metals are chromium, manganese, iron, cobalt, nickel, and copper. The actinides contain thorium, uranium, and plutonium—familiar names to those concerned with nuclear energy, nuclear warfare, and related issues.

DETAILED LOOK AT ATOMIC STRUCTURE

So far, this chapter has covered some important aspects of atoms. These include the facts that an atom is made of three major subatomic particles, and consists of a very small, very dense, positively charged nucleus surrounded by a cloud of negatively charged electrons in constant, rapid motion. The first *20* elements have been discussed in some detail and placed in an abbreviated version of the periodic table. Important concepts introduced so far in this chapter include:

- Dalton's atomic theory
- Electron shells
- Inner shell electrons
- Octet rule
- Lewis symbols to represent outer e^-
- Significance of filled electron shells
- Outer shell electrons
- Abbreviated periodic table

The information presented about atoms so far in this chapter is adequate to meet the needs of many readers. These readers may choose to forego the details of atomic structure presented in the rest of this chapter without major harm to their understanding of chemistry. However, for those who wish to go into more detail, or who do not have a choice, the remainder of this chapter discusses in more detail the electronic structures of atoms as related to their chemical behaviour and introduces the quantum theory of electrons in atoms.

Electromagnetic Radiation

The *quantum theory* explains the unique behaviour of charged particles that are as small and move as rapidly as electrons. Because of its close relationship to electromagnetic radiation, an appreciation of quantum theory requires an understanding of the following important points related to electromagnetic radiation:

- Energy can be carried through space at the speed of light, 3.00×10^8 meters per second (m/s) in a vacuum, by *electromagnetic radiation*, which includes visible light, ultraviolet radiation, infrared radiation, microwaves, and radio waves.
- Electromagnetic radiation has a *wave character*. The waves move at the speed of light, c, and have characteristics of *wavelength* (λ), amplitude, and *frequency* (ν, Greek "*nu*") as illustrated below:

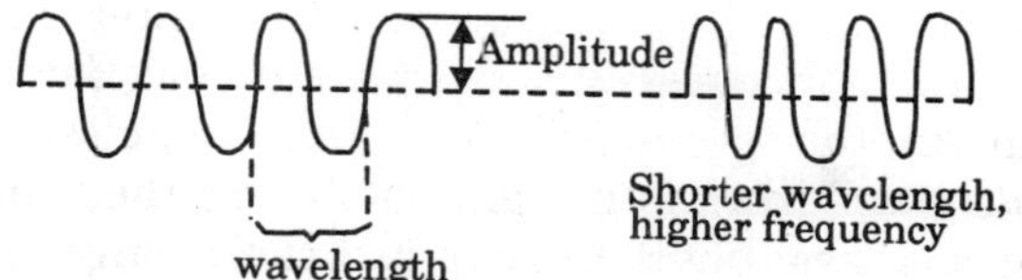

- The wavelength is the distance required for one complete cycle and the frequency is the number of cycles per unit time. They are related by the following equation:

$$\nu \lambda = c$$

where v is in units of cycles per second (s^{-1}, a unit called the *hertz*, Hz) and λ is in meters (m).

- In addition to behaving as a wave, electromagnetic radiation also has characteristics of particles.
- The dual wave/particle nature of electromagnetic radiation is the basis of the quantum theory of electromagnetic radiation, which states that radiant energy may be absorbed or emitted only in discrete packets called *quanta* or *photons*. The energy, E, of each photon is given by

$$E = h\nu$$

where h is Planck's constant, 6.63×10^{-34} J-s (joule × second).

- From the preceding, it is seen that *the energy of a photon is higher when the frequency of the associated wave is higher* (and the wavelength shorter).

MODELS OF ELECTRONS IN ATOMS

The *quantum theory* introduced in the preceding section provided the key concepts needed to explain the energies and behaviour of electrons in atoms. One of the best clues to this behaviour, and one that ties the nature of electrons in atoms to the properties of electromagnetic radiation, is the emission of light by energized atoms. This is easiest to explain for the simplest atom of all, that of hydrogen, which consists of only one electron moving around a nucleus with a single positive charge. Energy added to hydrogen atoms, such as by an electrical discharge through hydrogen gas, is re-emitted in the form of light at very specific wavelengths (656, 486, 434, 410 nm in the visible region).

The highly energized atoms that can emit this light are said to be "excited" by the excess energy originally put into them and to be in an excited state. The reason for this is that the electrons in the excited atoms are forced farther from the nuclei of the atoms and, when they return to a lower energy state, energy is emitted in the form of light. The fact that very specific wavelengths of light are emitted in this process means that electrons can be present only in specified states at highly specific energy levels. Therefore, the transition from one energy state to a lower one involves the emission of a specific energy of electromagnetic radiation (light). Consider the equation

$$E = h\nu \qquad ...(2.37)$$

that relates energy to frequency, ν, of electromagnetic radiation. If a transition of an electron from one excited state to a lower one involves a specific amount of energy, E, a corresponding value of ν is observed. This is reflected by a specific wavelength of light according to the following relationship:

$$\lambda = \frac{c}{\nu} \quad ...(2.38)$$

The first accepted explanation of the behaviour outlined above was the Bohr theory advanced by the Danish physicist Niels Bohr in 1913. Although this theory has been shown to be too simplistic, it had some features that are still pertinent to atomic structure. The *Bohr theory* visualized an electron orbiting the nucleus (a proton) of the hydrogen atom in orbits. Only specific orbits called *quantum states* were allowed. When energy was added to a hydrogen atom, its electron could jump to a higher orbit. When the atom lost its energy as the electron returned to a lower orbit, the energy lost was emitted in the form of electromagnetic radiation as shown in Fig. 2.19.

Because the two energy levels are of a definite magnitude according to quantum theory, the energy lost by the electron must also be of a definite energy, $E = h\nu$. Therefore, the electromagnetic radiation (light) emitted is of a specific frequency and wavelength.

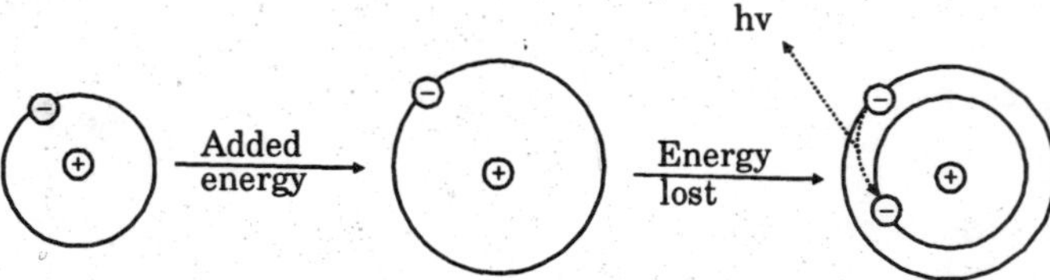

Fig. 2.19. According to the Bohr model, adding energy to the hydrogen atom promotes an electron to a higher energy level. When the electron fails back to a lower energy level, excess energy is emitted in the form of electromagnetic radiation of a specific energy, E = hν.

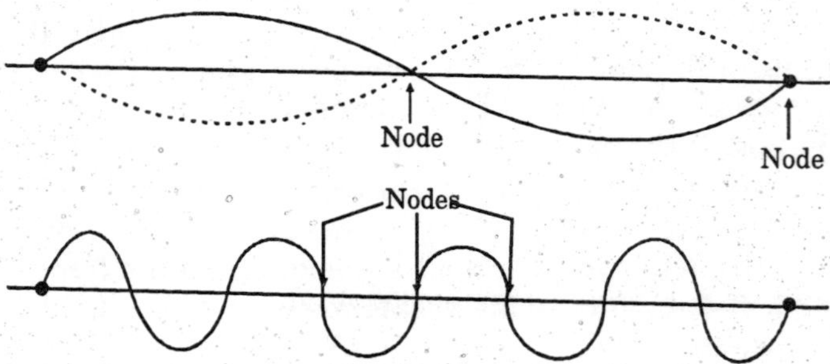

Fig. 2.20. A string anchored at both ends (such as on a stringed musical instrument) can only vibrate at multiples of waves or half-waves. The illustration above shows one wave (top) and four waves (bottom). Each point at which a wave intersects the horizontal lines is called a node.

Wave Mechanical Model

Though shown to have some serious flaws and long since abandoned, the Bohr model laid the groundwork for the more sophisticated theories of atomic structure that are accepted today and introduced the all-important concept that *only specific energy states are allowed for an electron in an atom.* Like electromagetic radiation, electrons in atoms are now visualized as having a dual wave/ particle nature. They are treated theoretically by the *wave mechanical model* as *standing waves* around the nucleus of an atom. The idea of a standing wave can be visualized for the string of a musical instrument as represented in Fig. 2.20.

Such a wave does not move along the length of a string because both ends are anchored, which is why it is called a *standing wave.* Each wave has *nodes*, which are points of zero displacement. Because there must be a node on each end where the string is anchored, the standing waves can only exist as multiples of *half-wavelengths.* According to the *wave mechanical* or *quantum mechanical* model of electrons in atoms, the known *quantization* of electron energy in atoms occurs because only specific multiples of the standing wave associated with an electron's movement are allowed.

Such a phenomenon is treated mathematically with the *Schrodinger equation* resulting from the work of Erwin Schrodinger, first published in 1926. Even for the one-electron hydrogen atom, the mathematics is quite complicated, and no attempt will be made to go into it here. The Schrodinger equation is represented as

$$H\psi = E\psi \qquad ...(3.39)$$

where ψ is the Greek psi. Specifically, ψ is the *wave function*, a function of the electron's energy and the coordinates in space where it may be found. The term H is an *operator* consisting of a set of mathematical instructions, and E is the sum of the kinetic energy due to the motion of the electron and the potential energy of the mutual attraction between the electron and the nucleus.

Schrodinger Equation

With its movement governed by the laws of quantum mechanics, it is not possible to know where an electron is relative to an atom's nucleus at any specific instant. However, the Schrodinger equation permits calculation of the probability of an electron being in a specified region. Solution of the equation gives numerous wave functions, each of which corresponds to a definite energy level and to the probability of finding an electron at various locations and distances relative to the nucleus.

The wave function describes *orbitals, each of which has a characteristic energy and region around the nucleus where the electron has certain probabilities of being found.* The term orbital is used because quantum mechanical calculations do not give specific orbits consisting of defined paths for electrons, like planets going around the sun. The square of the wave function, ψ^2, calculated and integrated for a small segment of volume around a nucleus is proportional to the probability of finding an electron in that small volume, which may be regarded as part of an *electron cloud.* With this view, the electron is more likely to be found in a region where the cloud is relatively more dense. The cloud has no definite outer limits, but fades away with increasing distance from the nucleus to regions in which there is essentially no probability of finding the electron.

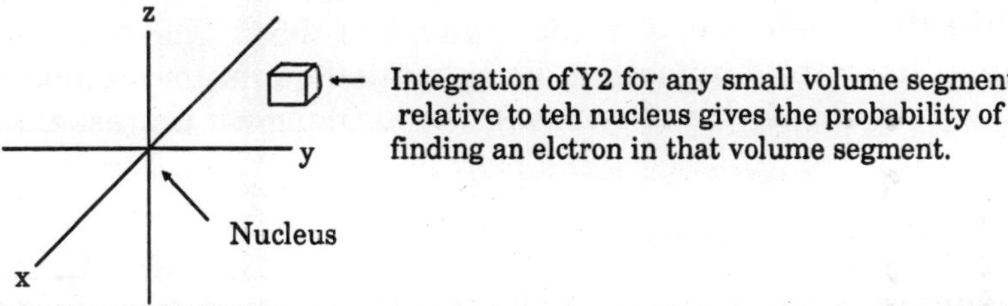

Fig. 2.21. The *xyz* coordinate system used to describe the orientations in space of orbitals around the nucleus of an atom.

Multielectron Atoms and Quantum Numbers

Wave mechanical calculations describe various ***energy levels*** for electrons in an atom. Each energy level has at least one orbital. *A single orbital may contain a maximum of only two electrons.* Although the Schrodinger approach was originally applied to the simplest atom, hydrogen, it has been extended to atoms with many electrons (multielectron atoms). Electrons in atoms are described by four *quantum numbers*, which are defined and described briefly below. With these quantum numbers and a knowledge of the rules governing their use, it is possible to specify the orbitals allowed in a particular atom. *An electron in an atom has its own unique set of quantum numbers; no two electrons may have exactly identical quantum numbers.*

The Principal Quantum Number, n

Main energy levels corresponding to *electron shells* discussed earlier in this chapter are designated by a *principal quantum number, n*. Both the size of orbitals and the magnitude of the average energy of electrons contained therein increase with increasing n. Permitted values of n are 1, 2, 3, ..., extending through 7 for the known elements.

The Azimuthal Quantum Number, l

Within each main energy level (shell) represented by a principal quantum number, there are *sublevels* (subshells). Each sublevel is denoted by an *azimuthal quantum number, l*. For any shell with a principal quantum number of n, the possible values of l are 0, 1, 2, 3, ..., $(n - 1)$. This gives the following:

- For $n = 1$, there is only 1 possible subshell, $l = 0$.
- For $n = 2$, there are 2 possible subshells, $l = 0, 1$.
- For $n = 3$, there are 3 possible subshells, $l = 0, 1, 2$.
- For $n = 4$, there are 4 possible subshells, $l = 0, 1, 2, 3$.

From the above it is seen that the maximum number of sublevels within a principal energy level designated by n is equal to n. Within a main energy level, sublevels denoted by different values of l have slightly different energies. Furthermore, l designates *different shapes of orbitals.*

The italicized letters *s, p, d,* and *f*, corresponding to l values of 0, 1, 2, and 3, respectively, are frequently used to designate sublevels. The number of electrons present in a given sublevel is limited to 2, 6, 10, and 14 for *s, p, d,* and *f* sublevels, respectively. It follows that, since each orbital may be occupied by a maximum of 2 electrons, there is only 1 orbital in the *s* sublevel, 3 orbitals in the *p* sublevel, 5 in the *d* and 7 in the *f*. The value of the principal quantum number and the letter designating the azimuthal quantum number are written in sequence to designate both the shell and subshell. For example, 4*d* represents the *d* subshell of the fourth shell.

The Magnetic Quantum Number, m_l

The *magnetic quantum number*, m_l, is also known as the *orientational quantum number.* It designates the orientation of orbitals in space relative to each other and distinguishes orbitals within a subshell from each other. It is called the magnetic quantum number because the presence of a magnetic field can result in the appearance of additional lines among those emitted by electronically excited atoms (that is, in atomic emission spectra). The possible values of m_l in a subshell with azimuthal quantum number l are given by $m_l = +l, +(l-1),..., 0, ..., -(l-1), -l$. As examples, for $l = 0$, the only possible value of m_l is 0, and for $l = 3$, m_l, may have values of 3, 2, 1, 0, –1, –2, –3.

Spin Quantum Number, m_s

The fourth and final quantum number to be considered for electrons in atoms is the *spin quantum number* m_s, which may have values of only +½ or –½. It results from the fact that an electron spins in either of two directions and generates a tiny magnetic field with an associated magnetic moment. *Two electrons may occupy the same orbital only if they have opposite spins so that their magnetic moments cancel each other.*

For each electron in an atom, the orbital that it .occupies and the direction of its spin are specified by values of n, l, m_l, and m_s, assigned to it. These values are unique for each electron; *no two electrons in the same atom may have identical values of all four quantum numbers.* This rule is known as the *Pauli exclusion principle.*

ENERGY LEVELS OF ATOMIC ORBITALS

Electrons in atoms occupy the lowest energy levels available to them. Fig. 2.22 is an energy level diagram that shows the relative energy levels of electrons in various atomic orbitals. Each dash, —, in the figure represents one orbital that is a potential slot for two electrons with opposing spins. The electron energies represented in Fig. 2.22 can be visualized as those required to remove an electron from a particular orbital to a location completely away from the atom. As indicated by its lowest position in the diagram, the most energy would be needed to remove an electron in the 1*s* orbital; comparatively little energy is required to remove electrons from orbitals having higher n values, such as 5, 6, or 7.

Furthermore, the diagram shows decreasing separation of the energy levels (values of n) with increasing n. It also shows that some sublevels with a particular value of n have lower energies than sublevels for which the principal quantum number is $n - 1$. This is first seen from the placement of the 4*s* sublevel below that of the 3*d* sublevel. As a consequence, with increasing atomic number, the 4*s* orbital acquires 2 electrons before any electrons are placed in the 3*d* orbitals. The value of the principal quantum number, n, of an orbital is a measure of the relative distance of the maximum electron density from the nucleus. An electron with a lower value of n, being on the average closer to the nucleus, is more strongly attracted to the nucleus than an electron with a higher n value that is at a relatively greater distance from the nucleus.

Hund's Rule of Maximum Multiplicity

Orbitals that are in the same sublevel, but that have different values of m_l, have the same energy. This is first seen in the 2*p* sublevel where the three orbitals with m_l of –1, 0, +1 (shown as three dashes on the same level in Fig. 2.22) all have the same energies. The order in which electrons go into such a sublevel follows *Hund's rule of maximum multiplicity*, which states that *electrons in a sublevel are distributed to give the maximum number of unpaired electrons* (that is, those with parallel spins having the same sign of m_s). Therefore, the first three electrons to be placed in the 2*p* sublevel would occupy the three separate available orbitals and would have the same spins. Not until the fourth electron out of a maximum number of 6 is added to this sublevel are two electrons placed in the same orbital.

Table 2.6. Quantum Numbers for Electrons in Atoms

n, Principal Quantum Number Denoting an Energy Level, or Shell	*l, Azimuthal Quantum Number of a Sublevel or Subshell*[a]		*m_l, Magnetic Quantum Number Designating Orbital (m_s in paraentheses)*[b]	
1	0	(1*s*)	0(m_s = + ½ or – ½)	1 orbital
	0	(2*s*)	0(m_s = + ½ or – ½)	1 orbital
2			–1(m_s = + ½ or – ½)	
	1	(2*p*)	0(m_s = + ½ or – ½)	3 orbitals
			+1(m_s = + ½ or – ½)	
	0	(3*s*)	0(m_s = + ½ or – ½)	1 orbital
			–1(m_s = + ½ or – ½)	
	1	(3*p*)	0(m_s = + ½ or – ½)	3 orbitals
3			+1(m_s = + ½ or – ½)	
			–2(m_s = + ½ or – ½)	
			–1(m_s = + ½ or – ½)	
	2	(3*d*)	0(m_s = + ½ or – ½)	5 orbital
			+1(m_s = + ½ or – ½)	
			+2(m_s = + ½ or – ½)	
	0	(4*s*)	0(m_s = + ½ or – ½)	1 orbital
			–1(m_s = + ½ or – ½)	
	1	(4*p*)	0(m_s = + ½ or – ½)	3 orbitals
			+1(m_s = + ½ or – ½)	
			–2(m_s = + ½ or – ½)	
			–1(m_s = + ½ or – ½)	
	2	(4*s*)	0(m_s = + ½ or – ½)	5 orbitals
4			+1(m_s = + ½ or – ½)	
			+2(m_s = + ½ or – ½)	
			–3(m_s = + ½ or – ½)	
			–2(m_s = + ½ or – ½)	
			–1(m_s = + ½ or – ½)	
	3	(4*f*)	0(m_s = + ½ or – ½)	7 orbitals
			+1(m_s + ½ or – ½)	
			+2(m_s = + ½ or – ½)	
			+3(m_s = + ½ or – ½)	

[a]Subshell designation in parentheses.

[b]Each entry in this column designates an orbital capable of holding 2 electrons with opposing spins.

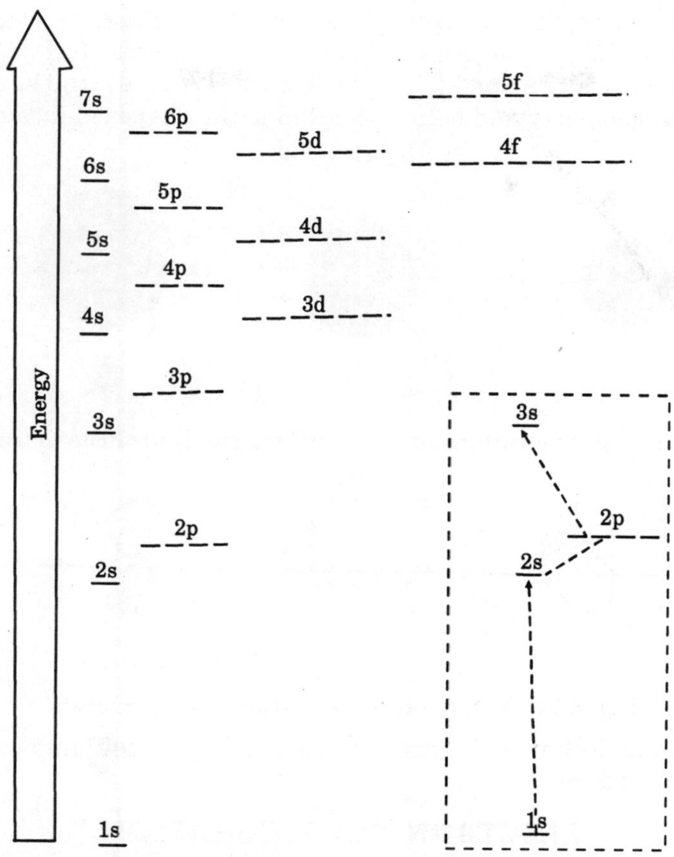

Fig. 2.22. Energy levels of atomic orbitals. Each orbital capable of containing 2 electrons is shown with a dash,—. The order of placing electrons in orbitals is from the lowest-lying orbitals up, as shown in the inset.

SHAPES OF ATOMIC ORBITALS

In trying to represent different shapes of orbitals, it is important to keep in mind that they do not contain electrons within finite volumes, because there is no outer boundary of an orbital at which the probability of finding an electron drops to exactly 0. However, an orbital can be drawn as a figure ("fuzzy cloud") around the atom nucleus within which there is a relatively high probability (typically 90%) of finding an electron within the orbital. These figures are called *contour representations of orbitals* and are as close as one can come to visualizing the shapes of orbitals.

Each point on the surface of a contour representation of an orbital has the same value of ψ^2 (square of the wave function of the Schrodinger equation), and the entire surface encloses the volume within which an electron spends 90% of its time. The two most important aspects of an orbital are its size and shape, both of which are rather well illustrated by a contour representation. Fig. 2.23 show the contour representations of the first three *s* orbitals. These are seen to be spherically shaped and to increase markedly in size with increasing principal quantum number, reflecting increased average distance of the electron from the nucleus. Fig. 2.24 shows contour representations of the three $2p$ orbitals. These are seen to have different orientations in space; they have directional properties. In fact, s orbitals are the only ones that are spherically symmetrical. As discussed already, the shapes of orbitals are involved in molecular geometry and help to determine the shapes of molecules.

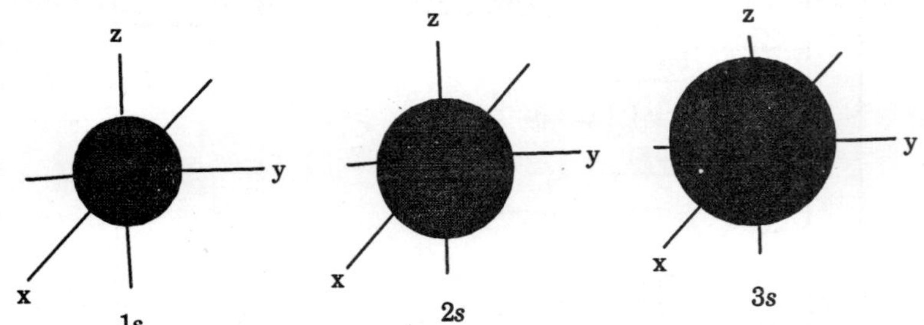

Fig. 2.23. Contour representations of s orbitals for the first three main energy levels.

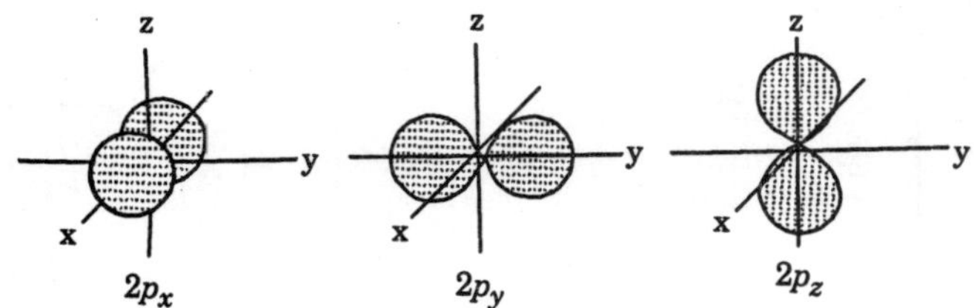

Fig. 2.24. Contour representations of $2p$ orbitals.

The shapes of d and f orbitals are more complex and variable than those of p orbitals and are not discussed in this book.

ELECTRON CONFIGURATION

Electron configuration is a means of stating which kinds of orbitals contain electrons and the numbers of electrons in each kind of orbital of an atom. It is expressed by the number and letter representing each kind of orbital and superscript numbers telling how many electrons are in each sublevel. The hydrogen atom's one electron in the $1s$ orbital is shown by the following notation:

Sublevel notation

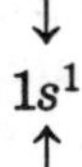

↓

$1s^1$

↑

Number of electrons in sublevel

Nitrogen, atomic number 7, has 7 electrons, of which 2 are paired in the $1s$ orbital, 2 are paired in the $2s$ orbital and 3 occupy singly each of the 3 available $2p$ orbitals. This electron configuration is designated as $1s^2 2s^2 2p^3$.

The *orbital diagram* is an alternative way of expressing electron configurations in which each separate orbital is represented by a box. Individual electrons in the orbitals are shown as arrows pointing up or down to represent opposing spins (m_s = + ½ or – ½). The orbital diagram for nitrogen is the following:

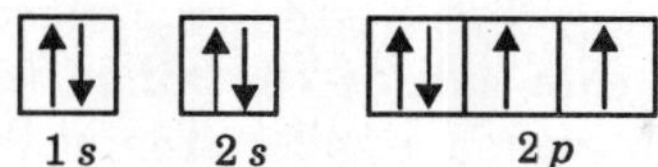

This gives all the information contained in the notation $1s^22s^22p^3$, but emphasizes that the three electrons in the three available 2*p* orbitals each occupy separate orbitals, a condition of Hund's rule of maximum multiplicity. Most of the remainder of this chapter is devoted to a discussion of the placement of electrons in atoms with increasing atomic number, how this affects the chemical behaviour of the atoms, and how it leads to a systematic organization of the elements in the periodic table.

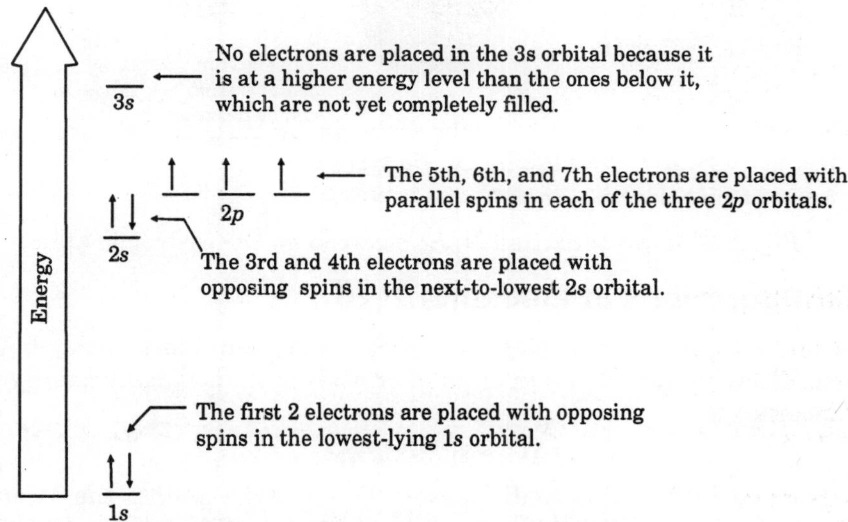

Fig. 2.25. Placement of electrons in orbitals for the 7-electron nitrogen atom according to the energy level diagram.

ELECTRONS IN THE FIRST 20 ELEMENTS

The most meaningful way to place electrons in the orbitals of atoms is on the basis of the periodic table. This enables relating electron configurations to chemical properties and the properties of elements in groups and periods of the periodic table. In this section, electron configurations are deduced for the first 20 elements and given in an abbreviated version of the periodic table.

Electron Configuration of Hydrogen

The 1 electron in the hydrogen atom goes into its lowest-lying 1s orbital. Fig. 2.26 summarizes all the information available about this electron and its configuration.

Electron Configuration of Helium

An atom of helium, atomic number 2, has 2 electrons, both contained in the 1*s* orbital and having quantum numbers n = 1, l = 0, m_l = 0, and m_s = +1/2 and −1/2. Both electrons have the same set of quantum numbers except for m_s. The electron configuration of helium can be represented as $1s^2$, showing that there are 2 electrons in the 1*s* orbital. Two is the maximum number of electrons that can be contained in the first principal energy level; additional electrons in atoms with atomic number greater than 2 must go into principal energy levels with *n* greater than 1. There are several other noble gas elements in the periodic table, but helium is the only one with a filled shell of 2 electrons—the rest have stable outer shells of 8 electrons.

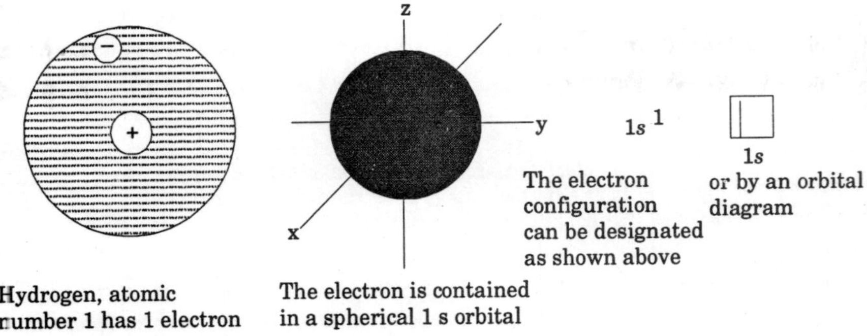

Fig. 2.26. Representation of the electron on the hydrogen atom.

Electron Configurations of Elements 2-20

The electron configurations of elements with atomic numbers through 20 are very straightforward. Electrons are placed in order of orbitals with increasing energy as shown in figure 2.23. This order is

$$1s^2 2s^2 2p^6 3s^2 3p^6 4s^2$$

It should be noted that in this configuration the electrons go into the 4*s* orbital before the 3*d* orbital, which lies at a slightly higher energy level. In filling the *p* orbitals, it should also be kept in mind that 1 electron goes into each of three *p* orbitals before pairing occurs. In order to follow the discussion of electron configurations for elements through 20, it is useful to refer to the abbreviated periodic table in figure 2.30.

Lithium

For lithium, atomic number 3, two electrons are placed in the 1*s* orbital, leaving the third electron for the 2*s* orbital. This gives an electron configuration of $1s2s^1$. The two 1*s* electrons in lithium are in the stable noble gas electron configuration of helium and are very difficult to remove. These are lithium's *inner electrons* and, along with the nucleus, constitute the *core* of the lithium atom. The electron configuration *of the core* of the lithium atom is $1s^2$, the same as that of helium. Therefore, the lithium atom is said to have a *helium core. The core of any atom consists of its nucleus plus its inner electrons, those with the same electron configuration as the noble gas immediately preceding the element in the periodic table.*

Valence Electrons

Lithium's lone 2*s* electron is an outer electron contained in the *outer shell* of the atom. Outer shell electrons are also called *valence electrons*, and are the electrons that can be shared in covalent bonding or lost to form cations in ionic compounds.

Beryllium

Beryllium, atomic number 4, has 2 inner electrons in the 1*s* orbital and 2 outer electrons in the 2*s* orbital. Therefore, beryllium has a *helium core,* plus 2 *valence electrons.* Its electron configuration is $1s^22s^2$.

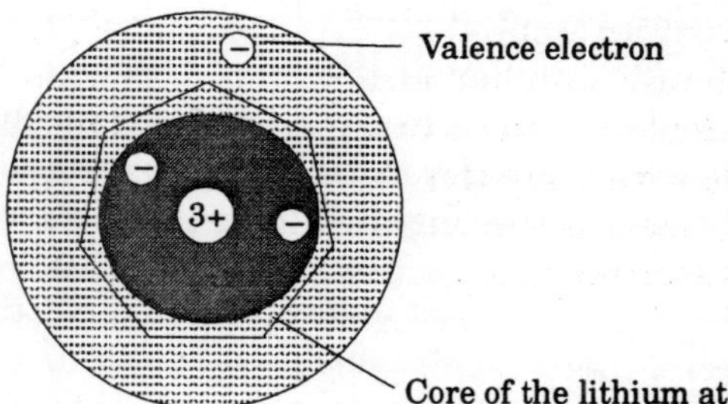

Fig. 2.27. Core (kernel) and valence electrons of the lithium atom.

Filling the 2*p* Orbitals

Boron, atomic number 5, is the first element containing an electron in a 2*p* orbital. Its electron cofiguration is $1s^22s^22p^1$. Two of the five electrons in the boron atom are contained within the spherical orbital closest to the nucleus, and 2 more are in the larger spherical 2*s* orbital. The lone 2*p* electron is in an approximately dumbbell-shaped orbital in which the average distance of the electron from the nucleus is about the same as that of the 2*s* electrons. This electron could have any one of the three orientations in space shown for *p* orbitals.

The electron configuration of carbon, atomic number 6, is $1s^22s^22p^2$. The four outer, valence electrons are shown by the Lewis symbol

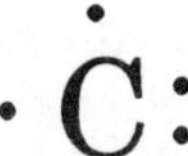

Two of the four valence electrons are represented by a pair of dots, :, to indicate that these electrons are paired in the same orbital. These are the two 2*s* electrons. The other two are shown as individual dots to represent two unpaired 2*p* electrons in separate orbitals.

The 7 electrons in nitrogen, *N*, are in a $1s^22s^22p^1$ electron configuration. Nitrogen is the first element to have at least one electron in each of 3 possible *p* orbitals (see figure 2.28).

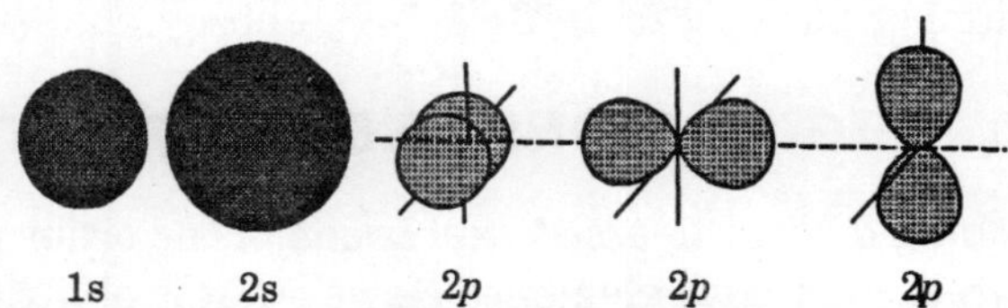

Fig. 2.28. The N atom contains 2 electrons in the 1*s* orbital, 2 in the 2*s* orbital, and 1 in each of three separate 2*p* orbitals oriented in different directions in space.

The next element to be considered is oxygen, atomic number 8. Its electron configuration is $1s^22s^22p^1$. It is the first element in which it is necessary for 2 electrons to occupy the same p orbital as shown by the following orbital diagrams:

The electron configuration of fluorine, atomic number 9, is $1s^22s^22p^5$. Its Lewis symbol,

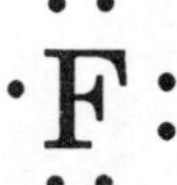

shows that it has only one unpaired electron in its valence shell. In its chemical reactions, fluorine seeks to obtain another electron to give a stable *octet.*

Neon, atomic number 10, is at the end of the second period of the periodic table and is a noble gas as shown by its Lewis symbol,

$$:\ddot{\underset{\cdot\cdot}{\mathrm{Ne}}}:$$

denoting a filled octet of electrons. Its electron configuration is $1s^22s^22p^6$. In this configuration the $2s^22p^6$ portion stands for the outer electrons. Like neon, all other elements with the outer electron configuration ns^2np^6 are noble gases located in the far right of the periodic table (for example, argon, with the electron configuration $1s^22s^22p^23s^23p^6$).

Filling the 3*s*, 3*p*, and 4*s* Orbitals

The 3*s* and 3*p* orbitals are filled in going across the third period of the periodic table from sodium through argon. Atoms of these elements, like all atoms beyond neon, have their 10 innermost electrons in the neon electron configuration of $1s^22s^22p^6$. Therefore, these atoms have a *neon core,* which may be designated {Ne}.

{Ne} stands for $1s^22s^22p^6$

With this notation, the electron configuration of element number 11, sodium, may be shown as {Ne}$3s^1$, which is an abbreviation for $1s^22s^22p^63s^1$. The former notation has some advantage in simplicity, while showing the outer electrons specifically. In the example just cited it is easy to see that sodium has 1 outer shell 3*s* electron, which it can lose to form the Na^+ ion with its stable noble gas neon electron configuration.

At the end of the third period is located the noble gas argon, atomic number 18, with the electron configuration $1s^22s^22p^63s^23p^6$. For elements beyond argon, that portion of the electron configuration identical to argon's may be represented simply as {Ar}. Therefore, the electron configuration of potassium, atomic number 19 is $1s^22s^22p^63s^23p^64s^1$, abbreviated {Ar}$4s^1$, and that of calcium, atomic number 20 is $1s^22s^22p^63s^23p^64s^1$, abbreviated {Ar}$4s^2$.

ELECTRON CONFIGURATIONS AND THE PERIODIC TABLE

There are several learning devices to assist expression of the order in which atomic orbitals are filled (electron configuration for each element). However, it is of little significance to express electron configurations without an understanding of the meaning of the configurations. By far the most meaningful way to understand this important aspect of chemistry is within the context of the periodic table as outlined in figure 2.30. To avoid clutter, only atomic numbers of key elements are shown in this table. The double-pointed arrows drawn horizontally across the periods are labeled with the kind of orbital being filled in that period. The first step in using the periodic table to figure out electron configurations is to note that the periods are numbered 1 through 7 from top to bottom along the left side of the table.

These numbers correspond to the principal quantum numbers (n values) for the orbitals that become filled across the period for, both 5 and *p* orbitals. Therefore, in the first group of elements—those with atomic numbers 1, 3, 11, 19, 37, 55, and 87—the last electron added is in the ns orbital; for example, 5*s* for Rb, atomic number 37. The last electron added to each of the second group of elements—those with atomic numbers 4, 12, 20, 38, 56, and 88—is the second electron going into the as orbital. For example, in the third period, Mg, atomic number

12, has a filled 3s orbital containing 2 electrons. For the group of elements in which the *p* orbitals start to be filled—those in the column with atomic numbers 5, 13, 31, 49, 81—the last electron added to each atom is the first one to enter an *np* orbital.

1 H $1s^1$							2 H $1s^2$
3 Li {He}$2s^1$	4 Be {He}$2s^2$	5 B {He} $2s^22p^1$	6 C {He} $2s^22p^2$	7 N {He} $2s^22p^3$	8 O {He} $2s^22p^4$	9 F {He} $2s^22p^5$	10 Ne {He} $2s^22p^6$
11 Na {Na}$3s^1$	12 Mg {Na}$3s^2$	13 Al {Ne}$3s^23p^1$	14 Si {Ne}$3s^23p^2$	15 P {Ne}$3s^23p^3$	16 S {Ne}$3s^23p^4$	17 Cl {Ne}$3s^23p^5$	18 Ar {Ne}$3s^23p^6$
19 K {Ar} $4s^1$	20 Ca {Ar} $4s^2$						

Fig. 2.29. Abbreviated periodic table showing the electron configurations of the first 20 elements.

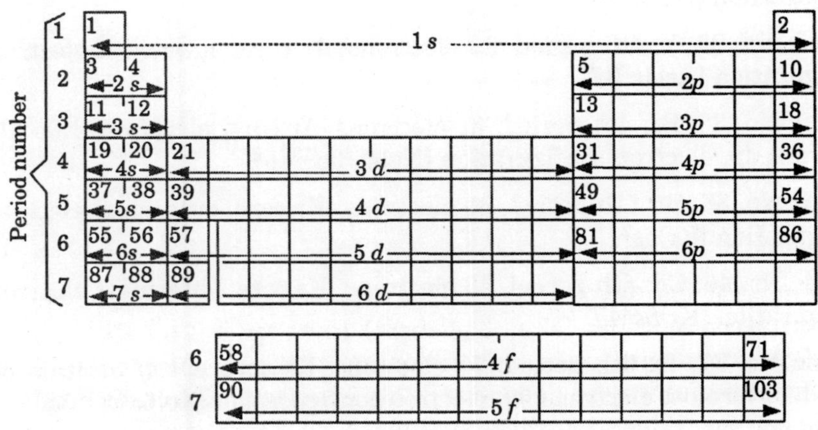

Fig. 2.30. Outline of the periodic table labeled to show the filling of atomic orbitals. The type of element being filled with increasing atomic number is shown by the labeled horizontal lines across the periods. The left point of each arrow, ←, marks an element in which the filling of a new kind of orbital starts, and the right point of each arrow, →, marks an element for which the filling of a new kind of orbital is completed. The atomic numbers are given for each element at which the filling of a kind of orbital begins or is completed.

For example for element 31, Ga, which is contained in the 4th period, the outermost electron is in the 4*p* orbital. In going to the right across each period in this part of the periodic table, the outermost electrons are, successively, np^2, np^3, np^4, np^5, and np^6. Therefore, for elements with atomic numbers 6, 7, 8, 9, and 10 in the 2nd period, the outermost electrons are $2p^2$, $2p^3$, $2p^4$, $2p^5$, and $2p^6$. Each of the noble gases beyond helium has filled *np* orbitals with a total of 6 outermost p electrons in the three *np* orbitals. Each set of five *d* orbitals becomes filled for the *transition metals* in the three horizontal periods beginning with atomic numbers 21, 39, and 57 and ending with, successively, atomic numbers 30, 48, and 80. *For each of these orbitals the value of n is* 1 *less than the period number in which the orbitals become filled.*

The first *d* orbitals to become filled are the lowest-lying ones possible, the 3*d* orbitals, which become filled in the fourth period. Across the 5th period, where the 5s and 5*p* orbitals become filled for elements 37-38 and 49-54, respectively, the 4*d* orbitals (*n* 1 *less than the period number*) become filled for the transition metals, atomic numbers 39-48. Shown below the main body of the periodic table, the *inner transition elements* consist of two rows of elements that are actually parts of the 6th and 7th periods, respectively. The principal quantum numbers of their *f* orbitals are 2 *less than* their period numbers. Fig. 2.31 shows that the first *f* orbitals begin to fill with the first of the *lanthanides,* element number 58. These are 4*f* orbitals of which there are 7 that are filled completely with element number 71. The 5*f* orbitals are filled in the 6*th* period. The 5*f* orbitals are filled with the actinide elements, atomic numbers 90-103.

With Fig. 2.30 in mind, it is possible to write the expected electron configurations of any of the elements. Consider the following examples:

(*i*) Atomic No. 16, S: 3rd period, 16 electrons, Ne core, outermost electrons 3*p*, electron configuration {Ne}$3s^23p^4$

(*ii*) Atomic No. 23, V: 4th period, 23 electrons, Ar core, outermost electrons 3*d*, electron configuration {Ar}$4s^23d^3$

(*iii*) Atomic No. 35, Br: 4th period, 35 electrons, Ar core, all 3*d* orbitals filled, outermost electrons 4*p*, electron configuration {Ne}$4s^23d^{10}4p^5$

(*iv*) Atomic No. 38, Sr: 5th period, 38 electrons, Kr core, outermost electrons 5*s*, electron configuration {Kr}$5s^2$

(*v*) Atomic No. 46, Cd: 5th period, 48 electrons, Kr core, outermost electrons 4*d*, electron configuration {Kr}$5s^24d^{10}$

(*vi*) Atomic No. 77, Ir: 6th period, 77 electrons, Xe core, all 4*f* orbitals filled in the 6th period, outermost electrons 5*d*, electron configuration {Xe}$6s^24f^{14}5d^7$

The actual electron configurations are given in Table 2.7. In some cases these vary slightly from those calculated according to the rules outlined above These exceptions occur because of the relatively higher stabilities of two half-filled sets of outermost orbitals, or orbitals in which one is half-filled and one entirely filled. The examples below illustrate this point:

- Cr, atomic number 24. Rules predict {Ar}$4s^23d^4$. However the actual electron configuration is {Ar}$4s^13d^5$ because this gives the slightly more stable electron configuration with *half-filled* 4*s* and 3*d* orbitals.
- Cu, atomic number 29. Rules predict {Ar}$4s^23d^9$. However the actual electron configuration is {Ar}$4s^13d^{10}$ because this gives *half-filled* 4*s* and *filled* 3*d* orbitals.

Table 2.7. Electron Configurations of the Elements in the Ground (Unexcited) State.

Atomic Number	*Symbol*	*Configuration of Ground State Atoms*	*Atomic Number*	*Symbol*	*Configuration of Ground State Atoms*
1	H	$1s^1$	55	Cs	{Xw}$6s^1$
2	He	$1s^2$	56	Ba	{Xe}$6s^2$
3	Li	{He}$2s^1$	57	La	{Xe}$6s^2 5d^1$
4	Be	{He}$2s^2$	58	Ce	{Xe}$6s^2 4f^1 5d^1$
5	B	{He}$2s^2 2p^1$	59	Pr	{Xe}$6s^2 4f^3$
6	C	{He}$2s^2 2p^2$	60	Nd	{Xe}$6s^2 4s^4$
7	N	{He}$2s^2 2p^3$	61	Pm	{Xe{$6s^2 4f^5$
8	O	{He}$2s^2 2p^4$	62	Sm	{Xe}$6s^2 4f^6$
9	F	{He}$2s^2 2p^5$	63	Eu	{Xe}$6s^2 4f^7$
10	Ne	{He}$2s^2 2p^6$	64	Gd	{Xe}$6s^2 4f^7 5d^1$
11	Na	{Ne}$3s^1$	65	Tb	{X}$6s^2 4f^9$
12	Mg	{Ne}$3s^2$	66	Dy	{Xe}$6s^2 4f^{10}$
13	Al	{Ne}$3s^2 3p^1$	67	Ho	{Xe}$6s^{24} 4f^{11}$
14	Si	{Ne}$3s^2 3p^2$	68	Er	{Xe}$6s^2 4f^{12}$
15	P	{Ne}$3s^2 3p^3$	69	Tm	{Xe}$6s^2 4f^{13}$
16	S	{Ne}$3s^2 3p^4$	70	Yb	{Xe}$6s^2 4f^{14}$
17	Cl	{Ne}$3s^2 3p5$	70	Lu	{Xe}$6s^2 4f^{14} 5d^1$
18	Ar	{Ne}$3s^2 3p^6$	72	Hf	{Xe}$6s^2 4f^{14} 5d^2$
19	K	{Ar}$4s^1$	73	Ta	{Xe}$6s^2 4f^{14} 5d^3$
20	Ca	{A}$4s^2$	74	W	{Xe}$6s^2 4f^{14} 5d^4$
21	Sc	{Ar}$4s^2 3d^1$	75	Re	{Xe}$6s^2 4f^{14} 5d^5$
22	Ti	{Ar}$4s^2 3d^2$	76	Os	{Xe}$6s^2 4f^{14} 5d^6$
23	V	{Ar}$4s^2 3d^3$	77	Ir	{Xe}$6s^2 4f^{14} 5d^7$
24	Cr	{Ar}$4s^1 3d^5$	78	Pt	{Xe}$6s^1 4f^1 4d^9$
25	Mn	{Ar}$4s^2 3d^5$	79	Au	{Xe}$6s^1 4f^{14} 5d^{10}$
26	Fe	{Ae}$4s^2 3d^6$	80	Hg	{Xe$6s^2 4f^{14} 5d^{10}$
27	Co	{Ae}$4s^2 3d^7$	81	Ti	{Xe}$6s^2 4f^{14} 5d^{10} 6p^1$
28	Ni	{Ar}$4s^2 3d^8$	82	Pb	{Xe}$6s^2 4f^{14} 5f^{14} 5d^{10} 6p^2$
29	Cu	{Ar}$4s^1 3d^{10}$	83	Bi	{Xe}$6s^2 4f^{14} 5d^{10} 6p^3$
30	Zn	{Ar}$4s^2 3d^{10}$	84	Po	{Xe}$6s^2 4f^{14} 5d^{10} 6p^4$
31	Ga	{Ar}$4s^2 3d^{10} 4p^1$	85	At	{Ae}$6s^2 4f^{14} 5d^{10} 6p^5$
32	Ge	{Ar}$4s^2 3d^{10} 4p^2$	86	Rn	{Xe}$6s^2 4f^{14} 5d^{10} 6p^6$
33	As	{Ar}$4s^{23} 3d^{10} 4p^3$	87	Fr	{Rn}$7s^1$
34	Se	{Ar}$4s^2 3d^{10} 4p^4$	88	Ra	{Rn}$7s^2$

Atomic Number	*Symbol*	*Configuration of Ground State Atoms*	*Atomic Number*	*Symbol*	*Configuration of Ground State Atoms*
35	Br	$\{Ar\}4s^23d^{10}4p^5$	89	Ac	$\{Rn\}7s^26d^1$
36	Kr	$\{Ar\}4s^23d^{10}4p^6$	90	Th	$\{Rn\}7s^26d^2$
37	Rb	$\{Kr\}5s^1$	91	Pa	$\{Rn\}7s^25f^26d^1$
38	Sr	$\{Kr\}5s^2$	92	U	$\{Rn\}7s^25f^36d^1$
39	Y	$\{Kr\}5s^24d^1$	93	Np	$\{Rn\}7s^25f^46d^1$
40	Zr	$\{Kr\}5s^24d^2$	94	Pu	$\{Rn\}7s^25f^6$
41	Nb	$\{Kr\}5s^14d^4$	95	Am	$\{Rn\}7s^25f^7$
42	Mo	$\{Kr\}5s^14d^5$	96	Cm	$\{Rn\}7s^25f^76d^1$
43	Tc	$\{K\}5s^24d^5$	97	Bk	$\{Rn\}7s^25f^9$
44	Ru	$\{Kr\}5s^14d^7$	98	Cf	$\{Rn\}7s^25f^{10}$
45	Rh	$\{Kr\}5s^14d^8$	99	Es	$\{Rn\}7s^25f^{11}$
46	Pd	$\{Kr\}4d^{10}$	100	Fm	$\{Rm\}7s^25f^{12}$
47	{Ag}	$\{Kr\}5s^14d^{10}$	101	Md	$\{Rn\}7s^25f^{13}$
48	Cd	$\{Kr\}5s^24d^{10}$	102	No	$\{Rn\}7s^25f^{14}$
49	In	$\{Kr\}5s^24d^{10}5p^1$	103	Lr	$\{Rn\}7s^25f^{14}6d^1$
50	Sn	$\{Kr\}5s^24d^{10}5p^2$	104	Rf	$\{Rn\}7s^25f^{14}6d^2$
51	Sb	$\{Kr\}5s^24d^{10}5p^3$	105	Ha	$\{Rn\}7s^25f^{14}6d^3$
52	Te	$\{Kr\}5s^24d^{10}5p^4$	106	Unh	$\{Rn\}7s^25f^{14}6d^4$
53	I	$\{Kr\}5s^24d^{10}5p^5$	107	Uns	$\{Rn\}7s^25f^{14}6d^5$
54	Xe	$\{Kr\}5s^24d^{10}5p^6$	109	Une	$\{Rn\}7s^25f^{14}6d^7$

3 Metals

Increasing technologic use of metals is one measure of man's progress since his emergence from the Stone Age. This has posed hazards to health from the time metals were fashioned into spears to present-day exposures to space-age metals, alloys, or salts. High natural concentrations of metals in food or water could have led to the first exposures. Metal leached from eating utensils or metallic cookware increased the risk of exposure. Intentional use of compounds containing toxic metals as pesticides or as therapeutic agents increased the opportunity for hazardous exposures. Although some metals have been known for centuries as industrial poisons, the coming of the industrial age led to more widespread occurrence of occupational disease related to exposure to a variety of toxic metals.

In recent years a justifiable concern has arisen in regard to pollution of our environment by toxic metals. Metals or their salts are also used therapeutically but their use has declined, particularly in the treatment of infectious diseases, with the advent of more effective organic drugs. Certain metal salts such as the mercurial diuretics still have a place in therapy. Recently lithium carbonate has been introduced for use in the treatment of manic-depressive psychosis, and one of the major uses of bismuth is in the manufacture of over-the-counter medications for the treatment of gastrointestinal distress. Some metals are essential for life. Others have no known biologic function, but are not serious toxic hazards. Still other metals have the potential to produce disease.

Metals that are essential nutrients can also exert toxic action if the homeostatic mechanism maintaining them within physiologic limits is unbalanced. Iron, for example, may be purposely included in the diet or given as supplements to correct symptoms of deficiency. Excessive intake of iron, however, is a common cause of accidental poisoning. Iron may also produce industrial disease and, in special cases (*e.g.*, in the Bantus of Africa), it may produce maladies classified as environmental or at least nonoccupational diseases. The metals having the greatest potential for causing disease are those which accumulate in the body. The daily intake (largely from food) and the body burden of a variety of metals are shown in Table 3.1.

Concentrations of aluminum, vanadium, titanium, chromium, strontium, tin, lead, and cadmium in the lung increase up to age 40; these increases are due to the accumulation of insoluble particles. Levels of nickel, tin, strontium, cadmium, lead, and perhaps barium are increased in other tissues as well as in the lungs due to inhalation and relocation. The tolerance levels for metals in drinking water established by the U.S. Public Health Service along with the results of a survey of concentrations of metals found in drinking water samples are shown in Table 3.2. The survey suggests that significant numbers of people are exposed to the hazards of excess metals in the municipal water supplies.

Table 3.1. Body Burden and Human Daily Intake and Content in the Earth's Crust of Selected Elements*

Element	Human Body Burden (Mg / 70 Kg)	Daily Intake (Mg)	Earth's Crust (Ppm)
Aluminum	100	36.4	81,300
Antimony	<90		0.2
Arsenic	<100	0.7	2
Barium	16	16	400
Boron	<10	0.01-0.02	16
Cadmium	30	0.018-0.02	0.2
Calcium	1,050,000		36,300
Cesium	<0.01		1
Chromium	<6	0.06	200
Cobalt	1	0.3	23
Copper	100	3.2	45
Germanium	Trace	1.5	1
Gold	<1		0.005
Iron	4,100	15	50,000
Lead	120	0.3	15
Lithium	Trace	2	30
Magnesium	20,000	500	20,900
Manganese	20	5	1,000
Mercury	Trace	0.02	0.5
Molybdenum	9	0.35	1
Nickel	<10	0.45	80
Niobium	100	0.60	24
Potassium	140,000		25,900
Rubidium	1,200	10	120
Selenium	15	0.06-0.15	0.09
Silver	<1		0.1
Sodium	105,000		28,300
Strontium	140	2	450
Tellurium	600	0.6	0.002
Tin	30	17	3
Titanium	<15	0.3	4,400
Uranium	0.02		2
Vanadium	30	2.5	110
Zinc	2,300	12	65
Zirconium	250	3.5	70

*Data derived largely from Schroeder, 1965*b*.

Although excessive concentrations of metals may occur in water, air, or soil as a result of natural deposits, technologic use of these non-biodegradable materials can lead to their accumulation in the environment. Vanadium may be released into the atmosphere from the combustion of oil. Many metals, including mercury, may be released from the combustion of coal. The use of leaded gasoline has added to the levels of lead in the environment. Table 3.3 shows figures for the concentration of various metals in the atmosphere. Metals released to the environment may be bioconcentrated and thus enter the food chain.

Mercury compounds released with industrial wastes may be converted by microbial systems in aquatic bottom mud to the highly toxic methyl mercury, which is then taken up by fish living in the contaminated waters. Such an incident led to deaths and tragic disabling disease among residents of Minimata, including infants of exposed mothers. In industrial situations, inhalation is the most important route of exposure. The background of long experience has led to the recommendation of concentrations in the air of the workplace that are deemed safe for eight-hour exposures. These values, which were adapted as Standards by the Occupational Safety and Health Administration (OSHA), are shown in Table 3.4. In some instances (alkyl lead compounds and thallium) the hazards of skin absorption have been taken into consideration as well in establishing the safe level. Other metals, such as nickel, beryllium, and arsenic, include skin changes as part of their spectrum of toxicity.

Topical exposure to certain occupational metals may result in irritation of the skin and eyes or sensitization reactions and provide a route of absorption resulting in systemic toxicity. Contact with abraded rather than intact skin can produce serious symptoms of toxicity. While parenteral exposure is generally limited to medicinal use, cases of metal splinters being embedded as the result of industrial use are not unknown.

FACTORS INFLUENCING TOXICITY

Before considering the toxic properties of individual metals, it is useful to call attention to certain general properties of this class of elements that have considerable impact on their toxicity. To begin with, they seldom interface with biologic systems in the elemental form. Rather, they occur as discrete compounds that vary considerably in the ease with which they pass across biologic membranes. Soluble salts of metals dissociate readily in the aqueous environment of biologic membranes, facilitating thereby their transport as metal ions. Conversely, insoluble salts are relatively poorly absorbed, particularly if they are presented to absorptive biologic surfaces in a polymeric state of aggregation. Even in the case of soluble metallic salts, certain factors modify their absorption. Thus, a soluble salt may interface with an organism in the presence of anions that favour the formation of insoluble salts. As a case in point, a high level of dietary phosphate reduces the gastrointestinal absorption of lead because the highly insoluble lead phosphate salt is formed.

Foods have a high capacity for metal binding with consequent reduction of absorption. Thus, absorption is much greater when ingestion occurs during a period of fasting than on a full stomach. The matter of solubility is particularly important in determining the fate of metals deposited in the airways. The more insoluble the metal compound, the more likely it is to be cleared from the pulmonary bed by retrograde movement to the pharynx with subsequent swallowing. In the course of this retrograde movement, systemic absorption is minimal.

Table 3.2. Tolerance Levels for Metals in Drinking Water and Results of Sampling of Community Water Supplies (969)

	Limits in Mg/Liter			*Number of Samples of A Total of 2,595 Exceeding*	
Element	*Mandatory Upper*	*Desirable Upper*	*Maximum Concentrations Found*	*Mandatory*	*Desirable*
Arsenic	0.05	0.01	0.10	5	10
Barium	1.0		1.55	2†	
Boron	5.0	1.0	3.28	0	20
Cadmium	0.01		3.94	4	
Chromium ($Cr^{6\dagger}$)	0.05		0.79 ‡	5	
Copper		1.0	8.35		42
Iron		0.3	26.0		223
Lead	0.05		0.64	37	
Manganese		0.05	1.32		211
Selenium	0.01		0.07	10	
Silver	0.05		0.03	0	
Uranium (uranyl)§		5.0	Not included		
Zinc		5.0	13.0		8

† Not measured in all samples.
‡ Total chromium measured.
§ Proposed.

Table 3.3. Urban Air Metal Particle Concentration in the United States, 1964-1965*

	Concentration (Mg/M^3)	
Pollutant	*Average*	*Maximum*
Antimony	0.001	0.160
Arsenic	0.02	
Beryllium	<0.0005	0.010
Bismuth	<0.0005	0.064
Cadmium	0.002	0.420
Chromium	0.015	0.33
Cobalt	<0.0005	0.060
Copper	0.09	10.0
Iron	1.58	22.0
Lead	0.79	8.60
Manganese	0.10	9.98
Molybdenum	<0.005	0.78
Nickel	0.034	0.460
Tin	0.02	0.50
Titanium	0.04	1.10
Vanadium	0.050	2.20
Zinc	0.67	58.0
Barium†	0.09	
Samarium†	0.07	

† 1970 values.

Table 3.4. Acceptable Average Concentrations (μg/M³) of Occupational Exposure Based on Eight-hour Exposures*

Antimony and compounds (as Sb)	500	
Stibine (SbH_3)	500	(0.1 ppm)
Arsenic and compounds (as As)	500	
Arsine (AsH_3)	200	(0.05 ppm)
Arsenate, calcium	1,000	
Arsenate, lead	150	
Barium (soluble compounds)	500	
Beryllium and compounds	2	(5†)
Boron oxide	15,000	
Boron trifluoride	3,000†	
Diborane	100	
Pentaborane		(0.005 ppm)
Decaborane (skin)	300	
Cadmium fume	100	(3,000†)
Cadmium dust	200	(600†)
Chromic acid and chromates	100†	
Chromium, soluble salts	500	
Chromium, metal and insoluble salts	1,000	
Cobalt, metal fume and dust	100	
Copper fume	100	
Copper, dusts and mists	1,000	
Hafnium	500	
Iron oxide fume	10,000	
Ferbam (ferric dimethyldithiocarbamate)	15,000	
Ferrovanadram dust (FeV)	1,000	
Lead and its inorganic compounds	200	
Lead arsenate	150	
Lead, tetraethyl (as Pb—skin)	75	
Lead, tetramethyl (as Pb—skin)	75	
Lithium hydride	25	
Magnesium oxide fume	15,000	
Manganese	5,000†	
Mercury	100†	
Mercury (organo alkyl)	10	(40†)
Molybdenum (soluble compounds)	5,000	

Molybdenum (insoluble compounds)	15,000	
Nickel, metal and soluble compounds as Ni	1,000	
Nickel carbonyl	7	(0.001 ppm)
Osmium tetroxide	2	
Platinum (soluble salts as Pt)	2	
Rhodium, metal fume and dust as Rh	100	
Rhodium (soluble salts)	1	
Selenium compounds as Se	20	
Selenium, hexafluoride	400	
Silver, metals and soluble compounds	10	
Tantalum	5,000	
Tellurium	100	
Tellurium hexafluoride	200	(0.02 ppm)
Thallium (soluble compound—skin as Tl)	100	
Tin (inorganic compounds except oxides)	2,000	
Tin (organic compounds)	100	
Titanium dioxide	15,000	
Uranium (soluble compounds)	50	
Uranium (insoluble compounds)	250	
Vanadium (V_2O_5 dust)	500†	
Vanadium (V_2O_5 fume)	100†	
Yttrium	1,000	
Zinc chloride fume	1,000	
Zinc oxide fume	5,000	
Zirconium compounds as Zr	5,000	

† Ceilings.

Some metals occur in the environment as alkyl compounds, in which case the metal is firmly bonded to carbon. These alkyl compounds remain largely intact in the biologic environment. They are lipid soluble and pass readily across biologic membranes unaltered by the surrounding medium. Even after their absorption, they are only slowly dealkylated and are, therefore, distributed in the body in accordance with their lipid-soluble characteristics.

The most notable examples of this type of organometallic compound are methyl mercury and tetraethyl lead. Their toxicologic properties are quite different from those of inorganic forms, attesting to their integrity in the body as distinct molecular entities. The strong attraction between metal ions and organic ligands has implications beyond the matter of their availability for absorption. It also influences the disposition of metals in the body and their rate of excretion. Most lexicologically important metals bind strongly to tissues and therefore are only slowly excreted. Consequently, with continuing intake they tend to accumulate to a high degree. Tissue affinities of the various metals are, however, quite dissimilar. Thus, lead and radium have a strong affinity for osseous tissue, whereas cadmium and mercury localize mainly in the kidney.

METAL CHELATION

Because metals persist so strongly in the body, a major therapeutic objective in poisoning is the administration of drugs that enhance their excretion. The concept of enhancing metal excretion by administration of readily excreted complexing agents should probably be credited to Seymour Kety. Starting with the observation that citrate has a powerful solvent action on lead phosphate, he determined by a potentiometric method the high affinity of citrate for ionic lead and suggested the use of citrate as a means of promoting lead excretion. The concept was further defined with the development of dimercaptopropanol (British Anti-Lewisite; BAL) as an antidote for arsenic poisoning. Starting with the observation that arsenic has an affinity for sulfhydryl-containing substances, a series of low-molecular-weight sulfhydryl compounds was synthesized and tested for efficacy.

Dithiols were found to be more protective than monothiols and, among the dithiols, those with sulfhydryl groups on adjacent carbon atoms were best able to reverse the toxic effects of arsenic. This led to the conclusion that the simultaneous binding of arsenic to two sulfur atoms on adjacent carbon atoms was required to compete successfully with the critical binding site responsible for the toxic effects (Stocken and Thompson, 1946). Further, these observations led to the

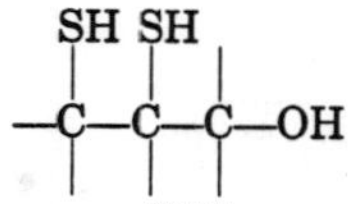

SH SH
—C—C—C—OH
BAL

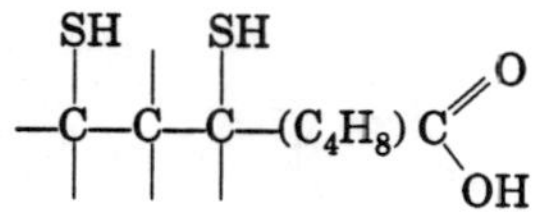

SH SH
—C—C—C—(C_4H_8) C(=O)OH

6, 8-dithiooctanoic acid
(α-lipoic acid)

prediction that the "biochemical lesion" of arsenic poisoning would prove to be a dithiol with sulfhydryl groups separated by one or more intervening carbon atoms. This prediction was borne out a few years later with the discovery that arsenic interferes with the function of 6, 8-dithiooctanoic acid in biologic oxidation. Since these initial studies, the concept of metal-binding agents as drugs has received considerable attention.

Certain principles have emerged and a few more effective agents have been developed. When two or more ligands (*e.g.*, sulfhydryl groups) in a molecule simultaneously form bonds with a metal atom, the donor molecule is properly referred to as a chelating agent. This term is derived from the Greek *chela,* for claw. Multidentate complexes of this type generally are more stable than simple unidentate complexes. The classic ligands are anions of oxo acids, *e.g.*, R—C(=O)—O— or neutral molecules in which the donor atom is nitrogen, *e.g.*, $R—NH_2$.

Thiols, such as in dimercaptopropanol, resemble oxo acids in that they coordinate with metal ions by giving up protons. The usual role of these ligands is that they are electron donors. In the process of forming metal chelates, the formation of five or six-membered rings generally is required for optimal stability. Chelating agents are generally nonspecific in regard to their affinity for metals. To varying degrees, they will mobilize and enhance the excretion of a rather wide range of metals, including essential metals such as calcium and zinc. Their efficacy depends not solely on their affinity for the metal of interest, but also on their affinity for endogenous metals, mainly Ca, which compete in accordance with their own affinities for the chelator. The affinity constant $\left(K\frac{M}{ML}\right)$ for a metal (*M*) and a ligand donor chelating agent (*L*) is defined by

$$K\frac{M}{ML} = \frac{[KM]}{[K][M]}$$

The net affinity of the chelator (K'_M) for the metal *M* is approximated by

$$K'_M = \frac{K^M_{ML[L]}}{a_L + K^{C_a}_{C_{aL}[Ca^{2+}]}}$$

in which a_L is calculated from the affinity of the ligand groups for hydrogen ions at the biologic pH. K'_M is an oversimplified figure of merit. It holds only for 1:1 metal-chelator complexes. For 1:2 complexes the product of the successive formation constants K^M_{ML} and K^M_{ML2} is substituted for K^M_{ML}, giving the net affinity constant:

$$K''_M = \frac{K^M_L\, K^M_{L2}\,[L]^2}{(a_L + K^{Ca}_{CaL}\,[Ca^{2+}])^2}$$

Further, it is obvious that calcium is only one of many metals that compete with toxic metals in the body. However, other metals in the body either form less stable metal chelates or are present in much lower concentrations than calcium.

The most thoroughly studied group of chelating agents is the family of polyaminocarboxylic acids. The most widely used member of this group is ethylenediaminetetraacetic acid (EDTA):

```
    O                      O
     \\                   //
HO—C—C                 C—C—OH
      \    |   |      /
       N—C—C—N
      /    |   |      \
HO—C—C                 C—C—OH
   //                    \\
  O                        O
```

It has been used extensively in the treatment of lead poisoning. The association between metal and chelator probably is usually quadridentate in nature:

```
     O=C—O—M—O—C=O
       |   /\ \  |
      —C—         —C—
       |           |
   O                   O
    \\                //
⁻O—C—C—N—C—C—N—C—C—O⁻
```

Its congener, diethylenetriaminepentaacetic acid (DTPA), has a slightly higher K'_M. As would be predicted, its lead-mobilizing efficacy is somewhat higher than for EDTA

Table 3.5. Net Affinity Constants* for Some Selected Metal Ions and Chelating Agents

Metal ION	*EDTA‡*	*DTPA§*	*DFOA‖*	*PA¶*
Be^{2+}	−2.3	−1.1		
Fe^{3+}	13.4	15.6	20.7	
Cu^{2+}	7.1	9.6	4.2	8.9 (10.1)
Zn^{2+}	4.6	6.5	1.2	2.4
Ce^{3+}	4.3	8.5		
Hg^{2+}	10.1	14.8		9.9 (11.9)
Pb^{2+}	6.3	7.2		4.8 (5.7)

* Numbers = log K'_M; numbers in parentheses = log K''_M

‡ Ethylenediamine tetraacctate.

§ Diethylenetriamine pentaacetate.

‖ Desferrioxamine.

¶ D-Penicillamine.

K^M_{ML}'*s,* however, are not always reliable determinants of *in vivo* metal-mobilizing efficacy. Thus, K^M_{ML}'S for mercury suggest that DTPA would be a better mobilizing agent than D-penicillamine (PA) whereas the reverse actually is true (Catsch and Harmuth-Hoene, 1975). Factors other than relative affinities obviously are involved in *in vivo* metal mobilization. At best, affinity characteristics serve only as a guide to the investigator searching for potential metal mobilizers.

Some chelating agents may not be useful because they are rapidly metabolized to inactive forms in the body. Others may bind tightly to a toxic metal but may at the same time remain immobilized at the site of metal complexation by formation of ternary complexes with fixed ligand acceptors in the tissues. See Catsch and Harmuth-Hoene (1975) for an up-to-date review of chelating agents and metal antidotes.

TOXIC EFFECTS

There is very little by way of generalizing principles concerning the mechanisms of action of toxic metals. The very elemental nature of the metals and their diverse affinities for organic ligands in biologic structures are characteristics that discourage any Unitarian concepts of toxic actions. Those which have been most thoroughly studied produce a bewildering array of biologic effects. Their toxic actions similarly involve a multiplicity of target organs and systems. In no case can the multiple manifestations of toxicity be assigned to the inhibition of a single enzyme or a single biochemical process. It is perhaps because of this multiplicity of effects that the concepts of "critical organ" and "critical dose" have evolved in connection with the metals rather than with some other class of poisons.

The term "critical" is used to denote "most sensitive." Thus, the critical organ is the one showing adverse effects at the lowest dose. There is no implication involved as to severity of effect. Other organs and systems may be much more severely affected, but only at higher doses. The concept assumes considerable importance in regard to toxic agents for which a tolerance limit value greater than zero seems necessary because of technologic or economic imperatives. Needless to say, a ranking of effects in accordance with dose has practical meaning only insofar as it can be defined in man or other species for which benefits are an issue. A ranking in laboratory animal species is useful, but only as a first step toward assessment in the species of ultimate concern.

ACQUIRED TOLERANCE

The field of biology owes much of its fascination to the intricacies of adaptation, whether it be on an evolutionary scale or in the moment-to-moment responses of the organism to a changing environment. Tolerance may be viewed as a special form of adaptation in which continued exposure to a chemical agent results in an increased resistance to the noxious consequences of the exposure. In the fields of pharmacology and toxicology, the best-known mechanism of acquired tolerance is induction of the mixed-function oxidase system, wherein certain organic compounds stimulate their own detoxification and that of other organic compounds. The concept of acquired tolerance to metals has its origins in anecdotal form.

Natives of the Austrian Tyrol reputedly acquired a tolerance to arsenic by continuous exposure. No evidence has ever been presented confirming the existence of acquired tolerance to arsenic. Nonetheless, in the case of lead and cadmium at least, there is fragmentary evidence for acquired tolerance. In both cases continued exposure to low doses of the metals results in the elaboration of proteins that strongly bind the metals. The implications of this phenomenon are discussed in more detail under the separate headings for these two metals. The various aspects of metal toxicology that have been discussed above are not equally understood for all metals.

As a matter of fact, the body of knowledge concerning the individual metals is extremely spotty. For a few, notably lead, mercury, and cadmium, a fairly complete picture emerges. For one reason or another, these have been studied most intensively. They serve as models. Admittedly, current and future problems in the toxicology of metals will only be solved by the integration of knowledge drawn from the broad spectrum of biomedical sciences. None-theless, the search for knowledge in this area clearly requires an understanding of concepts and phenomena that have evolved in the immediate field of interest. With that thought in mind, the metals whose toxicology is best known will be considered first.

LEAD

Introduction

It is fitting that consideration of individual metals in this chapter should begin with lead. No metal has been more intensively studied from a toxicologic point of view and no metal has presented a broader range of problems, both in regard to the multiplicity of routes of entry, and in regard to the spectrum of organs and systems affected in man, and in domesticated and wild animals as well. The highest level of exposure occurs principally among people working in lead smelters. The various processes involved in refining lead result in generation of metal fumes and deposition of lead oxide dust in the workers' occupational environment.

Conditions are only somewhat better in storage battery factories, where lead oxide dust is a constituent of the battery grids and is likewise an inevitable by-product of grid preparation.

Other manufacturing operations, too numerous to describe here, result in varying degrees of lead exposure.

In the general population, the major hazard is for young children who chew and swallow objects contaminated with lead-containing paint, *e.g.*, flaking paint on walls and woodwork or weathered lead paint dust and flakes leaching from the exterior of residential and commercial structures into adjacent soil and dust. For those who are interested in the broad aspects of the environmental significance of lead as a pollutant, there are two good recent reviews.

Metabolism

For all practical purposes, there are two forms of lead. The first is inorganic lead, in which the various salts and oxides are considered to act identically once absorbed into the systemic circulation. The second form is alkyl lead, notably tetraethyl lead and tetramethyl lead. These are clearly different from inorganic forms of lead, as to both absorption and disposition in the body. They will therefore be discussed separately. Metabolism of organic forms of lead has been studied extensively, not only in animal models but also in man.

Distinctions are not generally made regarding the disposition of the various inorganic compounds. It is assumed that lead ions dissociate to some degree and are absorbed and distributed in the body in the same manner, regardless of environmental origin. The validity of this assumption has not been tested rigorously, but there is no reason to suspect that lead salts retain their identity as a molecular species during the processes of absorption and subsequent distribution and excretion.

Absorption

The major routes of lead absorption are the gastrointestinal tract and the respiratory system. Small amounts of lead may also be absorbed from the intact or abraded skin when applied in high concentrations. So far as the general population is concerned, dermal contact with concentrated aqueous solutions of lead is infrequent. This is in contrast to the continual tngestion and inhalation of lead that are experienced to varying degrees by everyone. In contrast to the relative insignificance of skin as a route of inorganic lead absorption, alkyl lead compounds are absorbed to such a degree that toxicity among handlers of these compounds in the blending of leaded gasoline has been attributed to skin absorption.

The average dermal lethal dose of tetraethyl lead is 700 mg/kg Pb in rabbits, only approximately six times the oral lethal dose. The absorption of lead from the gastro-intestinal tract is greatly influenced by con-current dietary levels of numerous substances, notably calcium, iron, fats, and proteins. Absorption also is considerably greater in infants than in adults and during the fasting state than with meals.

The overall absorption of lead taken as inadvertent contaminants of foods and beverages by adults is approximately 8 per cent, an estimate arrived at years ago by long-term balance studies in volunteers and more recently confirmed by tracer studies using the stable isotope ^{204}PB. The reason for the substantially greater absorption of dietary lead in children, approximately 40 per cent as contrasted to 8 per cent in adults, is not known. It is doubtful that the difference is due to differences in the nature of the diet, since the same phenomenon has been observed in experimental animals. Little is known regarding the nature of lead transport across the gastrointestinal mucosa. It has been speculated that a competition exists between lead and calcium for a common transport mechanism, based on the known reciprocal relationship between dietary calcium levels and lead absorption. The reason for this is not known at present.

Active calcium transport by the everted gut sac, however, is not influenced by lead. The absorption of lead by inhalation has in many instances been responsible for documented cases of lead poisoning. Further, there is concern over the contribution that lead in ambient air makes to total exposure from all sources in the general population. The sources of lead in air range from large particle-size dusts to highly respirable aerosols. The chemical composition is equally variable. Lead oxide is the dominant form in smelters and storage battery plants, while in ambient air a complex mixture exists of halides, carbonates, oxides, phosphates, and sulfates. Since both particle size and chemical form have a substantial influence on the regional deposition and clearance of aerosols, it is not surprising that fractional absorption of inhaled lead is inadequately understood.

A few elegant studies have been reported concerning the deposition and clearance kinetics for lead aerosols using radioactive lead tracers. Such studies are of limited value, at least theoretically, because they can only be conducted on a very limited number of volunteers under very restricted circumstances as to ventilatory patterns and physico chemical aerosol characteristics.

Distribution

The kinetics of lead distribution are complex. The fate of a single dose of lead in experimental animals has been reported by several investigators. The major characteristics are (1) rapid, profound transfer to bone, and (2) progressively decreasing rate of excretion (Fig. 3.1). It is seen from the figure that within the two-month period, of observation, the fractional excretion of the residual amount of lead in the body at any point in time becomes progressively smaller. This is probably because the residual fraction of the single original dose of lead becomes progressively more deeply buried in the bone matrix with the passage of time and, consequently, becomes progressively less accessible to the circulating body fluids (blood, lymph, canalicular transudate).

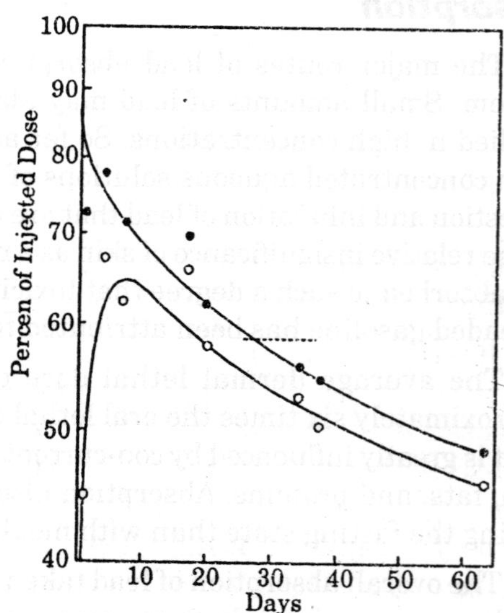

Fig. 3.1. The fate of a single dose of lead in rats. • = total amount in body; O = amount in skeleton. (From Hammond, P. B.: The effect of chelating agents on the tissue distribution and excretion of lead.

The simplest mathematical description of excretion kinetics for a single dose of lead is one wherein a linear relationship is evident for In residual fraction of dose vs. In time. If these interpretations of animal studies are correct, it would be expected that continued administration of lead would result in continued accumulation, even over very long periods of time. Autopsy studies of lead in people of different ages in fact do indicate that lead accumulates throughout life, at least in bone. In contrast to bone, lead in most soft tissues seems to rise only to early adulthood or middle life. An approximate steady state of intake vs. outgo is reached within variable but finite periods of time. Equilibration in blood is quite rapid, in contrast to other organs and systems. In a group of volunteers exposed continuously to an elevated concentration of lead in air, virutal equilibration of blood lead with the new exposure level occurred within approximately 100 days. It is not to be inferred that this plateau indicates that lead in blood has equilibrated completely with all tissues.

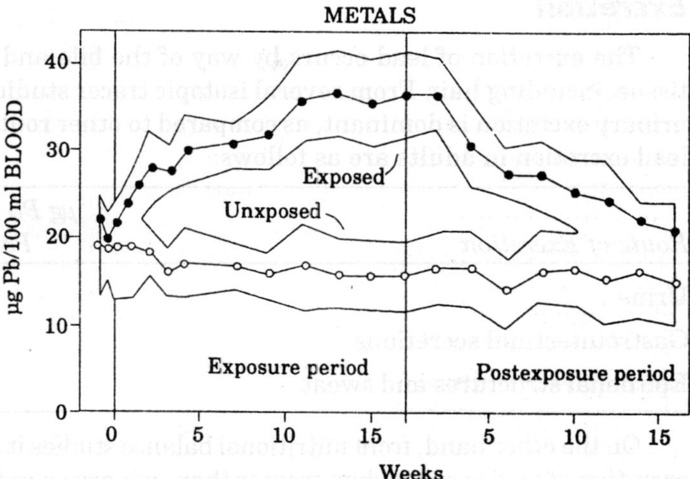

Fig. 3.2. Effect of continuous air exposure to 10.9 μg Pb/M³ on the concentration of lead in the blood of human volunteers. (From Cole, J. F., and Lynam, D. R.: ILZRO's research to define lead's impact on man. *Proceedings of the International Symposium, Environmental Health Aspects of Lead.*

It is known that the rate of return of blood lead to normal following cessation of long-term occupational exposure is much slower than is depicted in the post-exposure period of Fig. 3.2. Approximately 90 per cent of the total body burden of lead with long-term exposure is in the bones. No other strikingly high affinities exist, although liver and kidney have somewhat higher-than-average concentrations. It has been observed that high lead exposure in both man and animals results in the appearance of dense, intranuclear inclusion bodies, principally in the proximal tubular cells of the kidney, but in some other organs as well. These bodies are composed of protein and also have an extremely high concentration of lead. Their formation, which is stimulated even by a single dose of lead, is dependent on *de novo* protein synthesis.

It has been speculated that this is a defense mechanism whereby lead is sequestered away in a relatively nonavailable form. The concentration of lead in the blood is of unique importance. It is used as an index of recent lead exposure, both in the lead-using industries and in childhood lead-screening programs. By convention, the concentration of lead in the blood is expressed as micrograms per deciliter and is referred, to generally in those terms by the symbol PbB. It is the only index of exposure that has been used extensively in linking external dose (*e.g.*, concentration of lead in the air) to internal dose, or amount of lead in the body.

It is also the only index of internal dose that has been linked to the various biologic effects of lead. Thus, in the development of dose-response and dose-effect relationships, "dose" is almost always expressed as PbB. The toxic effects of alkyl lead compounds are quite different from those of inorganic lead, as is their distribution in the body. Tetraethyl and tetramethyl lead are rapidly dealkylated by the liver to the trialkyl metabolites that are responsible for toxicity. The trialkyl metabolites are then only slowly metabolized to inorganic lead.

Excretion

The excretion of lead occurs by way of the bile and urine and by exfoliation of epithelial tissue, including hair. From several isotopic tracer studies conducted on men, it was found that urinary excretion is dominant, as compared to other routes. Approximate contributions to daily lead excretion in adults are as follows:

Route of Excretion	*μg Pb Excreted Per Day*	*%*
Urine	36	76
Gastrointestinal secretions	8	16
Epithelial structures and sweat	4	8

On the other hand, from nutritional balance studies it appears that in infants gastrointestinal excretion of lead is somewhat greater than urinary excretion. The pattern of excretion observed in experimental animals resembles that in human infants. The mechanism of urinary lead excretion is not fully known, but the most likely process is glomerular filtration with variable, degrees of tubular reabsorption, depending oh the filtered load.

Biologic Effects

Many organs and systems are adversely affected by lead. Some effects have been observed in both man and experimental animals. Others have been observed only in experimental animals. The implications of effects reported only in animal studies often are not known because relevant human studies have not been conducted. The four major target organs and systems are the central nervous system, the peripheral nerves, the kidney, and the hematopoietic system. In all four cases the effects have been observed in man and have been studied extensively.

Central Nervous System

There are numerous reports of a severe, often fatal condition commonly referred to as lead encephalopathy, occurring as a result of chronic or subchronic exposure to high doses of inorganic lead. The major features are dullness, restlessness, irritability, headaches, muscular tremor, ataxia, and loss of memory. These signs and symptoms may progress to convulsions, coma, and death. A high incidence of residual damage is seen, including epilepsy, hydrocephalus, and idiocy. These sequelae are similar to those seen following infectious or traumatic injury to the brain.

The pathogenic mechanism leading to these effects is not well understood. Although varying degrees of cerebral vascular damage are common in fatal cases, along with demyelination and axonal damage in neurons, these are not constant findings. A major concern today is subtle behavioural effects, particularly in children, at levels of exposure below those causing encephalopathy. Epidemiologic studies suggest that only moderately elevated lead exposure in infants and young children (PbB = 40 to 80) may cause deficits as reflected in psychometric performance tests and in certain neurologic tests. The toxic effects of alkyl lead compounds on the central nervous system are more of a psychic nature, compared to inorganic lead.

Hallucinations, delusions, and excitement are the most common effects. These progress to delirium in fatal cases. There have been numerous studies utilizing experimental animal models regarding the effects of inorganic lead on the central nervous system. These studies have mainly been concerned with possible effects of lead on certain performance tasks that might reflect effects on cognitive function (learning and memory) or sensorimotor function in the infant animal exposed to lead very early in life or *in utero*. These studies have not as yet yielded any conclusive, consistent information, but they do tend to confirm observations in children, suggesting subtle effects not accompanied by overt signs of lead poisoning. The mechanism whereby functional disturbances of the central nervous system occur is poorly understood.

Some investigators have studied effects of lead on the action of neurotransmitters using isolated peripheral nerve preparations. Both cholinergic and adrenergic synaptic evoked transmitter release is inhibited by lead. This effect is inhibited by calcium. The significance of these observations in regard to the brain is highly uncertain at this time.

Peripheral Nervous System

The older literature cites the frequent occurrence of lead palsy among workers in the lead trades. The major manifestation of lead palsy is weakness of the extensor muscles. Sensory disturbances also occur, *e.g.*, hyperaesthesia and analgesia. This peripheral neuropathy has been studied in some detail experimentally. The anatomic lesion is characterized by segmental demyelination and by axonal degeneration. Functionally, nerve conduction velocity is slowed, even in the absence of palsy, an effect seen in both children and adults even with no discernible impairment of myoneural function.

Kidney

Two distinct types of renal effect have been observed in man. In the first type, the effects are manifestations of damage to the proximal tubules. Tubular reabsorption of glucose, amino acids, and phosphate is depressed. These effects are readily reversible with chelation therapy.

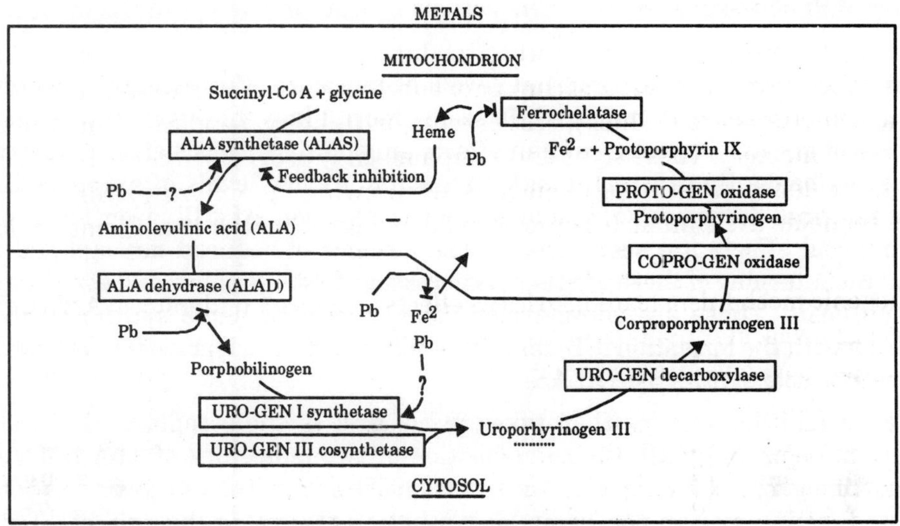

Fig. 3.3. Effects of lead on heme metabolism.

The other type of renal effect occurs with prolonged high lead exposure. It is a progressive disease characterized by interstitial fibrosis, sclerosis of vessels, and glomerular atrophy. It occurs mainly among heavy consumers of moon-shine whiskey and in workers with high, long-term industrial exposure. Death may ensue due to renal failure. A dramatic example of the slow, progressive nature of this condition was reported from Queensland, Australia. As early as 1897 it was observed that deaths from chronic nephritis in early adulthood were much more common than elsewhere. This excessive death rate was traced to childhood lead poisoning. The most remarkable aspect of this was that the clinical disease occurred many years after termination of the exposure.

No comparable incidents have been reported elsewhere, suggesting that unknown ancillary factors were involved. Another interesting feature of this "Queensland disease" was the occurrence of gout, presumably due to reduced renal excretion of uric acid. This condition is occasionally reported in lead poisoning due to moonshine whiskey, but not in occupational lead poisoning. Most of the toxic effects seen in man have also been observed in experimental animals. There are exceptions. Gout has never been reported in experimental lead poisoning; on the other hand, renal tumors are observed with prolonged lead exposure in rats and mice but not in man. As a matter of fact, excess deaths in industrial lead exposure have been shown only for the categories of cerebrovascular disease and chronic nephritis.

Hematopoiesis and Heme Synthesis

The multiple effects of lead on organs and systems are most vividly apparent here. It has long been known that anemia is one of the early manifestations of lead poisoning. It results from reduction of the lifespan of circulating erythrocytes as well as from inhibition of synthesis of hemoglobin. The shortened lifespan of erythrocytes is inconstant, occurring only in some cases of lead-induced anemia. Erythrocytes exposed to lead *in vitro* show increased osmotic resistance but also show increased mechanical fragility. In addition, it has been shown *in vivo* that, even in moderate lead exposure, erythrocyte Na-K-ATPase is somewhat inhibited, suggesting a loss of cell membrane integrity. This may account for the shortened lifespan of erythrocytes that sometimes occurs. The actions of lead on the synthesis of hemoglobin are complex.

Various effects are seen at different levels of exposure. For example, in some cases of poisoning globin synthesis is impaired. It seems more likely, however, that effects on heme synthesis are of greater importance than effects on globin synthesis since they occur even at exposure levels below those which result in anemia. At low levels of exposure these effects result in a marginal decrement of hemoglobin concentration. At still lower levels of exposure, some biochemical effects are seen, even in the absence of reduced hemoglobin. In order to understand the interplay of these effects, it is necessary first to understand some major features of heme synthesis. These are summarized, The process of heme synthesis begins in the mitochondrion with the formation of S-aminolevulinic acid (ALA), a process requiring the enzyme δ-aminolevulinic acid synthetase (ALAS).

A series of additional steps then takes place, first in the cytoplasm, then again in the mitochondrion, beginning with the condensation of two moleeules of ALA to form a pyrrole ring, porphobilinogen, and ending with the insertion of iron into the tetrapyrrole, protoporphyrin IX. The rate-limiting step in the heme biosynthetic pathway is the rate of ALA formation, which, in turn, is dependent on the rate of synthesis of the enzyme ALAS. The end product, heme, regulates ALAS synthesis by negative feedback inhibition. When heme concentration falls, compensatory derepression of ALAS synthesis occurs, with a consequent increase in ALAS-generated ALA synthesis. The several biochemical effects of lead depicted at the bottom of figure 3.3 are best explained by its known inhibitory effects on the incorporation of iron into protoporphyrin IX. This is either due to its direct inhibition of the enzyme ferrochelatase or to its interference with the entry of iron into the mitochondrion. Both mechanisms are compatible with the rise in erythrocytic protoporphyrin that occurs at a level of lead exposure below the level associated with reduced circulating hemoglobin. The consequent reduction in heme concentration probably triggers the derepression of ALAS, which in turn may explain the rise in urinary ALA and coproporphyrin, effects that occur only at a level of lead exposure greater than is necessary to inhibit conversion of protoporphyrin IX to heme.

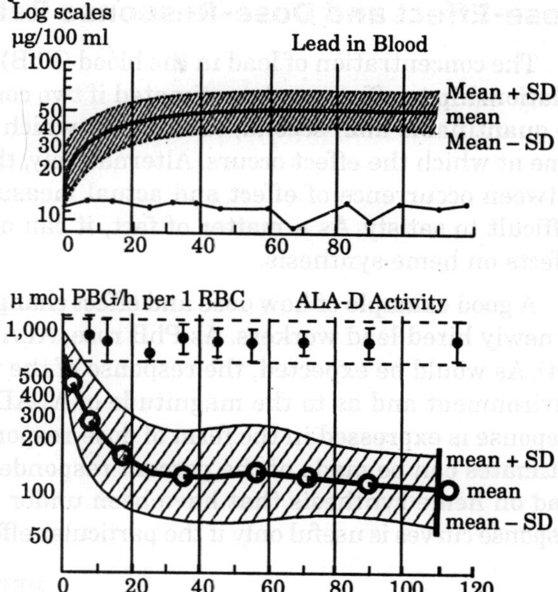

Fig. 3.4. Smoothed average blood lead concentrations and ALAD activities plotted against time in semilogarithmic scales. ALAD activities in the control group are presented together with the ALAD curve of the exposed subjects, and the blood lead concentrations in one control subject are presented together with the blood lead curve of the exposed subjects. Shadowed area indicates the standard deviation. (From Tola, S.; Hernberg, S.; indicative of absorption and biological effect in new lead exposure.

The evidence in support of this summary view is admittedly not totally satisfactory. The most puzzling aspect is with regard to inhibition of ALAD. This seems to occur at levels of lead exposure considerably below those involving elevation of its substrate ALA in plasma or urine. The concept of "reserve enzyme" usually is invoked to explain this. A discussion of the effects of lead on heme synthesis would not be complete without calling attention to the fact that hemoglobin is only one of many hemoproteins essential to normal body function. Others include the cytochromes cytochrome *c* oxidase and hydroperoxidases, all of which are part of the electron transfer systems' requiring heme. The effect of lead on these hemo-proteins is poorly understood.

Other Effects

The toxicologic effects of lead are not limited to the systems discussed above. Thus, lead has long been known to cause colic in cases of poisoning. The mechanism is not understood. Lead also causes chromosomal aberrations and, perhaps, abnormal sperm morphology in man. The significance of these effects is at present uncertain.

Dose-Effect and Dose-Response Relationships

The concentration of lead in the blood (PbB) is the best indicator of the dose in the body. Its relationship to effect can be estimated if two conditions can be satisfied. First, the effect must be quantifiable and, second, the time at which the dose is measured must correspond to the time at which the effect occurs. Alternatively, the dose must not change during the lag period between occurrence of effect and actual measurement of effect. These conditions are rather difficult to satisfy. As a matter of fact, it can only be done satisfactorily with regard to lead effects on heme synthesis.

A good example of how dose and effect change together was provided in a prospective study of newly hired lead workers. As PbB rose, ALAD activity fell more or less con-currently (Fig. 3.4). As would be expected, the response of the workers varied both as to the PbB in the work environment and as to the magnitude of ALAD inhibition. The interindividual variability in response is expressed in the form of dose-response curves. For lead effects on heme synthesis, estimates can be made of the percent respondents. This is illustrated for the major effects of lead on heme synthesis that have been under discussion (Fig. 3.5). The elaboration of dose-response curves is useful only if the particular effect (*e.g.*, FEP > 80) has some health significance.

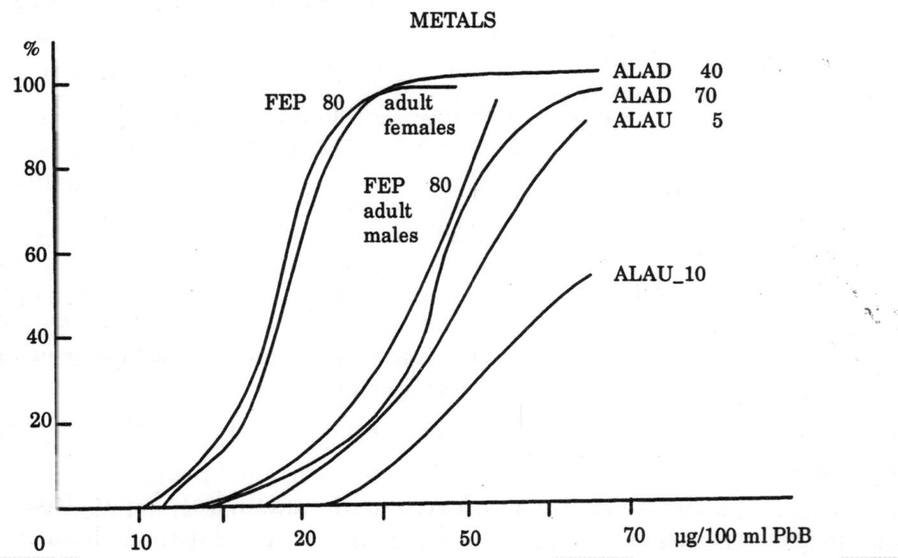

Fig. 3.5. Dose-response relationships for effects of lead on heme intermediates. The ordinate represents percent respondents for the various effects.
Effects:
FEP = Concentration of erythrocytic porphyrin concentration, expressed as pg/dl packed cells.
ALAD = Aminolevulinic acid dehydrase activity in blood, expressed as % inhibition.
ALAU = Concentration of aminolevulinic acid in urine, expressed as mg/l.

Treatment of Lead Poisoning

It is extremely difficult to evaluate the benefits of therapeutic regimens for the treatment of lead poisoning, or other metallic poisons for that matter. To begin with, the incidence of frank poisoning and the corresponding amount of clinical experience is low compared to many other diseases, *e.g.,* hypertension or the common headache. Further, the wide spectrum of manifestations of illness among cases makes it extremely difficult to compare responses to various regimens. In adults and children both, the major specific therapeutic objective is removal of lead from the body using chelating agents.

In adults the most widely accepted procedure is intravenous infusion of the calcium salt of disodium ethylenediamine tetraacetate (CaEDTA), 1 to 2 g per day, for 4 to 5 consecutive days. The lead chelate formed by exchange of Ca for Pb is excreted promptly in the urine. Curiously, the major source of lead mobilized in this manner is the bones. The treatment of lead poisoning in children also entails a course of CaEDTA therapy, either alone or in combination with dimercaptopropanol (BAL). Combined therapy has been found to be more effective than therapy with either drug alone.

MERCURY

Introduction

It is "the hottest, the coldest, a true healer, a wicked murderer, a precious medicine, and a deadly poison, a friend that can flatter and lie."

There has always been an aura of magic surrounding mercury. Even the name, shared by a Roman god and a distant planet, and the lustrous liquid appearance of this metal suggest magic. Before the time of Christ and even lo this day, magic properties have been ascribed to mercury. It occupied a central role among alchemists in the transmutation of base metals into gold. It was carried about as amulets to ward off disease and other evils.

At various times down through the centuries it has been used to treat almost every ailment known to man. Even lo this day it is used to a limited extent for therapeutic purposes. Its toxic properties, however, were not unappreciated. It has long been widely condemned as being a drug with no reasonable margin of safely, even at a time when it was being used widely, notably for the treatment of syphilis. Even the characteristics of occupational intoxications were described in the Middle Ages. Man is apparently a poor student of history. Mercury poisoning still occurs to some extent in certain occupations, principally from inhalation of mercury vapor. There also have occurred episodes of environmental contamination with organic forms of mercury, principally methyl mercury.

The most widely known such episode occurred in Minamata Bay, Japan, from 1953 into the early 1960s. This was followed by a similar episode in Niigata, Japan. In both cases the cause was consumption by the local inhabitants of fish that were contaminated with mercury from industrial waste. In all, 1,200 cases of poisoning were reported. Even more extensive episodes have resulted from contamination of bread made from cereal grains treated with alkyl-mercury fungicides. The largest of these episodes occurred in Iraq, 1971-1972. It involved some 6,000 cases and 500 deaths. At the time that outbreaks of methyl mercury poisoning in Japan were under active investigation, Swedish scientists found high concentrations of methyl mercury in freshwater fish.

The sources of mercury were mainly chloralkali plants, drainage of fields in which cereal grain seeds had been treated with mercury, and wood pulp plants. Methylation of mercury occurred due to the action of aquatic organisms, leading lo transfer and bioconcentration up the food chain to the large carnivorous fish. For more details concerning the environmental significance and lexicology of mercury, there are two good recent reviews.

Metabolism

The chemical form of mercury has a profound influence on its disposition. For all practical purposes there are three general forms of mercury.

1. Elemental mercury—Hg^0: This form is of considerable importance texicologically because it has a high vapor pressure. A saturated atmosphere at 24° C contains approximately 18 mg/ M^3. Of equal significance is the fact that the vapor exists in a monoatomic state. It is therefore distributed primarily to the alveolar bed upon inhalation. Finally, metallic mercury has limited but toxicologically significant solubility in water (20 μg/1) and organic solvents (2.7 mg/1 in pentane).

2. Inorganic Mercury—Hg^{1+} and Hg^{2+}: Of these two oxidation states, Hg^{2+} is the more reactive, readily forming complexes with organic ligands, notably sulfhydryl groups. In contrast to $HgCl_2$, which is both highly soluble in water and highly toxic, HgCl is highly insoluble and correspondingly less toxic.

3. Organic Mercury: These compounds are of diverse chemical structure. As the term is used here, organic mercury refers to all compounds in which mercury forms a bond with one carbon atom. For all practical purposes, the group is limited to methyl and ethyl mercury, phenyl mercury, and the family of alkoxyalkyl mercury diuretics.

$$CH_3Hg^+ ; C_2H_5Hg^+ ; C_6H_5Hg^+ : RCH_2CORCH_2Hg^+—$$

These organic cations form salts with inorganic and organic acids, *e.g.*, chlorides and acetates. They also react readily with biologically important ligands, notably sulfhydryl groups. Finally, they pass readily across biologic membranes since they are lipid soluble. The major difference among these various organomercury cations is that the stability of the carbon-mercury bonds *in vivo* varies considerably. Thus, the alkyl mercury compounds are considerably more resistant to biodegradation than either phenyl mercury or the alkoxyalkyl mercury compounds.

Absorption

For elemental mercury the most important route of absorption is the respiratory tract. As would be expected from the mono-atomic nature and lipid solubility of mercury vapor, percent deposition and retention are quite high, of the order of 80 per cent in man. Although confirmatory data are not available, monoalkyl mercurials (*e.g.*, methyl mercury) probably also are deposited and retained to a high degree, since they have high vapor pressures and high lipid solubility. Elemenlal mercury is very poorly absorbed from the gastrointestinal tract, probably less than 0.01 per cent. This may be because, unlike in the lungs, mercury is not in a monoatomic state but, rather, occurs as large globular particles.

Inorganic mercury in food is absorbed to the extent of about 7 per cent, and organic mercury compounds are very efficiently absorbed, owing to their lipid solubility. For example, the absorption of methyl mercury, even mixed with food, is about 95 per cent in adults. As with all metals, the degree of skin absorption in man is not known with any precision. Systemic absorption of alkyl mercurials probably is substantial. People have been poisoned as a result of dermal application of methyl mercury ointments. Some absorption of elemental mercury and even of inorganic salts of mercury occurs. Thus, in experimental animals, 5 per cent of an aqueous solution of mercuric chloride was absorbed through the skin of guinea pigs within five hours.

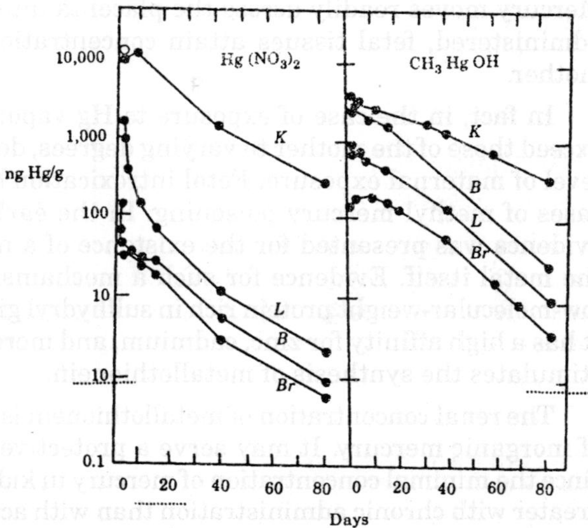

Fig. 3.6. The fate of single doses of inorganic mercury and methyl mercury in rats. (Modified from Swensson, Å., and Ulfvarson, U.: Distribution and excretion of mercury compounds in rats over a long period following a single injection.

Distribution and Metabolism

The distribution of mercury varies considerably, depending on the chemical form and, to a lesser extent, on the route of administration. Elemental mercury is rapidly oxidized to Hg^{2+} and organic mercury compounds are also to varying degrees metabolized to yield Hg^{2+}. As far as tissue affinity is concerned, the most outstanding single characteristic of mercury is its affinity for the kidney. This is true both for Hg^{2+} and for the organic mercury cation $R\text{-}Hg^{+}$. The disposition of inhaled elemental mercury vapor is of special interest because of the importance of intoxication by this route.

It has long been known that Hg° is rapidly oxidized in the erythrocytes to Hg^{2+}, even *in vitro*. Yet, while chronic mercury poisoning due to intake of Hg^{2+} is essentially a renal problem, chronic mercury poisoning due to inhalation of Hg° is a disease of the central nervous system. This apparent paradox is explained by the fact that transfer of the lipid-soluble Hg° from the blood to the brain is sufficiently rapid to result in a toxicologically significant differential distribution to that organ. The subsequent oxidation of Hg° in the brain serves to trap it there.

A similar selective distribution occurs in the fetus. The oxidative process is enzyme mediated, with the catalase complex being the most likely site of oxidation. The disposition of organic mercury compounds is, in general, quite unlike that of Hg^{2+}. This is particularly true of the short-chain alkyl mercury compounds, notably methyl mercury Fig. 3.6. Compares the fate of single equal subcutaneous doses of Hg as Hg^{2+} and as methyl Hg. Although both forms of mercury distribute preferentially to the kidney, the concentration in the brain and blood is substantially higher in the case of methyl mercury. Toxic manifestations of inorganic mercury are renal whereas those for methyl mercury poisoning are neurologic. The disposition of phenyl mercury is essentially the same as for inorganic mercury. This is because the carbon-mercury bond is rapidly cleaved in *vivo,* yielding benzene and Hg^{2+}.

The benzene is subsequently oxidized to phenol, conjugated, and excreted. Alkyl and alkoxyalkyl mercury compounds also are subject to some carbon-mercury bond cleavage, but the mechanism whereby this occurs is poorly understood. In general, cleavage occurs much more slowly than with phenyl mercury. Rate of clearance from the body varies a great deal. In some cases, as with alkoxyalkyl mercurials used as diuretics, clearance is much faster than for Hg^{2+}. In other cases, as with methyl mercury, clearance is considerably slower than for Hg^{2+}.

Mercury moves readily across the placenta into fetal tissue. Regardless of the chemical form administered, fetal tissues attain concentrations of mercury at least equal to those of the mother.

In fact, in the case of exposure to Hg vapor and to methyl mercury, fetal concentrations exceed those of the mother to varying degrees, depending on the species, duration, and absolute level of maternal exposure. Fetal introxication by way of the mother has been documented in cases of methyl mercury poisoning. In the earlier section of this chapter dealing with lead, evidence was presented for the existence of a metal-sequestering mechanism, stimulated by the metal itself. Evidence for such a mechanism also exists for mercury. Metallothionein, a low-molecular-weight protein rich in sulfhydryl groups, was originally isolated from horse kidney. It has a high affinity for zinc, cadmium, and mercury. Administration of any one of these metals stimulates the synthesis of metallothionein.

The renal concentration of metallothionein is increased as much as sixfold by administration of inorganic mercury. It may serve a protective role for the kidney by sequestering mercury, since the minimal concentration of mercury in kidney associated with toxic effects is considerably greater with chronic administration than with acute administration. The role of metallothionein in toxicology is discussed further in the section on cadmium. As with lead, the concentration of mercury in the blood has been used as a biologic indicator of exposure. The distribution between blood cells and plasma is dependent on the chemical form. Thus, with methyl mercury exposure the concentration ratio, whole blood/plasma, is approximately 20, while with exposure to mercury vapor the ratio is only slightly above unity.

Measurement of mercury in blood has been extremely useful in assessing safe steady-state levels of methyl mercury exposure in man. The kinetics of methyl mercury disposition are fairly straightforward and simple. Elimination rates as determined in human volunteers receiving-single radioactive tracer doses indicate that the clearance of the dose from the body is adequately described by a single exponential elimination constant. The clearance half-time from these studies was found to be about 70 days. This rate of elimination seems to hold over a range of input rates. Thus, the rate of disappearance of mercury from the blood or hair of people who consumed moderate or large amounts of fish or bread contaminated with methyl mercury for long periods of time was quite consistent with the clearance half-time for the amount administered as a tracer dose to volunteers.

For other forms of mercury the biologic half-life is not so well described. Limited data suggest, however, that the biologic half-life of inorganic mercury is only about 40 days in man, as contrasted to 70 days for methyl mercury.

Excretion

The relative contribution of urine and feces to the total elimination of mercury is, as with other features of mercury metabolism, quite variable depending on the particular form of— mercury in the body. Upon prolonged inhalation of mercury, urinary excretion somewhat exceeds fecal excretion. The same probably applies to mercury administered as Hg^{2+}. The rate of urinary excretion for any one individual fluctuates considerably from day to day even under steady-state exposure conditions but has been found in industry to be roughly proportional to the level of air exposure.

The mechanism of renal excretion of mercury is complex. The weight of evidence suggests that glomerular filtration contributes little to the renal excretion of any form of mercury. The mechanisms whereby the tubules release mercury into the lumen of the nephron are not well understood. At nephrotoxic doses of inorganic mercury, however, a substantial excretion occurs by exfoliation of renal cells. In contrast to the excretion of inorganic mercury, methyl mercury is excreted mainly in the feces.

Two separate processes are involved: biliary excretion of methyl mercury and excretion by exfoliation of intestinal epithelial cells move mercury into the intestinal lumen. Intestinal reabsorption of the mercury, however, substantially cancels the biliary contribution to net excretion. A polythiol resin has been developed that short-circuits this enterohepatic recirculation. When the resin is given orally it traps the mercury excreted into the bile and carries a substantial fraction of it out into the feces. This procedure has been demonstrated to accelerate mercury excretion in human cases of poisoning.

Biologic Effects

As in the case of lead, mercury has toxic effects involving numerous organs and systems. Some have been clearly shown to occur in man under current and recent circumstances of exposure. Others either are of historic interest or are phenomena that have been demonstrated only in animal models. The relevance of these to the human condition is in many cases uncertain. Regardless of all that, our present perceptions suggest that, irrespective of the chemical form of mercury, the major target organs are the central nervous system and the kidney. As will be seen, there is no sound basis for concluding that this is because the toxic actions of the various chemical forms can be attributed to a single common metabolite *e.g.*, Hg_2^+.

Central Nervous System

The most consistent and pronounced effects of exposure to both elemental mercury vapor and to short-chain alkyl mercury compounds are on the central nervous system. There are similarities but there also are distinct differences. The effects of mercury vapor exposure are strikingly neuropsychiatric in nature whereas those resulting from methyl mercury exposure are largely of a sensorimotor nature. Certain effects are similar in both types of exposure, however, notably tremor. In mercury vapor exposure the tremor progresses in severity with duration of exposure. Initially it involves only the hands but later may spread to other parts of the body.

Tremors are triggered by voluntary use of the affected muscles (intentinal tremor). Neuropsychiatric signs also occur at relatively low levels of exposure, notably excessive shyness, insomnia, and emotional instablility with depressive moods and irritability most frequently reported. This neuropsychiatric complex is known as "erethism." Tremor also occurs in methyl mercury intoxication, but other effects not seen in mercury vapor exposure occur more consistently and at lower exposure levels. These are sensory in nature.

The earliest signs are paresthesias and constriction of the visual field. At somewhat higher levels of exposure other sensory effects occur, such as loss of hearing, of vestibular function, and of the senses of smell and taste. These effects are not known to occur in elemental mercury intoxication. Numerous other neurologic effects occur in methyl mercury intoxication. These are motor effects such as incoordination, paralysis, and abnormal reflexes. It is not clearly known whether motor neurons are uniquely involved. Some of these motor effects, *e.g.*, incoordination, could result from defects in sensory input. Neuropsychiatric effects, which are so prominent in elemental mercury exposure, are also reported to occur in methyl mercury poisoning, but not so consistently.

Further, the effects seem somewhat different from those observed in elemental mercury poisoning. Thus, shyness and irritability are not observed in methyl mercury poisoning but are very prominent in elemental mercury poisoning. On the other hand, spontaneous fits of laughing and crying and intellectual deterioration occur only in methyl mercury poisoning. The pathogenetic mechanism of neurotoxicity has been investigated intensively in recent years, particularly concerning methyl mercury.

The various approaches used reflect the diversity of toxic phenomena exhibited. Thus, the problem has been viewed from the neuroanatomic standpoint as well as from the biochemical, neurophysiologic, and pharmacologic standpoints. Reviewed these studies and has proposed a tentative pathogenetic mechanism, which applies primarily to methyl mercury because relatively few of the studies have dealt with inorganic mercury or have directly compared effects of the inorganic and alkyl forms. Both inorganic and alkyl mercury disrupt the integrity of the blood-brain barrier as manifested by extravasation of plasma protein into adjacent cerebral tissue. Since the blood-brain barrier acts to regulate the uptake of amino acids and other metabolites, it is possible that brain metabolism is affected at this point of interface with the circulation.

The actions of mercury are probably not limited to the blood-brain barrier, however. Both forms of mercury are widely distributed elsewhere in the central nervous system, but there are quantitative differences. For example, while both forms of mercury localize to a high degree in dorsal root ganglion neurons and nerve fibers, inorganic mercury has a lesser degree of localization in neurons of the calcarine cortex than does methyl mercury. Degenerative changes are widespread in both forms of poisoning, although the nature of the changes differs in certain respects. In general, sensory neurons have been found to be more severely affected than motor neurons. Glial elements actually proliferate in areas of neuronal damage, perhaps to provide metabolic and structural support to the injured neurons. Damage is not limited to the cell bodies. Nerve fibers also are affected. Degeneration of both the axoplasm and the myelin sheath is common. Biochemical explanations for the various degenerative changes abound.

Enzymes along the glycolytic pathway and protein synthesis are inhibited. Inhibition of protein synthesis precedes both changes in anaerobic and aerobic glycolysis as well as neurologic effects. This "silent period" of several days between inhibition of protein synthesis and neurologic effects is seen even with single doses of mercury. It is not certain whether the inhibition is due to reduced amino acid up-take across the blood-brain barrier or whether the inhibition occurs by some other means. Various neurophysiologic parameters also are altered by mercury. Spike potentials of sensory ganglionic neurons of animals receiving methyl mercury are prolonged, indicating retardation of repolarization.

It has also been found that mercury blocks synaptic and neuromuscular transmission. The latter finding is of particular interest since a similar effect occurs in some cases of methyl mercury intoxication in man. It is not possible at this time to say whether all the varied effects that have been noted experimentally occur as a result of independent actions of mercury or as stages in a chain of related events. Unitarian hypotheses are always attractive. There is none at present that adequately accounts for all the diverse neurologic phenomena that have been observed clinically and experimentally in mercury intoxication.

Kidney

By far the highest concentration of mercury occurs in the kidney, regardless of the chemical form absorbed. Yet, for all that, the kidney is the primary target organ only in the case of inorganic mercury (Hg^{2+}). At least this is so with respect to toxic effects. It is true that the kidney also is the primary target organ for non-toxic effects in the special case of the alkoxyalkyl mercury compounds used therapeutically as diuretics. It is also true that certain morphologic and functional effects are seen with sublethal doses of methyl mercury, but these are of uncertain significance in comparison to effects on the nervous system. Massive oral doses of inorganic mercury, such as may be taken with suicidal intent, initiate a train of events beginning with anuria, progressing to polyuria, and finally to a recovery of normal renal function. The phase of anuria is the most life-threatening and may last for many days.

The pathogenetic mechanism is not well understood. Experimental evidence from animal studies suggests that several factors are involved. These are tubular obstruction, increased back diffusion of tubular filtrate, and preglomerular vasoconstriction. The phase of polyuria is characterized by decreased renal concentrating capacity. It probably results mainly from a substantial inhibition of proximal tubular sodium reabsorption. In severe poisoning, disturbances in tubular function may persist for several months after poisoning. Acute inorganic mercury poisoning, as described above, is relatively rare.

The more usual form of mercury nephrotoxicity occurs with chronic industrial exposure and is characterized by proteinuria. If severe, the nephrotic syndrome is observed, wherein the loss of plasma protein is sufficiently great to cause hypoproteinemia with edema of dependent parts, *e.g.*, the ankles. Surprisingly, very little is known about the nature of the proteins that are excreted in cases of inorganic mercury poisoning. In one recent suicide attempt the proteinuria was mixed. Excretion of both albumin and low-molecular-weight proteins rose concurrently, suggesting both glomerular and tubular damage Exposure to inorganic mercury at levels sufficient to cause proteinuria does not result in increased amino acid excretion.

It appears, therefore, that the renal tubular transport system for reabsorption of α-amino acids is less sensitive than the transport system for absorption of low-molecular-weight proteins. Information concerning the effects of methyl mercury on renal function in man is nonexistent, except for statements to the effect that even in the presence of neurologic effects the kidney is seldom affected. This is consistent with experimental studies in rats in which subchronic exposure to methyl mercury caused moderate proteinuria only at a level of mercury administration that actually killed a substantial number of the animals.

Other Effect

Although the nervous system and the kidney are the usual major targets for effects of mercury, a variety of other toxic phenomena occur. Some of these are well-known accompaniments of clinical poisoning. Others are effects seen in experimental animals that may prove in the future to have significance for man. In the case of mercury vapor inhalation, the neuropsychiatric problem resulting from chronic exposure is accompanied by stomatitis, gingivitis, and sometimes excessive salivation and a metallic taste. These tend to occur mainly in people whose oral hygiene is poor.

A peculiar discoloration of the anterior surface of the lens in the eye also is frequent ("mercurialentis"). When people are exposed to very high concentrations of mercury vapor, pneumonitis occurs as a result of direct irritation of the lung. In the case of poisoning with inorganic mercury salts by mouth, severe inflammation of the mouth, esophagus, stomach, and small intestine occurs. Following this initial contact inflammation, a secondary inflammatory effect may occur wherein absorbed mercury localizes in the intestinal mucosa.

The colon is peculiarly sensitive to this secondary inflammatory action. There also is a disease of infants known as acrodynia or "pink disease" in which inorganic mercury seems to play a role. It is characterized by neuropsychiatric disturbances, peripheral vascular effects, disturbances of sensation of the extremities, stomatitis, and other vague, nonspecific signs. The disease is not uniquely associated with excessive mercury exposure, nor does it occur regularly among children exposed to mercdry. See Bidstrup (1964) for a detailed discussion of the role of mercury in this disease.

Dose-Effect and Dose-Response Relationships

It may be recalled that, with lead, it is possible to estimate both dose-effect and dose-response relationships in man. The critical system, the hematopoietic system, is peculiarly amenable to quantitative estimates of effect. Thus, the activity of the enzyme ALAD in blood and the concentration of heme intermediates in blood and urine are readily quantifiable and exhibit clear relationships to the internal dose of lead, as reflected in the concentration of lead in the blood.

The situation with mercury is not nearly so simple. To begin with, the critical effects both in the case of poisoning with elemental mercury and with alkyl mercury compounds are neurologic. Such phenomena as tremor, shyness, irritability in the case of elemental mercury and paresthesias and constriction of the visual field in the case of methyl mercury are not readily measured in quantitative terms. It is therefore not possible to construct dose-effect curves. So far as inorganic salts of mercury are concerned, the kidney is the critical organ. It is true that renal function is amenable to quantitative description, but even here there is a problem. The renal effect that seems to be the most sensitive index of mercury toxkity is elevated excretion of protein in the urine, an effect that has not been studied adequately to allow for definition of either dose-effect or dose-response relationships. For these reasons, it is necessary to forego any thought of estimating dose-effect relationships for any form of mercury in man.

As a matter of fact, even dose-response relationships can only be estimated for elemental mercury and methyl mercury exposures. In the case of elemental mercury vapor exposure the measurement of dose that seems to correlate best with response is the actual concentration of mercury breathed by the subjects, as contrasted to the concentration of mercury in the blood or urine of the exposed subjects. As the time-weighted average (TWA) air concentration rises above 100 μg Hg/M^3, the frequency of occurrence of classic signs of erethism among workers increases. On this basis it has been recommended that the TWA limit for workers should be 50 fig Hg/M^3 in order to provide some margin of safety. In the case of methyl mercury exposure, dose-response relationships have been calculated from the data obtained in the recent large-scale out-break of poisoning in Iraq, described by Bakir *et al.* Again, as with elemental mercury exposure, the critical effects measured are those affecting the central nervous system.

A hierarchy of response is seen with reference to sensitivity of the central nervous system. The effect occurring at the lowest level of exposure is paresthesia (Fig. 3.7). Both the estimated body burden of mercury (Fig. 3.7) and the concentration of mercury in the blood (Fig. 3.8) are adequate expressions of dose, at least under conditions of long-term intake (months).

Treatment of Mercury Poisoning

As was pointed out in the discussion of the treatment of lead poisoning, data concerning the relative efficacy of various approaches to the treatment of metallic poisoning in general are grossly inadequate. The treatment of mercury poisoning is no exception. Studies of therapeutic regimens conducted on man generally are limited to clinical reports on one or, at most, a few individual cases with one or another selected drug. The reader, and the investigator himself for that matter, are always left to ponder questions concerning what would have happened had the patients not been treated at all, or concerning what would have, happened if some alternate drug or regimen had been used. Even when a substantial number of cases is available to a single investigator, ethical considerations severely limit the options.

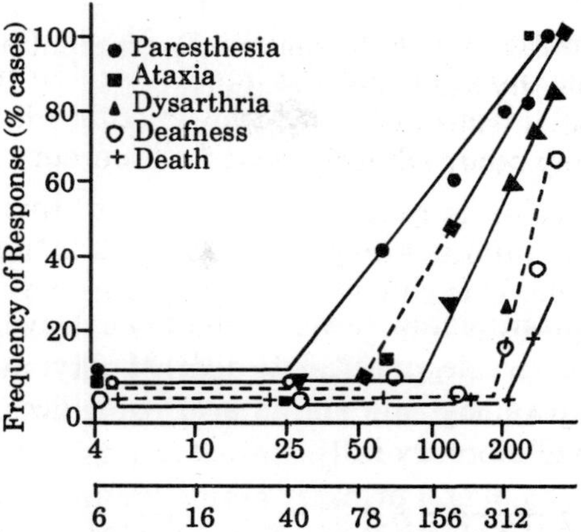

Fig. 3.7. Dose-response relationships for methyl mercury. The upper scale of estimated body burden of mercury was based on the authors' actual estimate of intake. The lower scale is based on the body burden, which was calculated based on the concentration of mercury in the blood and its relationship to intake derived from radioisotopic studies of methyl mercury kinetics in human volunteers.

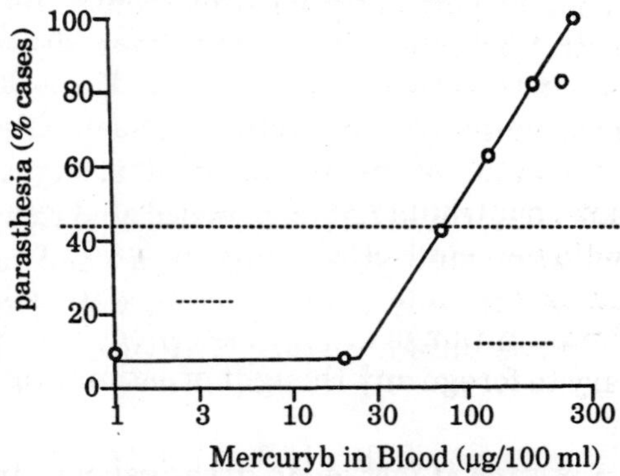

Fig. 3.8. Dose-response relationship for methyl mercury, using concentration of mercury in the blood as dose and paresthesia as response.

Untreated controls cannot be included in the study if the condition is even moderately severe. Reliance on data from experimental animals is equally unsatisfactory for various reasons. The experimental design may not correspond to the usual human exposure situation or the animal model selected may not behave like man. Chelating agents are relied upon, as in the case of lead poisoning, to remove mercury from the body. For many years the only chelating agent known to promote the excretion of mercury was dimercaptopropanol (BAL). It has been shown repeatedly that it enhances the excretion of mercury administered experimentally to animals in the inorganic form.

It has also been shown to be beneficial in the treatment of inorganic mercury poisoning in man, enhancing the excretion of mercury and effecting substantial improvements in the patients' clinical status. Its efficacy has been attributed to the presence of thiol groups on adjacent carbon atoms, with the presumed participation of both in the binding of divalent mercury as a stable, readily excreted chelate ring.

```
  |     |  |
 -C-----C--C-OH
  |     |  |
  S     S
   \   /
    Hg
```

Dimercaptopropanol, however, had two drawbacks. It was ineffective when given orally and it was ineffective as an antidote against alkyl mercury compounds. The next major advance was the discovery that the amino acid penicillamine was an effective chelator for mercury, even when administered orally. Its activity as a chelating agent is attributable to ring formation in which the thiol and primary amine groups probably serve as coordination sites. More recently, the N-acetyl derivative N-acetyl-D, L-penicillamine (NAP) has been discovered to be even more effective and less toxic than D,L-penicillamine. The toxicity of D, L-penicillamine

```
        CH3
        |      |
   CH3--C------C--COOH
        |      |
        S      N
         \    / \
           Hg
```

is attributable to the L-isomer, which inhibits a number of pyridoxine-dependent reactions, probably by reacting with pyridoxal-5-phosphate to form a thiazolidine. By contrast, the L-isomer of NAP is relatively nontoxic. The LD50 of D, L-NAP is considerably greater than for D, L-penicillamine or even than for D-penicillamine. The concept of ring formation, with the attendant stability of the metal complex, suggests that compounds of the type R-Hg^{+}, such as methyl mercury, would not form metal chelates with BAL or NAP. Surprisingly, NAP enhances the excretion of methyl mercury to a very substantial degree. It may be speculated that this effect is due only to interaction with Hg^{2+} formed by *in vivo* dealkylation of methyl mercury. This is unlikely, since NAP also removes methyl mercury bound to serum albumin *in vitro*.

An alternate possibility is that NAP first causes a cleavage of the carbon-mercury bond and subsequently reacts with Hg^{2+}. It has been demonstrated that sulfhydryl reagents such as glutathione enhance the cleavage of the carbon-mercury bond of phenyl mercury *in vitro* in the presence of γ-globulin. It is not inconceivable, therefore, that NAP might facilitate a similar cleaving action on the carbon-mercury bond of methyl mercury. In spite of some impressive animal data concerning the efficacy of the penicillamines as mercury chelators, their clinical efficacy in the treatment of methyl mercury poisoning is not impressive.

To a large extent at least, this is because the neurologic effects of methyl mercury result from irreversible damage to neurons. Even from the more limited point of view of metal-mobilizing efficacy, the results have been equivocal. In some cases there has been essentially no suggestion of metal mobilization, at least insofar as decrease in blood mercury levels is concerned.

CADMIUM

Introduction

Cadmium ranks close to lead and mercury as a metal of current toxicologic concern. It is a relative newcomer, having been identified as a distinct element only in 1817. It occurs in nature in association with zinc and lead. Extraction involves separation and recovery of either zinc and cadmium or of all three metals in a single refinery plant. It quickly found application as an alloy, in electroplating of other metals, and as a pigment. Later it came to be used extensively in the manufacture of alkali storage batteries and plastics. Most early recorded cases of cadmium poisoning were due to inhalation of cadmium fumes or dusts. These cases were generally due to industrial exposure, although the first recorded cases, reported in 1858, were not industrial in nature.

Three servants became severely ill from inhaling the dried film of cadmium carbonate that they had applied to silverware as a polishing agent and had subsequently brushed off. It was not until the second or third decade of the present century that the respiratory effects of acute and chronic inhalation of cadmium became recognized as a significant occupational disease entity. Later it was recognized that the kidney also is quite sensitive to cadmium. Friberg first reported in 1948, the coexistence of renal and lung damage in men working in an alkali storage battery plant. An event occurred shortly after World War II that stimulated interest in cadmium because it involved relatively low-level exposure of the general population through contamination of food.

In 1946, Dr. Hagino, a general practitioner, returned to Fuchu, Japan, from the army to re-open his medical practice. He was visited by numerous patients aged 40 to 70 who complained of severe rheumatic and myalgic pains. He gave this mysterious disease the name *Itai-itai* (pain-pain, or ouch-ouch). Gradually it came to be accepted that cadmium in the local rice played an etiologic role in this disease. The source of the cadmium was the effluent from a Pb-Zn-Cd mine upstream from the rice fields. Interest in cadmium was further stimulated when Schroeder published a provocative epidemiologic study in 1965 linking dietary cadmium to hypertension in the general population.

Because of the large number of people potentially affected by hypertension, a great deal of effort has since been expended investigating the effects of cadmium on the cardiovascular system. The usual sources of cadmium for the general population are mainly food and inhaled tobacco smoke. Among foods, the usual concentration is less than 0.1 μg/g wet weight. The only foods that accumulate excessively high concentrations of cadmium are shellfish, liver, and kidney. In these foods the concentration often exceeds 10 μg/g. Sources of cadmium in foods and other environmental media generated by man are not clearly defined.

It is difficult to point to any one or even several sources and to say that it or they are of outstanding significance so far as exposure of any broad segment of the general population is concerned. This is in contrast to mercury, where environmental transport from contaminated waters to fish stands out as a major identifiable concern. Similarly, with lead we know that residues of lead-base paint in old homes and fallout from the air surrounding smelters are major hazards to the general population of children. For those who have an interest in the broader aspects of cadmium as an environmental pollutant, there are two excellent recent reviews. Both of these are concerned primarily with man's exposure and with human health effects.

Metabolism

The specific chemical forms of lead and mercury have clearly been shown to exert considerable influence on their disposition in the body. Thus, the valence state of inorganic mercury profoundly influences the degree of absorption and even the pattern of distribution in the body. With both lead and mercury, alkyl compounds are lexicologically significant and exhibit unique toxicologic properties. In contrast, cadmium occurs only in one valence state, +2, and does not form stable alkyl compounds or other organometallic compounds of known toxicologic significance. The solubility of cadmium salts is highly variable. The halogen salts, sulfate, and nitrate are relatively soluble, while the oxide, hydroxide, and carbonate are insoluble in water. Cadmium has a relatively high vapor pressure. Thus, hazardous respirable air concentrations of CdO are readily attained in smelting and refining operations.

Absorption

As with any other aerosol, the pattern of deposition varies greatly with particle size; therefore, no generalizations can be made about fractional deposition. Similarly, no generalizations can be made concerning rate and route of clearance from the respiratory tract. A cadmium chloride aerosol would be cleared predominantly by direct absorption into the systemic circulation because of its high degree of water solubility. The insoluble cadmium oxide, dependent on particle size, would be cleared to a much greater extent by alveolar macrophages or via the mucociliary escalator system and subsequently swallowed. Absorption of cadmium from the gastrointestinal tract is relatively minor, ranging from 0.5 to 12 per cent in various species of animals.

Studies of cadmium absorption in man are limited but are consistent with animal studies. Tracer doses of radioactive cadmium administered with food to human volunteers were absorbed to the extent of 4.7 to 7 per cent. Numerous dietary factors enhance cadmium absorption, notably deficiencies in calcium, iron, and protein. Calcium is actively absorbed by the small intestine. This process requires a low-molecular-weight calcium-binding protein (CaBP), the synthesis of which is stimulated by dietary calcium deficiency. Recent evidence suggests that the increased CaBP activity induced by calcium deficiency may serve to enhance cadmium absorption as well as calcium absorption.

The enhancing effect of CaBP probably is counterbalanced by the influence of metallothionein, a low-molecular-weight protein whose synthesis is induced by cadmium, zinc, and mercury. It has a represser role in limiting the intestinal absorption of zinc probably exerts a similar effect in regard to cadmium. As with lead, it has been shown that young animals absorb cadmium to a much greater degree than older animals. Absorption of cadmium through the intact skin has been investigated only to a very limited extent. Up to 4 per cent of the water-soluble chloride salt is absorbed in five hours through the skin of the guinea pig.

Distribution

Cadmium has a strong preferential affinity for the liver and the kidney over a wide range of exposure levels. In general, about 50 percent of the total body burden is found in these two organs. The concentration ratio between the two varies with the total amount in the body. At low doses the concentration in the kidney is approximately ten times greater than in the liver. With increasing levels of exposure the concentration ratio approaches one. In heavy industrial exposure the concentration in the liver may actually exceed the concentration in the kidney.

Cadmium is widely distributed elsewhere in the body at low concentrations relative to the liver and kidney. Within the kidney there is a substantial gradation in concentration, progressing downward from outer cortex to inner medulla. Thus, average cortical concentration is about 1.5 times medullary concentration. Cadmium is highly cumulative. This is suggested by human autopsy data showing concentration rising to a peak in kidney at age 50 and in other organs somewhat earlier in life. This progressive accumulation of cadmium in the body over a major fraction of the lifespan has also been demonstrated experimentally in animals. Beyond age 50 a decrease in the amount of cadmium in the kidney has been reported by several investigators.

Steady-state concentrations of cadmium in blood occur relatively early in life under reasonably constant conditions of exposure. In workers newly exposed to a cadmium-contaminated environment, a new steady-state blood cadmium concentration was attained within one year. The biologic half-life of cadmium in man has been difficult to determine with any degree of precision because of limited opportunity to *d6* the appropriate experiments on volunteers. From indirect evidence, notably from rates of daily cadmium intake and amounts of cadmium in the body at age 50, the biologic half-life has been estimated to be anywhere from 19 to 38 years. In experimental animals, such as rats, mice, and dogs, the biologic half-life determined by observing the rate of loss of single tracer doses of radioactive cadmium has been calculated to be of the order of several hundred days.

From the standpoint of fraction of lifespan, these estimates in animals correspond roughly to the estimates in man, arrived at using more indirect approaches. The great disparity in rate and degree of accumulation of cadmium in various organs and systems is not readily explained. Speculation abounds concerning the role of the cadmium-binding protein metallothionein as a determinant of the distribution and retention characteristics of cadmium. This protein, originally isolated from equine kidney by Margoshes and Vallee (1957), was found to contain high concentrations of zinc (2.2 per cent) and cadmium (5.9 per cent), hence the name metallothionein. It actually consists of two low-molecular-weight proteins (MW ca. 6,500), which differ somewhat as to amino acid composition.

Major distinctive features are (1) the absence of aromatic amino acids; (2) high cysteine content; and (3) high affinity for certain metals, notably Cd, Zn, Hg, Ag, and Sn. The strong metal affinity is due to the high sulfhydryl content attributable to cysteine. It is widely distributed in the body of both man and animals, having been isolated from kidney, liver, spleen, intestine, heart, brain, lung, and skin. Perhaps the most intriguing property of metallothionein is the fact that cadmium induces synthesis of the protein, as do zinc and mercury. Pretreatment of animals with either cadmium or zinc in moderate doses increases tolerance to otherwise lethal single doses of cadmium given later. This effect is accompanied by an increase in hepatic metallothionein.

There is some question, however, as to the exact mechanism whereby this protective effect is exerted. Thus, it has been demonstrated that this protective effect is much more transient than the inductive effect and that the metallothionein content of organs other than the liver and perhaps the kidney does not necessarily increase in conjunction with this protective effect. The role of metallothionein as a transport protein is uncertain. During chronic exposure to cadmium more than 90 per cent of the circulating cadmium is in the blood cells, partially bound to metallothionein and partially bound to hemoglobin. Cadmium in plasma, however, is not normally bound to metallothionein; rather, it is bound to high-molecular-weight proteins.

Excretion

The major route of cadmium excretion in man is generally stated to be the urine. This conclusion is based on a study of five volunteers who received single oral doses of radioactive cadmium. In animals, fecal excretion appears to be greater than urinary excretion. Miscellaneous routes of excretion, principally hair and exfoliated epithelium, have not been evaluated in any meaningful fashion. The urinary excretion of cadmium is of special interest because there is need for a biologic monitoring process whereby the degree of body contamination with cadmium can be assessed. It has already been pointed out that the concentration of cadmium in the blood fails in this regard.

Blood cadmium equilibrates with any given level of exposure within a year, while the total amount in the body keeps accumulating over a period of decades. Much the same situation has been noted in regard to the urinary excretion of cadmium. In the general population, there is practically no increase in cadmium excretion with age in spite of strong evidence that the body burden, notably in the kidneys, increases markedly to age 50. By contrast, in heavily exposed population groups urinary excretion does increase with age. A reasonable explanation for this difference has been proposed. At low levels of exposure, excretion is not proportional to intake because the binding capacity of the accumulative organs is not saturated and may actually increase progressively due to increasing metallothionein levels. At higher levels of exposure the binding capacity more nearly approaches saturation and metallothionein levels may be maximal. Under such circumstances excretion would be greater and more nearly proportional to intake.

A substantial rise in the urinary excretion of cadmium is a danger signal. It is associated with renal damage. From animal studies, it has been suggested that the rise in cadmium excretion associated with renal damage is due to presentation of cadmium to the kidney as the metallothionein complex. Indeed, cadmium metallothionein administered parenterally is both much more nephrotoxic and much more readily excreted in the urine than cadmium administered as an inorganic salt. It its passage through the kidney, the metallothionein is degraded, and the cadmium released thereby binds to high-molecular-weight renal proteins. Thus, there occurs a somewhat paradoxic situation wherein metaltothionein reduces cadmium toxicity under some circumstances and enhances toxicity under other circumstances.

Biologic Effects

In keeping with the policy adhered to in the discussion of lead and mercury, the major focus of this section will be on effects known or suspected to occur in man. Animal data will be cited where they are deemed helpful in explaining or confirming observations in man. As was indicated in the introduction to the subject of cadmium, the major toxic phenomena in man are respiratory and renal toxicity, seen principally in industrial workers; the *itai-itai* disease complex reported from Japan, involving primarily elderly multiparous women; and finally, the highly controversial hypertensive effect suspected to involve a large number of people in the general population. These are all chronic effects. This is not to suggest that there is no such thing as acute cadmium poisoning. Nor is it to suggest that acute poisoning does not deserve a certain amount of attention. Numerous cases of acute cadmium poisoning have been documented. Many of these have been cases of inhalation of very high concentrations of cadmium. Most other recorded cases have resulted from consumption of beverages and foods contaminated by cadmium transferred from the container.

Acute lethal doses by inhalation in man can only be estimated rather crudely. Further, they vary considerably depending on the chemical form, particle size, and period of time over which inhalation occurred. For inhalations occurring within a span of eight hours, the lethal dose of cadmium oxide fumes has been estimated to be approximately 2,600 mg/M^3 for one minute, that is to say that inhalation of 2,600 mg/M^3 for one minute would be fatal, as would any other product of minutes and concentration equaling 2,600, *e.g.*, 26 mg/M^3 for 100 minutes. For cadmium oxide dust the lethal product of time and concentration would be substantially larger and for a cadmium chloride aerosol the product would be substantially lower. Precise numbers are not available, however.

The minimal toxic dose for an eight-hour inhalation is probably 1 to 3 mg/M^3 depending on aerosol particle size. The principal toxic effects of cadmium inhalation are attributable to local irritation of the respiratory tract. Death is usually due to massive pulmonary edema. Signs and symptoms are delayed a few hours and consist mainly of irritation of the upper respiratory tract, chest pains, nausea, and dizziness. Gastrointestinal effects, *e.g.*, nausea and diarrhea, may also occur. Long-term or permanent lung damage may occur, taking the form of emphysema and peribronchial and perivascular fibrosis. Oral lethal doses in man, again, are difficult to estimate and vary considerably depending on the chemical form. Thus, in experimental animals the acute oral LD50 varies from approximately 100 mg/kg for soluble salts of cadmium to several thousand mg/kg for metallic cadmium powder or for the insoluble selenide and sulfide. Estimated lethal doses in man range from 350 to 8,900 mg. The minimal acute toxic dose is probably less than 10 mg. As with inhalation of cadmium, major toxic effects are referable to local irritant effects.

In the case of oral intake the manifestations are nausea, vomiting, salivation, diarrhea, and abdominal cramps. Death may occur within 24 hours due to shock and dehydration or may be delayed one or two weeks following onset of various systemic effects, notably renal and cardiopulmonary failure. Extensive damage to the liver also may occur. Turning to the health effects of long-term exposure, those observed in industrial workers are best understood and least controversial.

The principal target organs are the lungs and the kidneys. What follows concerning respiratory effects is documented purely on the basis of cadmium inhalation in workers. By contrast, the renal effects noted in workers are essentially the same as those in cases of *itai-itai* and in other Japanese persons affected as a result of excessive dietary intake of cadmium. The critical target organ in chronic cadmium exposure is generally held to be the kidneys. That is to say that adverse effects have been detected at lower levels of exposure compared to other organs and systems.

The greater sensitivity, however, may be more apparent than real. The techniques used to characterize abnormalities in renal function are unusually sensitive and precise in comparison to those used to study toxic effects in some other organs, notably the lungs. Thus, the specification of the critical organ in the case of cadmium exposure may be somewhat illusory. The overall consequence of excessive inhalation of cadmium fumes and dusts is loss of ventilatory capacity, with a corresponding increase in residual lung volume. Thus, the disease has the cardinal features of emphysema.

Subjective complaints of shortness of breath upon exercise are common. Standard indices of ventilatory capacity are reduced, such as vital capacity and maximal ventilatory capacity. Mean volume of residual air, expressed as percentage of total lung volume, is increased. The most sensitive indices of reduced functional capacity were found recently to be (1) forced vital capacity; (2) forced expiratory volume at one second; and (3) peak expiratory flowrate. These effects are largely irreversible and are clearly more pronounced with long-term exposure than with short-term exposure. Studies of the anatomic progression of the disease are limited to relatively short-term animal studies, but the pathogenesis as seen in animals is consistent with the functional features of the disease as seen in man. In rats exposed to a cadmium aerosol for 15 days, the initial reaction of the lungs is of an inflammatory nature. This is followed by the appearance of emphysema and, later, of fibrosis.

In the general population, cigarette smoking contributes to the development of emphysema. Although a substantial amount of cadmium is inhaled with cigarette smoke, approximately 3 μg from 40 cigarettes, this amount is quite a bit lower than even the minimal amounts inhaled by workers experiencing emphysema from cadmium. The contribution of this source of cadmium to emphysema therefore is questionable. The mechanism whereby cadmium causes emphysematous and fibrotic changes in the lungs is not well understood.

There have been a number of studies in animals demonstrating that cadmium inhibits pulmonary defense mechanisms against respiratory infection, but the relevance of these studies to the pathogenesis of emphysema and lung fibrosis is not at all clear. Of greater interest is the observation that cadmium inhibits serum α_1-antitrypsin. This effect is specific for cadmium as compared to several other metals studied. There is an association, between severe α_1-antitrypsin deficiency of genetic origin and emphysema in man. Proteolytic agents, including papain and trypsin, have been demonstrated experimentally to cause emphysema.

Inhibition of α_1-antitrypsin by cadmium may therefore play a role in the pathogenesis of cadmium-induced emphysema. The current view is that the kidney is the most cadmium-sensitive organ. Toxic effects are noted in man at levels of exposure below those affecting other organs and systems. It must be remembered, however, that the kidney attains much higher concentrations of cadmium than other organs. Thus, it is not really a matter of sensitivity as much as it is a matter of predilection for accumulation. Attention was first directed to the renal toxicity of cadmium in a report concerning industrial exposure.

It was noted that men exposed to cadmium oxide dust in an alkaline storage battery factory exhibited consistent proteinuria and a reduced ability to concentrate urine. It was also noted that the greater the duration of exposure, the greater the incidence of renal damage. We know today that the importance of duration of exposure is in large part due to the highly accumulative character of cadmium. Based on limited renal biopsy studies of affected and nonaffected workers, a threshold concentration for renal toxicity of 200 μg Cd/g kidney cortex has been proposed. This threshold has assumed considerable importance in the calculation of acceptable levels of daily cadmium intake.

Curiously, with the onset of renal damage the concentration of cadmium decreases somewhat. Perhaps this is due to breakdown in the metallothionein cadmium-sequestering capacity. Since the initial report of renal damage in cadmium workers, there has been a series of reports that expand on the initial findings. Most recent studies have focused on the nature of cadmium-induced proteinuria.

High-molecular-weight proteins are normally filtered but are not reabsorbed by the renal tubules. Thus, increased urinary excretion of these high-molecular-weight proteins, *e.g.*, albumin, is considered to reflect glomerular damage. On the other hand, low-molecular-weight proteins are normally filtered and largely reabsorbed in the course of their passage through the renal tubules. Increased urinary excretion of low-molecular-weight proteins is therefore considered to indicate injury to the tubular reabsorptive mechanism. The effect of cadmium is mixed, in that both low- and high-molecular-weight protein excretion is enhanced. There is no question as to the existence of tubular damage in cadmium poisoning.

Glycosuria, hypercalciuria, aminoaciduria, and increased uric acid excretion with hypouricemia have all been reported. Limited studies indicate that these effects, notably proteinuria do not disappear following removal from excessive exposure. This is not to be interpreted as indicating that the kidney is irreversibly damaged. Continued malfunction more likely is due to persistence of the cadmium burden in the kidney. Of all the various effects of cadmium on renal function, the most sensitive appears to be proteinuria. It occurs in a substantial number of cadmium workers who do not have other manifestations of renal damage such as glycosuria and aminoaciduria. Chronic cadmium poisoning as described above is of a special character in the sense that the route of intake has a considerable bearing on the nature of the effects.

The lungs are prominently affected upon inhalation exposure because they are the first sensitive tissues encountered in the passage of the toxicant from the external environment to the internal envi-ronment. *Itai-itai,* the cadmium-induced disease recorded as having occurred in the general population of Japan, presents us with the characteristics of cadmium poisoning—re-sulting from dietary intake. It too, however, is a disease with rather special characteristics. Among the people of the Jintsu River Valley who all consumed similar amounts of cadmium-contaminated rice, middle-aged to elderly multiparous women were mainly affected. The most prominent effects, osteomalacia with attendant spontaneous multiple bone fractures, have seldom been reported in cadmium poisoning of industrial origin. Despite this fact, the role of cadmium in the etiology of this disease is reasonably certain.

For one thing, cadmium in the food supply grown in the area was extremely high. Further, the disease featured, in common with the industrial disease, both proteinuria and glycosuria. Finally, the disease has been reproduced experimentally in rats by combining dietary excess of cadmium with calcium deficiency. Perhaps the most controversial issue of all concerning human health effects of cadmium is the suggestion, first put forth in 1965, that cadmium has a significant role in the etiology of hypertension in the general population. The initial study was purely epidemiologic. Persons dying from hypertension were found to have significantly higher concentrations of cadmium and higher cadmium-to-zinc ratios in their kidneys than people dying of other causes. Subsequently, Schroeder and his colleagues claimed to have reproduced cadmium hypertension in rats and to have reversed the effect by admmislration of a polyamino carboxylic acid chelating agent that removes cadmium from the body.

The role of cadmium in the etiology of hypertension is still not satisfactorily resolved. It has been pointed out that hypertension is not prominent either in industrial cadmium poisoning or in *itai-itai.* On the other hand, several investigators have demonstrated that cadmium has a biphasic effect on blood vessels, wherein at low concentrations it is vasoconstrictive and at higher concentrations produces vasodilation. Regardless of the ultimate outcome of this particular controversy, it should be pointed out to the student of toxicology that dose-response curves are not necessarily unidirectional. Numerous other effects of cadmium may be cited, but they are either relatively innocuous or have only been demonstrated to occur in experimental animals.

As examples of the former, cadmium causes anosmia and yellow staining of the teeth in heavy industrial exposure. So far as observations of effects in animals is concerned, these are of definite value, in that they suggest possible effects in man that require future study. Thus, cadmium causes cerebral and cerebellar damage to newborn animals, whereas adults are resistant to these effects. In addition, cadmium is toxic to the testes of rats and mice, probably as a result of toxicity to the vasculature. Cadmium also causes hyperglycemia and glucose intolerance in animals, possibly as a result of decreased secretory activity of pancreatic beta cells. These and numerous other effects in animals are described and relevant literature is cited in the review articles cited in the introduction to this section.

Dose-Effect and Dose-Response Relationships

As with mercury, it is at present not possible to specify dose-effect relationships in man, and even dose-response relationships are very difficult to develop. The critical organ is generally agreed to be the kidney and the critical effect is proteinuria. In order to develop a dose-effect relationship, it would be necessary to specify the degree of proteinuria, *e.g.*, 10 mg/day, normalized for variable body mass or surface area. For that matter, there is not even any agreement as to the relative contributions of low-molecular-weight proteins (*e.g.*, O_2-micro-globulin) and high-molecular-weight proteins (*e.g.*, albumin). The designation of proteinuria in the all-or-none sense for the development of dose-response relationships is beset with similar difficulties. One investigator's definition of proteinuria may be quite different from another investigator's definition. Dose is even more difficult to define than effect, at least so far as application for any practical human monitoring programs is concerned.

The concentration of cadmium in kidney cortex would be a logical index of dose, but it is not accessible for sampling except under very unusual circumstances. Furthermore, there is good evidence that the concentration increases with cumulative exposure only to the point at which renal damage occurs. With the onset of damage a reduction in kidney cadmium occurs coincident with a pronounced rise in urinary excretion of cadmium. Thus, even the concentration of cadmium in the kidney is difficult to interpret, at least following the onset of proteinuria. The concentration of cadmium in the blood is not a good index of internal dose to the kidney. It equilibrates rather rapidly with any given level of external exposure (less than one year), whereas the target organ of interest (the kidney) continues to accumulate cadmium for many years.

In a group of workers exposed to fairly constant levels of cadmium for many years, a relationship was found between the incidence of proteinuria and the duration of exposure (Fig. 3.9). This is analogous to a dose-response curve, since the cumulative dose to the kidney was probably somewhat proportional to the number of years of exposure. Numerous investigators have studied the relationship between urinary cadmium excretion and duration of cadmium exposure or age of the individuals. There does appear to be a correlation between urinary cadmium excretion and duration of cadmium exposure in workers, suggesting that urinary excretion corresponds roughly to the concentration of cadmium in the kidney.

Treatment of Cadmium Poisoning

There is surprisingly little information concerning the management of cadmium poisoning. In the case of *itai-itai,* large doses of vitamin D given over a period of months are effective in relieving painful symptoms and, in some instances at least, therapy reduces the incidence of spontaneous fractures. For a review of the Japanese literature. There is no therapeutic approach being utilized in the management of industrial cadmium poisoning. Chelating agents have been studied in regard to their efficacy in the mobilization of cadmium. Dimercaptopropanol (BAL) has been shown to increase the uptake of cadmium by the kidney and to increase its nephrotoxicity.

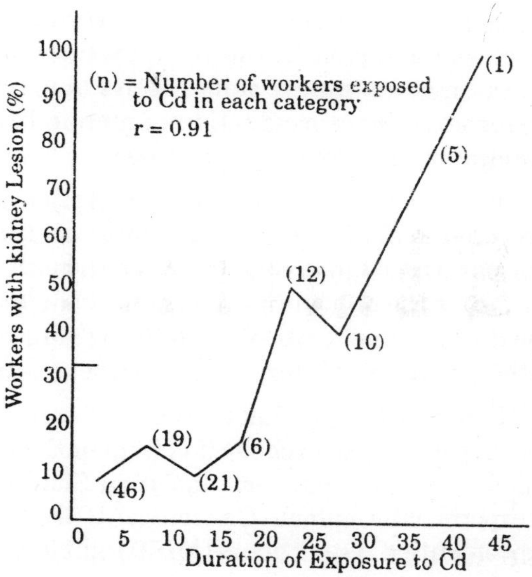

Fig. 3.9. The relationship between duration of cadmium exposure and renal response. The criterion for presence of renal lesion was the presence of an abnormal electrophoretic pattern of urinary proteins. Numbers in parentheses indicate number of men.

A similar effect has been noted regarding EDTA. It too increases both the concentration of cadmium in the kidney and the nephrotoxicity, in spite of an increased urinary excretion of cadmium. In one instance where EDTA was used in the treatment of a man who had swallowed about 5 g CdI_2, the patient died in spite of drug-induced enhancement of cadmium excretion in the urine. The therapeutic regimen did not prevent extensive damage to the kidney and liver. Based on the experience in animal studies, it may be that the therapeutic regimen actually increased damage to these organs.

ALUMINUM

Occurrence and Use

The principal ore of aluminum is bauxite. Aluminum is widely used as a building material and for other uses where light weight and corrosion resistance are important. Aluminum oxide has industrial uses as an abrasive and catalyst. Medically, various soluble salts of aluminum have been used as astringents, styptics, and antiseptics. The insoluble salts are used as antacids and as antidiarrheal agents. Inhalation of aluminum hydroxide is used as a preventive and curative agent for silicosis.

Absorption, Excretion, Toxicity

The normal blood level of aluminum is 17 μg Al/100 ml and most soft tissues contain between 0.2 and 0.6 ppm. The human body burden of aluminum is 50 to 150mg and is apparently unaffected by either normal daily intake levels, estimated to be approximately 10 to 100 mg, or considerably higher doses. The degree of absorption of ingested aluminum and its compounds is minimal.

Many aluminum salts are converted is minimal. Phosphate salt in the gastrointestinal tract and excreted in the feces as such. Milk is a secondary route of excretion. Parenteral injection of aluminum salts results in excretion in both the feces and urine as well as slight increase in the concentration of aluminum in the liver and the spleen. Massive oral doses of aluminum are reported to be toxic.

Interference with phosphate absorption occurs resulting in rickets. Gastrointestinal irritation also occurs following large oral doses of aluminum. The use of aluminum cooking utensils and cans is not enough to contribute significantly to either total body burden or toxic effects. Shaver's disease is the only aluminum-induced industrial disease. It may result from bauxite fume and the use of abrasive wheels containing aluminum. Exposure to the fume may produce weakness, fatigue, and respiratory distress. Chest x-ray may reveal extensive fibrosis with large blebs.

Spontaneous pneumothorax is a frequent complication. Silicon may also play a contributory role in the disease because it is frequently inhaled along with aluminum. Fibrosis has also been noted after aluminum dust inhalation. Similar changes can be reproduced by intratracheal injection in experimental animals. One source (AIHA, 1963) has recommended that the maximum atmospheric concentration (eight hours) be 50 million particles per cubic foot.

ANTIMONY

Occurrence and Use

The primary ore of antimony is stibnite (Sb_2S_3). The important uses of this metal are with lead alloys, in storage battery grids, in type alloys, pewter, bearing alloys, rubber, matches, ceramics, enamels, paints, lacquers, and textiles. Antimony may be present in food, resulting from the use of rubber, solders, and tinfoil for packaging. Leaching of antimony from cheap enameled vessels has caused some food contamination. Tarter emetic (antimony potassium tartrate) has been used as an insecticide. Antimony is a common pollutant in urban air. Antimony or its compounds were used medicinally as early as 4000 B.C. Their popularity in medicine has undergone several cycles of use and disuse.

At the present time their use is declining with the advent of newer parasiticides. The therapeutic activity involves reaction with the sulfhydryl groups in enzymes and a selective toxicity due to concentration in the parasite. Both trivalent and pentavalent organic compounds have been administered parenterally for parasiticidal effect; however, based on the hypothesis for the mechanism of effect, the trivalent forms would be expected to have greater efficacy. The probable *in vitro* conversion from the pentavalent to the trivalent forms may well account for the clinical effectiveness of the former. The trivalent compounds have been given orally as emetics and expectorants, but this use has been largely abandoned because of toxicity. The mechanism of emetic activity includes both a local and a central component.

Absorption, Excretion, Toxicity

Most of the information on the distribution and fate of antimony compounds arises from investigational results of therapeutic compounds. Antimony compounds are slowly absorbed from the gastro-intestinal tract and tend to produce vomiting. The distribution of antimony following intravenous or intramuscular administration is somewhat variable and cannot be fully accounted for solely on the basis of valence. The trivalent forms generally concentrate in red blood cells, while the pentavalent compounds are found in the plasma. Trivalent forms accumulate in the liver and are slowly excreted, principally in the feces. In experimental animals significantly high concentrations are found in the thyroid after administration of trivalent compounds.

The pentavalent forms tend to concentrate in the liver and spleen and are excreted in the urine. It is noteworthy that repeated dosages of labeled antimony tartar emetic were not accumulated in the body. Acute poisoning has resulted from accidental or intentional ingestion of antimonials. The symptoms are similar to those of arsenic poisoning and consist of vomiting, watery diarrhea, coltepse, irregular respiration, and lowered temperature. Vomiting increases the chance of recovery.

In fatal cases death occurs within a few hours after ingestion. Chronic incorporation of potassium antimony tartrate (5 ppm) into drinking water increased the mortality rate and decreased serum glucose levels in rats. The incidence of tumors was not increased; but there was evidence of antimony accumulation in the soft tissue, contrary to what has been reported by other investigators. Toxicity data have also been derived in connection with therapeutic use of antimonials. Cardiac effects, in a few cases atrial fibrillation due to a direct effect oh the heart and death, liver toxicity, characterized by jaundice and fatty degeneration, pulmonary congestion and edema, and papular skin eruptions have been reported.

Occupational poisoning by antimony is often difficult to establish since the antimony used in industry may contain some arsenic. The symptoms of toxicity of antimony and arsenic are similar. The signs ascribed to industrial antimony poisoning-include upper respiratory tract irritation, pneumonitis, dizziness, diarrhea, vomiting, and dermatitis. Antimony miners have developed disabling, but benign forms of silicosis. Some investigators have suggested a relationship between antimony and pulmonary carcinogenesis on the basis of possible antimony-containing abnormal enzyme systems. However, there is no positive evidence that diseased lung tissue contains excess amounts of antimony. Antimony and antimony compounds may generate stibine (antimony hydride) under reducing conditions. This has occurred during storage battery charging.

Although arsine is usually suspected of industrially induced hemolytic anemia, stibine may be involved more frequently than issuspected. The lethal concentration of stibine in the air for mice is about 100 ppm for 1.6 hours, while that of arsine is three hours. Stibine, like arsine, may be expected to cause rapid destruction of red blood cells, hemoglobinuria, and anuria. Subjective signs include headache, vomiting, nausea, and lumbar and epigastric pain.

ARSENIC

Occurrence and Use

Arsenic trioxide (As_2O_3) is used as the starting point in the manufacture of most arsenic compounds and is obtained from roasting arsenic-containing ores (FeAsS, As_2S3, and As_2S_2). The major use of arsenic has been in the form of its compounds whose toxicity makes them valuable as insecticides, weed killers, and wood preservatives. Lead-arsenic alloys are also used because they are more rigid than pure lead. Antifouling paints and materials to control sludge formation in lubricating oils also contain arsenic. Arsenic is ubiquitous in distribution, mostly pentavalent, in soil.

Arsenic is a natural constituent of food, although additional amounts may be added by contamination. Sea-foods, pork, liver, and salt may be exceptionally high in arsenic. In general, the naturally occurring arsenic is pentavalent while that added to the environment is trivalent. Medicinal uses of arsenic have ranged from treatment of leukemia to use as a tonic.

Arsenicals act locally and are slow corrosives; they have been used in the treatment of skin cancer. Both trivalent and pentavalent organic arsenic compounds have been used in the treatment of various parasitic diseases. In the United States there has been a decline in the use of arsenicals in human medicine but the decrease has not been as great in veterinarian or agricultural use. A certain amount of arsenic is added to the atmosphere by domestic coal use as well as by industrial contaminants. Arsenic is prevalent in small amounts in the water supply (Table 3.2). Some of this arsenic may come from phosphate fertilizers that contain arsenic.

Absorption, Excretion, Toxicity

Compounds of arsenic may be absorbed after ingestion or by inhalation. It is generally true that trivalent arsenic compounds are more toxic than pentavalent compounds and that natural oxidation favors the conversion of trivalent arsenic to the pentavalent form. It has been shown in some instances, however, that the arsenate is reabsorbed by the proximal renal tubule and excreted as the arsenite. Arsenate is the valence form most prevalent in nature and in this form tends to be rapidly excreted by the kidneys and probably does not accumulate. Arsenate can substitute for phosphate in some enzyme systems without adverse effects. Arsenites bind to tissue proteins and are concentrated in the leukocytes. They accumulate in the body primarily in the liver, muscles, hair, nails, and skin, perhaps because of combination with sulfhydryl groups.

Excretion is via the bile. Arsenite is not found in milk. The trivalent forms of arsenic are more toxic than the pentavalent forms. To illustrate these differences, single intraperitoneal injections of arsenic salts were administered to mice on days 6, 7, 8, 9, 10, 11, and 12 of gestation. Fetal deaths, resorption, exencephaly, and short jaws were observed among the fetuses. Those dams treated on day 8 of gestation were affected the most. At a dose of 25 mg/kg sodium arsenate was without effect, but at a level of 10 mg/kg sodium arsenite produced embryotoxicity and teratogenic effects. Chromosomal breaks in human leukocytes have been produced by arsenic *in vitro*. In rats fed sodium arsenate or arsenite at equal arsenic levels, those receiving the arsenate had less severe changes in bile duct enlargement.

Similar results in the rats were obtained when growth and survival were used as criteria. In rats, 62.5 ppm of arsenic as arsenite was without effect, while the no-effect level for arsenic as arsenate was 125 ppm. No dogs fed arsenic (125 ppm) as arsenite survived a two-year study. At the same level of arsenic as the arsenate only one of the six dogs died. In neither species was it evident that oral administration of arsenic might be carcinogenic. In man the symptom of acute inorganic arsenic poisoning occurring as a consequence of accidental or homicidal ingestion consist of burning and dryness of the oral and nasal cavities, gastrointestinal disturbance, and muscle spasms; vertigo, delirium, and coma may occur.

Edema of the face and about the eyelids may also be evident. Chronic arsenic intoxication is characterized by malaise and fatigue. Gastrointestinal disturbances, hyperpigmentation, and peripheral neuropathy may ultimately occur. Pale bands on the fingernails and toes may develop. Clinical pathology may reveal anemia (slightly hypochromic) and basophilic stippling.

Red cell disruption, decreased red cell production, and leukopenia are frequently observed. These signs disappear rapidly when exposure is halted, except for neuropathy, which may regress at a slower rate. Increased arsenic content of hair, nails, and urine is frequently present for long periods after exposure has been discontinued. Industrial poisoning generally follows the same pattern, although skin changes may occur more frequently than the hematologic changes. Nasal septum ulceration is seen after long industrial exposure. While there has been some controversy, the epidemiologic evidence indicates that industrial and agricultural exposure to arsenic is implicated in cancer of the skin and respiratory tract.

The individuals at greatest risk are smelter workers, although there is some suggestion that women residing near such operations incur a greater incidence of respiratory cancer. In experiments, the oral ingestion by animals has not suggested a carcinogenic potential by this route. Epidemiologic studies have implicated ingestion of arsenic as goitrogenic, and this possible effect has been confirmed in animals. Epidemiologic studies have suggested that arsenic in drinking water may be related to increased incidence of skin cancer. Extraordinarily high arsenic levels in soil and water have tentatively been linked with a severe form of peripheral arteriosclerosis (blackfoot disease) observed in Taiwan.

However, members of the same family, some with blackfoot disease and some without, have similarly high serum arsenic levels. Arsine, the hydride of arsenic, is one of the more toxic arsenic compounds. Arsine maybe generated when acids are combined with arsenic-containing metals. Poisoning by arsine is the principal source of industrial arsenic poisoning today and has been reported in connection with the refining or processing of tin, lead, and zinc.

Poisonings from this source have dire consequences because of the severe hemolytic effect and the inadequacy of available therapy. Arsine, a gas with a slight garlic-like smell detected only above safe levels, produces massive hemolysis and renal failure. Nausea, emesis, diarrhea, disturbance of the vascular bed, pulmonary edema, cyanosis, electrocardiogram abnormalities, hemoglobinuria, and liver dys-function may occur. There is generally some delay in the onset of symptoms. Exposures as low as 10 ppm have produced delirium, coma, and death. If the exposure is not fatal, the signs of chronic arsenic poisoning may appear. Urine may continue to contain arsenic for same time after poisoning.

BARIUM

Occurrence and Use

Barite ($BaSO_4$) and witherite ($BaCO_3$) are the more common mineral forms of barium. Barium is used in various alloys, in paints, soap, paper, and rubber, and in the manufacture of ceramics and glass. Barium fluorosilicate and carbonate have been used as insecticides. Barium sulfate, an insoluble compound, is used as a radiopaque aid to x-ray diagnosis. Barium is relatively abundant in nature and is found in plants and animal tissue. Plants accumulate barium from the soil. Brazil nuts have very high concentrations (3.000 to 4,000 ppm). Some water contains barium from natural deposits.

Absorption, Excretion, Toxicity

The soluble compounds of barium are absorbed and small amounts are retained in the body. Reports indicate the lung has an average concentration of 1 ppm (dry weight). The kidney, spleen, muscle, heart, brain, and liver concentrations are 0.10, 0.08, 0.05,. 0.04, 0.03, and 0.03, respectively-Some barium is also found in the skeleton. Studies suggest that barium may be an essential element inasmuch as rats and guinea pigs maintained on barium-free diets fail to grow normally. The soluble compounds once absorbed are transported by the plasma. The biologic half-life is short (less than 24 hours).

Feces appear to be the major excretion route of absorbed barium although some is lost through the kidney. The renal tubules reabsorb barium in the filtrate. The insoluble forms of barium, particularly barium sulfate, are not toxic by the oral route because of minimal absorption. However, the soluble barium compounds are highly toxic, in contrast to calcium and strontium, the other members of this group in the periodic table. Accidental poisoning from ingestion of soluble barium salts has resulted in gastroenteritis, muscular paralysis, decreased pulse rate, and ventricular fibrillation and extrasystoles. Potassium deficiency occurs in acute poisoning and the heroic measure, treatment with intravenous potassium, appears beneficial.

The digitalis-like toxicity, muscle stimulation, and central nervous system effects have been confirmed by experimental investigation. Baritosis, a benign pneumoconiosis, is an occupational disease arising from the inhalation of barium sulfate (barite) dust and barium carbonate. It is not incapacitating, but does produce radiologic changes in the lungs. The radiologic changes are reversible with cessation of exposure.

BERYLLIUM

Occurrence and Use

Beryl ($3BeO\text{-}Al_2O_3\text{-}6SiO_2$) is the chief ore of beryllium. Its industrial uses include the hardening of copper, the manufacture of nonsparking alloys for tools, and the manufacture of lightweight alloys and nuclear reactors. It is also used in the manufacture of ceramics and in the electronics industry (transistors, heat sinks, and x-ray and cathode tubes). Gas lantern mantles, when first ignited, volatilize most of the beryllium they contain. Formerly, beryllium was widely used in the manufacture of fluorescent lights and neon signs. The aerospace industry uses beryllium compounds as propellants, providing a possible source of environmental exposure; however, coal combustion is probably the largest source of environmental beryllium contamination.

Absorption, Excretion, Toxicity

Beryllium is not well absorbed when given by any route. Experimental animals may only absorb 1 per cent as a maximum. After inhalation exposure beryllium is retained in the lungs and mobilized slowly. In the bloodstream-a colloidal beryllium phosphate is formed. In addition, small amounts may be found in a soluble beryllium-citrate complex that is deposited in the bone or excreted via the urine. The colloidal portion is deposited in the liver, spleen, and bone marrow. The skin lesions are the most common sign of the industrial disease. Three distinct types of skin lesions have been described: dermatitis, ulceration, and granulomas. The dermatitis, sometimes accompanied by acute conjunctivitis and corneal ulceration, has been regarded as a hypersensitizing reaction.

The finding that victims of chronic beryllium lung disease react positively to patch tests supports this notion and indicates the possible immunologic component of the chronic disease. The ulcer is such that healing will take place only if the offending material is curetted. The granuloma is most frequently the result of broken fluorescent lamps embedding beryllium under the skin. The use of beryllium in this type of lamp has ceased. Chronic skin lesions sometimes appear after a long latent period in conjunction with the chronic pulmonary aspect of the disease. Sub-cutaneous granulomas have been produced in the pig using the beryllium phosphor. Short-term inhalation exposures to levels of soluble beryllium compounds in excess of 100 $\mu g/M^3$ result in acute lung distress. A latent period of one day to three weeks may exist.

The inflammatory response consists of nasopharyngitis, tracheobronchitis, and, in more severe cases, fulminating pneumonitis. When fulminating pneumonitis is present, chest radiographs show haziness to "snow-flurry" effect. Edema may become so severe that right-side heart failure occurs. The clinical course is four to six weeks in duration although x-ray changes may take longer. The acute pneumonitis is readily reproduced in experimental animals. Of 124 cases of chronic berylliosis, 11 per cent had a history of acute attacks. The symptoms of the chronic form of the disease are dyspnea, chronic cough, weight loss, weakness, fatigue, and chest pain or discomfort.

Latent periods of over 20 years have been observed, but a latent period of 10 to 19 years from the first exposure to five to nine years from the termination of exposure is the most common. In those cases in which histologic information from autopsy or biopsy is available, exposure was most frequently in connection with extraction and smelting operations or the manufacture of fluorescent lamps. The development of disease in persons exposed to minimal amounts of beryllium such as those handling a beryllium worker's clothes or living near a plant has been documented. The physical findings show a fine miliary nodulation (ground- glass appearance).

As the fibrosis increases, bleb formation is common and pneumothorax occurs. The striated muscles, liver, spleen, kidneys, and heart may be involved. Lung function and transfer of oxygen may be impaired. Other systemic aspects may include disturbances in nitrogen and calcium metabolism. Differential diagnosis of sarcoidosis and chronic beryllium disease is difficult. The use of the betylliutn patch test is helpful but may in fact sensitize the subject. *In vitro* tests utilizing immunologic indicators have been used. A history of beryllium exposure should be sought. While, histologically, granulomatosis pneumonitis is most characteristic of the histologic lung tissue from patients with chronic beryllium disease, 55 of 124 chronic cases had either indistinct or no granuloma formation. However, in many of these cases, the tissue may have been obtained by biopsy rather than at autopsy.

The degree of histologic abnormality did not correlate with the lung tissue beryllium content or the length of the latent period. There is some indication that early detection of the respiratory changes and reduction of the exposure levels can lead to reversal of the early radiographic abnormalities and to improvement of pulmonary gas exchange.

A high incidence of hyperuricemia seems to occur in cases of chronic beryllium disease, but the nature of this effect has not been ascertained. The human disease, chronic pulmonary granulomatosis in the diffuse form, has not been reproduced in experimental animals. Scattered granulomatous lesions have been produced in rats.

Rhesus monkeys react similarly to man when challenged by injection of beryllium oxide powder in the wall of the bronchus or inhalation of beryllium sulfate by development of widespread chronic beryllium pneumonitis with granulomata.

Osteogenic sarcoma has been produced in rabbits and bronchogenic carcinoma has been produced in rats and monkeys. Bony lesions, osteosclerosis in rats and rabbits, rickets in growing rats, and anemia are produced in experimental investigations but seem to be without counterparts in the human disease state.

BISMUTH

Occurrence and Use

Bismuth is obtained as a by-product of tin, lead, and copper ores. It is used in the-manufacture of type alloys, silvering of mirrors, low-melting solders (sometimes used in canning), and heat-sensitive devices such as automatic fire extinguishers. Bismuth telluride is used in the electronics industry as a semiconductor. Bismuth is one of the contaminants measured in urban air. Trivalent insoluble bismuth salts are used medicinally to control diarrhea and other types of gastrointestinal distress. Some of these preparations are available without prescription.

Various bismuth salts have been used externally for their astringent and slight antiseptic property. Bismuth salts have also been used as radio-contrast agents. Further self-exposure comes from the use of insoluble bismuth salts in cosmetics. Injections of soluble and insoluble salts, suspended in oil to maintain adequate blood levels, have been used to treat syphilis. Bismuth sodium thioglycollate, a water-soluble salt, was injected intramuscularly for malaria *(Plasmod-ium vivax)*. Bismuth glycolyarsanilate is one of the few pentavalent salts that have been used medicinally. This material was formerly used for treatment of amebiasis. Exposure to various bismuth salts for medicinal use has decreased with the advent of newer therapeutic agents.

Absorption, Excretion, Toxicity

Most bismuth compounds to which we are exposed are insoluble and poorly absorbed—either when taken orally or when applied to the skin, even if the skin is abraded or burned. Thus, most of the information on their distribution in the body is related to therapeutic use. Once the bismuth is absorbed from the site, tissue binding appears minimal. A diffusible equilibrium between tissues, blood, and urine is established. Tissue distribution, omitting injection depots, reveals the kidney as the site of the highest concentration.

The liver concentration is considerably lower at therapeutic levels but with massive doses in experimental animals (dogs), the kidney/liver ratio is decreased. Passage of bismuth into the amniotic fluid and into the fetus has been demonstrated. The urine is the major route of excretion. Traces of bismuth can be found in the milk and saliva. The total elimination of bismuth after injection is slow and dependent on mobilization from the injection site. There have been no reports of industrial poisoning from bismuth. Except for strong acidic salts such as bismuth trinitrate or violently reactive compounds such as bismuth tripentafluoride, the bismuth compounds do not present a hazard by dermal application, inhalation, or ingestion.

The dermal application of most bismuth compounds does not result in systemic toxicity. Oral administration of bismuth subnitrate produces poisoning through the formation of nitrites. Intravenous injections of bismuth salts were avoided because of toxicity; the soluble salts had a tendency to flocculate. Intramuscular injections tended to be painful, and if sufficient doses were employed, some necrosis was evident at the site of injection. In experimental animals renal and hepatic toxicity have been observed following the achievement of sufficient systemic bismuth levels. The symptoms of chronic toxicity in man consist of decreased appetite, weakness, rheumatic pain, diarrhea, fever, metal line on the gums, foul breath, gingivitis, and dermatitis.

Jaundice and conjunctival hemorrhage are rare, but have been reported. When nephritis does occur in man, albuminuria is used as a signal to discontinue administration. The renal effects include diuresis and are similar to the effects of mercury. Microscopically, lipid-carbohydrate-protein inclusions containing no bismuth are formed in the renal proximal convoluted tubular cell shortly after treatment with bismuth. They are apparently irreversible.

BORON

Occurrence and Use

While boron is, strictly speaking, a nonmetal, it is in the group IIIA metals and is of some toxicologic concern. Boron occurs regularly in natural water supplies (Table 3.2) and in plant and animal tissues. It is essential to plants but apparently not to animals. The average daily intake has been estimated at 10 to 20 mg. Borax ($Na_2B_4O_7$) is the most important mineral. Boric acid is useful medicinally as a mild antiseptic, especially as an eyewash. Borax is used in soldering and welding to remove oxide film, for softening water, in soaps, and in glass, pottery, and enamels.

Absorption, Excretion, Toxicity

Boron in the food, as sodium borate or boric acid, mostly in fruit and vegetables, is almost completely absorbed and is excreted in the urine. Treatment of large burned areas with boric acid results in systemic absorption. Large amounts of absorbed boron cause accumulation in the brain. In lambs, gastrointestinal and pulmonary disorders have been reported to result from grazing where pasture soils are high in boron content. Industrial poisoning has not been reported from exposure to boron salts except for the boranes. Death has been reported due to dermal application of boric acid for burns and cuts.

Central nervous system depression and gastrointestinal irritation are the mostisevere symptoms. Infants appear to be more susceptible to the toxic effects than adults. Skin irritation has occurred in in-fants following dermal application. The boranes diborane, decaborane, and pentaborane are used in high-energy fuels. Decabor-ane has also been used in vulcanizing rubber. All three are highly toxic. Pentaborane is the most hazardous. Diborane is an irritant to the lungs and kidneys. Decaborane and pentaborane are central nervous system poisons; however, the liver and kidneys may also be damaged if the exposure is severe.

CESIUM

Occurrence and Use

Cesium occurs in nature as pollucite, a hydrous cesium-aluminum silicate. Its main industrial uses are as a catalyst in the polymerization of resin-forming materials and in photoelectric cells. It is useful in this respect because the range of sensitivity is approximately that of the human eye. Radioactive cesium is a constituent of nuclear fallout.

Absorption, Excretion, Toxicity

Cesium is absorbed after oral administration and is bound within the cells of the soft tissues such as kidney and muscle. It is found in the red blood cells and may in some circumstances be able to replace potassium. The urine is the main route of excretion. Increased potassium levels facilitate cesium excretion.

The radioactive material is found in milk. No cases of industrial injury related to the chemical toxicity of cesium have been reported. It is likely that replacement of potassium by cesium would produce ill effects in man, probably neuromuscular in nature, as has been demonstrated in experimental animals.

CHROMIUM

Occurrence and Use

Chromite ($FeCr_2O_4$) is the most important chrome ore. Chromium plating is one of the major uses of this metal. Steel fabrication, paint and pigment manufacturing, and leather tanning constitute other major uses of chromium. The medicinal uses of chromium are limited to external application of chromium trioxide as a caustic and intravenous sodium radiochromate to evaluate the life-span of red cells.

Absorption, Excretion, Toxicity

Chromium exists in several valence states. Only the trivalent and hexavalent are biologically significant. While conversion from trivalent to hexavalent and other states is important chemically, the inner conversion from chromic to chromate does not apparently occur biologically. The conversion of hexavalent to trivalent does take place in the body. Trivalent chromium is an essential element in animals. It plays a role in glucose and lipid metabolism.

Chromium deficiency mimics diabetes mellitus and produces aortic plaques in rats. Chromium supplementation improves or normalizes glucose tolerance in diabetics, older people, and malnourished children. It has been suggested that chromium deficiency may be a basic factor in atherosclerosis. A deficiency of trivalent chromium apparently increases the toxicity of lead. The major environmental exposure to chromium occurs as a consequence of its presence in food. Brown sugar and animal fats, especially butter, are chromium-rich foods. Chromium is found in urban air (Table 3.7).

The concentration in natural water supplies is below 10 ppb; however, in municipal drinking water concentrations of 35 ppb have been reported (Table 3.2). The daily intake has been estimated at 60 μg (30 to 100 μg), 10 μg of which is due to water concentrations (Table 3.1). However, the absorption is limited to approximately 1 per cent. The occurrence of chromium in food or water has not been shown to produce any significant adverse effects in either man or experimental animals. The total chromium body burden of man has been estimated at less than 6 mg (Table 3.1).

Chromium is transported across the placenta and concentrated in the fetus. The tissue concentrations tend to decline rapidly with age except for the lung concentration, which tends to increase. The decline of chromium levels with age does not occur in rats. Wide geographic variations in tissue concentration, presumably due to differences in dietary intake and atmospheric concentration, have been reported. Water-soluble chromates disappear from the lungs into the circulatory system after intratracheal application, while the trivalent chromic chloride remains largely in the lungs. Oral administration of trivalent chromium results in little chromium absorption. The degree of absorption is slightly higher following administration of hexavalent compounds. Once absorbed, Cr^{3+} is bound to the plasma proteins.

Under normal conditions toe body contains stores of chromium in the skin, lungs, muscle, and fat. The bone contains chromium, but this is not due to selective deposition. The caudate nucleus has been reported to have high concentrations. Hexavalent chromium is reduced to the trivalent form in the skin. In the blood little hexavalent chromium can be detected. The reticuloendo-thelial system, liver, spleen, testes, and bone marrow have an affinity for chromite, possibly as the result of phagocytosis of colloidal particles formed at higher tissue concentrations.

On the other hand, chromates are bound largely to the red blood cells. Subcellular distribution studies have indicated that the nuclear fraction ductor. Bismuth is one of the contaminants measured in urban air. Trivalent insoluble bismuth salts are used medicinally to control diarrhea and other types of gastrointestinal distress. Some of these preparations are available without prescription. Various bismuth salts have been used externally for their astringent and slight antiseptic property.

Bismuth salts have also been used as radio-contrast agents. Further self-exposure comes from the use of insoluble bismuth salts in cosmetics. Injections of soluble and insoluble salts, suspended in oil to maintain adequate blood levels, have been used to treat syphilis. Bismuth sodium thioglycollate, a water-soluble salt, was injected intramuscularly for malaria *(Plasrriodium vivax).* Bismuth glycolyarsanilate is one of the few pentavalent salts that have been used medicinally. This material was formerly used for treatment of amebiasis. Exposure to various bismuth salts for medicinal use has decreased with the advent of newer therapeutic agents.

Absorption, Excretion, Toxicity

Most bismuth compounds to which we are exposed are insoluble and poorly absorbed—either when taken orally or when applied to the skin, even if the skin is abraded or burned. Thus, most of the information on their distribution in the body is related to therapeutic use. Once the bismuth is absorbed from the site, tissue binding appears minimal. A diffusible equilibrium between tissues, blood, and urine is established. Tissue distribution, omitting injection depots, reveals the kidney as the site of the highest concentration. The liver concentration is considerably lower at therapeutic levels out with massive doses in experimental animals (dogs), the kidney/liver ratio is decreased.

Passage of bismuth into the amniotic fluid and into the fetus has been demonstrated. The urine is the major route of excretion. Traces of bismuth can be found in the milk and saliva. The total elimination of bismuth after injection is slow and dependent on mobilization from the injection site. There have been no reports of industrial poisoning from bismuth. Except for strong acidic salts such as bismuth trinitrate or violently reactive compounds such as bismuth tripentafluoride, the bismuth compounds do not present a hazard by dermal application, inhalation, or ingestion. The dermal application of most bismuth compounds does not result in systemic toxicity.

Oral administration of bismuth subnitrate produces poisoning through the formation of nitrites. Intravenous injections of bismuth salts were avoided because of toxicity; the soluble salts had a tendency to flocculate. Intramuscular injections tended to be painful, and if sufficient doses were employed, some necrosis was evident at the site of injection. In experimental animals renal and hepatic toxicity have been observed following the achievement of sufficient systemic bismuth levels. The symptoms of chronic toxicity in man consist of decreased appetite, weakness, rheumatic pain, diarrhea, fever, metal line on the gums, foul breath, gingivitis, and dermatitis. Jaundice and conjunctiva hemorrhage are rare, but have been reported.

When nephritis does occur in man, albuminuria is used as a signal to discontinue administration. The renal effects include diuresis and are similar to the effects of mercury. Microscopically, lipidcarbohydrate-protein inclusions containing no bismuth are formed in the renal proximal convoluted tubular cell shortly after treatment with bismuth. They are apparently irreversible.

BORON

Occurrence and Use

While boron is, strictly speaking, a nonmetal, it is in the group (Table 3.2) and in IIIA metals and is of some toxicologic concern (Table 3.4). Boron occurs regularly in natural water supplies plant and animal tissues. It is essential to plants but apparently not to animals. The average daily intake has been estimated at 10 to 20 mg. Borax ($Na_2B_4O_7$) is the most important mineral. Boric acid is useful medicinally as a mild antiseptic, especially as an eyewash. Borax is used in soldering and welding to remove oxide film, for softening water, in soaps, and in glass, pottery, and enamels.

Absorption, Excretion, Toxicity

Boron in the food, as sodium borate or boric acid, mostly in fruit and vegetables, is almost completely absorbed and is excreted in the urine. Treatment of large burned areas with boric acid results in systemic absorption. Large amounts of absorbed boron cause accumulation in the brain. In lambs, gastrointestinal and pulmonary disorders have been reported to result from grazing where pasture soils are high in boron content. Industrial poisoning has not been reported from exposure to boron salts except for the boranes.

Death has been reported due to dermal application of boric acid for burns and cuts. Central nervous system depression and gastrointestinal irritation are the most severe symptoms. Infants appear to be more susceptible to the toxic effects than adults. Skin irritation has occurred in infants following dermal application. The boranes diborane, decaborane, and pentaborane are used in high-energy fuels. Decaborane has also been used in vulcanizing rubber. All three are highly toxic. Pentaborane is the most hazardous. Diborane is an irritant to the lungs and kidneys. Decaborane and pentaborane are central nervous system poisons; however, the liver and kidneys may also be damaged if the exposure is severe.

CESIUM

Occurrence and Use

Cesium occurs in nature as pollucite, a hydrous cesium-aluminum silicate. Its main industrial uses are as a catalyst in the polymerization of resin-forming materials and in photoelectric cells. It is useful in this respect because the range of sensitivity is approximately that of the human eye. Radioactive cesium is a constituent of nuclear fallout.

Absorption, Excretion, Toxicity

Cesium is absorbed after oral administration and is bound within the cells of the soft tissues such as kidney and muscle. It is found in the red blood cells and may in some circumstances be able to replace potassium. The urine is the main route of excretion. Increased potassium levels facilitate cesium excretion.

The radioactive material is found in milk. No cases of industrial injury related to the chemical toxicity of cesium have been reported. It is likely that replacement of potassium by cesium would produce ill effects in man, prob-ably neuromuscular in nature, as has been demonstrated in experimental animals.

CHROMIUM

Occurrence and Use

Chromite ($FeCr_2O_4$) is the most important chrome ore. Chromium plating is one of the major uses of this metal. Steel fabrication, paint and pigment manufacturing, and leather tanning constitute other major uses of chromium. The medicinal uses of chromium are limited to external application of chromium trioxide as a caustic and intravenous sodium radiochromate to evaluate the life-span of red cells.

Absorption, Excretion, Toxicity

Chromium exists in several valence states. Only the trivalent and hexavalent are biologically significant. While conversion from trivalent to hexavalent and other states is important chemically, the inner conversion from chromic to chromate does not apparently occur biologically. The conversion of hexavalent to trivalent does take place in the body. Trivalent chromium is an essential element in animals. It plays a role in glucose and lipid metabolism.

Chromium deficiency mimics diabetes mellitus and produces aortic plaques in rats. Chromium supplementation improves or normalizes glucose tolerance in diabetics, older people, and malnourished children. It has been suggested that chromium deficiency may be a basic factor in atherosclerosis. A deficiency of trivalent chromium apparently increases the toxicity of lead. The major environmental exposure to chromium occurs as a consequence of its presence in food. Brown sugar and animal fats, especially butter, are chromium-rich foods. Chromium is found in urban air.

The concentration in natural water supplies is below 10ppb; however, in municipal drinking water concentrations of 35 ppb have been reported. The daily intake has been estimated at 60 µg (30 to 100 µg), 10 µg of which is due to water concentrations. However, the absorption is limited to approximately 1 percent. The occurrence of chromium in food or water has not been shown to produce any significant adverse effects in either man or experimental animals. The total chromium body burden of man has been estimated at less than 6 mg.

Chromium is transported across the placenta and concentrated in the fetus. The tissue concentrations tend to decline rapidly with age except for the lung concentration, which tends to increase. The decline of chromium levels with age does not occur in rats. Wide geographic variations in tissue concentration, presumably due to differences in dietary intake and atmospheric concentration, have been reported. Water-soluble chromates disappear from the lungs into the circulatory system after intra-tracheal application, while the trivalent chromic chloride remains largely in the lungs. Oral administration of trivalent chromium results in little chromium absorption. The degree of absorption is slightly higher following administration of hexavalent compounds. Once absorbed, Cr^{3+} is bound to the plasma proteins.

Under normal conditions the body contains stores of chromium in the skin, lungs, muscle, and fat. The bone contains chromium, but this is not due to selective deposition. The caudate nucleus has been reported to have high concentrations. Hexavalent chromium is reduced to the trivalent form in the skin. In the blood little hexavalent chromium can be detected. The reticuloendothelial system, liver, spleen, testes, and bone marrow have an affinity for chromite, possibly as the result of phagocytosis of colloidal particles formed at higher tissue concentrations. On the other hand, chromates are bound largely to the red blood cells.

Subcellular distribution studies have indicated that the nuclear fraction contains almost one-half the intracellular chromium. Urinary excretion accounts for about 80 percent of injected chromium. However, elimination via the intestine may also play a role in chromium excretion. Milk is another secondary route of excretion. Average urinary and blood concentrations are 0.4 and 2.8 μg/100g, respectively. Occupational exposure to chromium com-pounds (Cr^{6+}) causes dermatitis, penetrating ulcers on the hands and forearms, perforation of the nasal septum, and inflammation of the larynx and liver.

The dermatitis is probably due to an allergenic response, although persons sensitive to Cr^{6+} also respond to large amounts of Cr^{3+}. The ulcers are believed to be due to chromate ion and not related to sensitization. Chromic acid, and, to a lesser extent, chromate, are presumably the causative agents in perforation of the nasal septum. Epidemiologic studies indicate that chromate is a carcinogen with bronchogenic carcinoma as the principal lesion. The latent period appears to be 10 to 15 years. The relative risk of chromate plant workers for respiratory cancer is 20 times greater than that of the general population.

Experimental studies have suggested that calcium chromate may be the specific carcinogenic agent. However, some investigators have produced cancer in experimental animals with injections of either the trivalent or hexavalent. Incorporation of hexavalent chromium (5 ppm) into the drinking water of mice over their lifetimes produced a slightly higher incidence of malignant tumors than in the controls. Trivalent chromium (chromium acetate) given to rats under similar conditions produced no such effect.

COBALT

Occurrence and Use

Cobalt is a relatively rare metal produced primarily as a by-product of other metals, chiefly copper. It is used in high-temperature alloys and in permanent magnets. Its salts are useful in paint driers, as catalysts, and in the production of numerous pigments. It is an essential element in that 1 μg of vitamin B_{12} contains 0.0434 μg of cobalt. Vitamin B_{12} is essential in the prevention of pernicious anemia. If other requirements exist, they are not well understood. Deficiency diseases of cattle and sheep caused by insufficient natural levels of cobalt are characterized by anemia and loss of weight or retarded growth.

Absorption, Excretion, Toxicity

Cobalt salts are generally well absorbed after oral ingestion, probably in the jejunum. Despite this fact, increased levels lend not to cause significant accumulation. About 80 per cent of the ingested cobalt is excreted in the urine. Of the remaining, about 15 per cent is excreted in the feces by an enterohepatic pathway, while the milk and sweat are other secondary routes of excretion. The total body burden has been estimated as 1.1 mg. The muscle contains the largest total fraction, but the fat has the highest concentration.

The liver, heart, and hair have significantly higher concentrations than other organs, but the concentration in these organs is relatively low. The normal levels in human urine and blood are about 98 and 0.18 μg/1, respectively. The blood level is largely in association with the red cells. Significant species differences have been observed in the excretion of radiocobalt. In rats and cattle 80 percent is eliminated in the feces. Polycythemia is the characteristic response of most mammals, including man, to ingestion of excessive amounts of cobalt.

Toxicity resulting from overzealous therapeutic administration has been reported to produce vomiting, diarrhea, and a sensation of warmth. Intravenous administration leads to flushing of the face, increased blood pressure, slowed respiration, giddiness, tinnitus, and deafness due to nerve damage. High levels of chronic oral administration may result in the production of goiter. Epidemiologic studies suggest that the incidence of goiter is higher in regions containing increased levels of cobalt in the water and soil. The goitrogenic effect has been elicited by the oral administration of 3 to 4 mg/kg to children in the course of sickle cell anemia therapy. Cardiomyopathy has been caused by excessive intake of cobalt, particularly in beer to which cobalt was added to enhance its foaming qualities.

The onset of the poisoning occurred about one month after cobalt was added in concentrations of 1 ppm. Why such a low concentration should produce this effect in the absence of any similar change when cobalt is used therapeutically is unknown. The signs and symptoms were those of congestive heart failure. Autopsy findings revealed a tenfold increase in the cardiac levels of cobalt. Alcohol may have served to potentiate the effect of the cobalt. Hyperglycemia due to alpha cell pancreatic damage has been reported after injection into rats. Reduction of blood pressure has also been observed in rats after injection and has led to some experimental use in man. Industrial exposure to cobalt salts leads to respiratory effects, although there is some question as to whether cobalt is the sole agent responsible for these effects.

Most industrial exposure comes from the cemented carbide industry where exposure to 1 to 2 mg/M^3 has produced pulmonary effects. Sensitization may be an important part of this effect. Experimental studies in animals, however, confirm the lung-irritant effect of the metal as used in this industry but not of other cobalt compounds. Skin and eye lesions similar to allergic dermatitis have also been reported. Skin tests show positive sensitization to cobalt but not to other material used in the cemented carbide industry. Gastric disturbances occurring shortly after daily exposure to cobalt acetate that progress to epigastric pain, pain in the limbs, hematuria, and occult blood in the stool have been reported. Recovery was complete in three weeks.

COPPER

Occurrence and Use

Copper occurs in several oxides, carbonate, and sulfide ores as well as native copper. It is widely used in industry because of its conductivity, malleability, and durability. Copper is widely distributed in nature and is an essential element. Copper deficiency is characterized by hypochromic, microcytic anemia resulting from defective hemoglobin synthesis. Oxidative enzymes, such as catalase, peroxidase, cytochrome oxides, and others, also require copper.

Medicinally, copper sulfate is used as an emetic. It has also been used for its astringent and caustic action and as an anthelmintic. Water purification processing has included the use of copper to remove algae. Copper sulfate mixed with lime has been used as a fungicide. Copper salts have also been used as a food additive to give a bright green colour to canned peas. Since copper proteins in the blood of many shellfish act as oxygen carriers, the concentration is frequently high.

Absorption, Excretion, Toxicity

The intestinal mucosa acts to some extent as a barrier to the absorption of ingested copper. Most cuprous salts are insoluble in water but they tend to oxidize to (he cupric form. Copper is initially bound to serum albumin and later more firmly bound to alpha-ceruloplasmin, where it is exchanged in the cupric form. The normal serum level of copper is 120 to 145 μg/1. The bile is the normal excretory pathway and plays a primary role in copper homeostasis. The liver and bone marrow are the storage organs for excess copper. The amount of copper in milk is not enough to maintain the copper levels in the liver, lung, and spleen of the newborn. The levels decline up to about ten years of age, remaining relatively constant thereafter. Brain levels, on the other hand, tend to almost double from infancy.

The ratios of newborn to adult liver copper show considerable species difference: man, 15:4; rat, 6:4; and rabbit, 1:6. Since urinary copper levels may be increased by soft water, under these conditions concentrations of approximately 60 μg/1 are not uncommon. While copper is an essential element in most organisms, the range between deficiency and toxicity is low in those without effective barriers to controlled absorption, for example, algae, fungi, and some invertebrates.

Fish are sensitive to copper, apparently because their gills do not provide an effective barrier against absorption. Ruminants are more sensitive to copper than monogastric mammals. Copper toxicity in sheep and cattle is characterized by excessive hepatic stores, hemolysis, and hemoglobinuria. Other mammals, including man, are less sensitive to copper, presumably because of a better-developed homeostatic mechanism. It is generally felt that excessive copper exposure in normal persons does not result in a chronic disease. While there is no increase in copper tissue stores with age, serum copper levels do increase. This has led to the speculation that increased serum copper levels may accelerate atherosclerosis. Industrial exposure to copper does not appear responsible for acute or chronic poisoning, except for "brass chills," which is another form of metal fume fever.

Whether the responses ascribed to copper are due to this metal or to some other factor is controversial. The increased incidence of lung cancer in coppersmiths is not generally accepted as being significant evidence of this metal's carcinogenicity. Acute poisoning resulting from ingestion of excessive amounts of oral copper salts, most frequently copper sulfate, may produce death.

The symptoms are vomiting, sometimes with a blue-green colour observed in the vomitus, hematemesis, hypotension, melena, coma, and jaundice. Autopsy findings have revealed centrilobular hepatic necrosis. Few cases of copper intoxication as a result of burn treatment with copper compounds have resulted in hemolytic anemia. Copper poisoning producing hemolytic anemia has also been reported as the result of using copper-containing dialysis equipment. Whether increased serum and liver copper concentrations, elicited after insertion of a copper wire into the uterus of rats will be indicative of difficulties as a result of using copper-containing intrauterine devices is open to speculation.

Table 3.6. Interrelationship Of Copper, Iron, And Lead In The Developing Erythrocyte

	Inron Deposition			
	Reticulocyte		*Erythroblast*	
Treatment	*Vesicles*	*Mitochondria*	*Vesicles*	*Mitochondria*
Copper supplemented	+	0	+	0
Copper deficient	+	0	+	0
Copper supplemented + Pb	+ + + +	+ + + +	+	0
Control diet + Pb	+ +	+ +	+	+
Copper deficient + Fe	+ + + +	0	++	0
Copper deficient + Pb	++	0	++	0
Copper deficient + Fe + Pb	+ +	0	+	0

A complex interrelationship of copper, iron, and lead exists in the developing erythrocyte. The results obtained in rats are summarized. These results indicate that excesses in copper and lead increase the iron content of the reticulocyte, but that copper deficiency in the presence of excess lead alters the intracellular distribution of iron.

GALLIUM

Occurrence and Use

Gallium is obtained as a by-product of copper, zinc, lead, and aluminum. It is used in high-temperature thermometers, as a substitute for mercury in arc lamps, in the manufacture of alloys, and as a seal for glass and vacuum equipment. The metal is a liquid at temperatures greater than 29.9° C. Radioactive gallium has been used as a diagnostic tool for the localization of bone lesions.

Absorption, Excretion, Toxicity

Gallium is not readily absorbed by the oral route, but occurs in bone at concentrations less than 1 ppm. Increasing intake produces slight increases in gallium levels in the liver, spleen, kidney, and bone. The urine is the major route of excretion. There are no reported adverse effects of gallium following industrial exposure. Therapeutic use of radiogallium has produced some effects, chiefly dermal and gastrointestinal in nature.

The bone marrow depression reported may be due largely to the radioactivity. In animals gallium acts as a neuromuscular poison and causes renal damage. Photophobia, blindness, and paralysis have been reported in rats. Renal damage ranging from cloudy swelling to necrosis has been reported. In dogs aplastic changes in the bone marrow have been observed.

GERMANIUM

Occurrence and Use

Germanium occurs in some mineral ores but is produced in the United States primarily as a by-product of zinc. It is a semiconductor and is used in electronics, in fine lenses, in certain aluminum alloys, and as a catalyst. However, only small amounts are actually used industrially; for example, only 7.5 tons were produced in the world in 1964. Germanium is present in most foods. Raw clams, tuna, baked beans, and tomato juice have significant amounts. The average daily intake has been estimated at 1,500 μg, but wide variations may occur. One would expect significant amounts of germanium to be in the urban atmosphere as a result of relatively high (1.6 to 7.5 per cent) germanium concentration in coal.

Absorption, Excretion, Toxicity

Sodium germanite is rapidly absorbed from the gastrointestinal tract—96 per cent in eight hours. It is transported at low serum levels unbound to plasma proteins. Some is also found in the red blood cells. Levels of 0.65 and 0.29 mg/ml have been reported as normal values for red blood cells and serum, respectively. In dogs 90 percent of radiogermanium oxide is excreted in the urine in 72 hours. The normal range of human urine concentration may be 0.40 to 2.16 μg/ml; milk and feces are secondary routes of excretion. In mice continuous feeding of germanium causes accumulation in the spleen; however, the same apparently does not hold true for the rat.

Inhalation exposure of rats to germanium and germanium oxide revealed rapid removal from the lung tissue. The toxicity of germanium and its compounds is low. The most widely studied is germanium oxide. The acute effects reported in animals are hypothermia, listlessness, diarrhea, respiratory and cardiac depression, edema, and hemorrhage in the lungs and gastrointestinal tract. The gaseous hydride, like other metal hydrides, is more toxic and causes hemolysis. Chronic administration to rats in food at 1000 ppm or in water at 100 ppm of germanium oxide caused growth inhibition and mortality.

The survivors appeared to develop a tolerance. The cause of the mortality was not apparent. Exposure to germanium is not considered an industrial hazard, nor has it been implicated in any chronic diseases of man.

GOLD

Occurrence and Use

Gold is rather widely distributed in small quantities, but the major economically-usable deposits occur as the free metal in quartz veins or alluvial gravel. It may be naturally alloyed as sylvanite $(AuAg)Te_2$. Sea water contains 3 to 4 mg per ton, and small amounts, 0.03 to 1 mg per cent, have been reported in many foods. Gold is used in jewelry, for other ornamental uses, and for special industrial purposes where its properties of electrical and heat conductivity, malleability, and ductility outweigh its expense. While gold and its salts have been used for a wide variety of medicinal purposes, their present uses are limited to the treatment of rheumatoid arthritis and rare skin diseases such as discoid lupus.

Absorption, Excretion, Toxicity

Gold salts are poorly absorbed from the gastrointestinal tract. The majority of the information we have concerning the distribution of gold salts Originates from its therapeutic use or through experimental studies. Normal urine and fecal excretions of 0.1 and 1 mg per day, respectively, have been reported. After injection of most of the soluble salts, gold is excreted via the urine, while the feces account for the major portion of insoluble compounds. Gold seems to have a long biologic half-life, and detectable blood levels can be demonstrated for ten months after the cessation of treatment. Most of the retained gold salts are found in the kidney; less is present in the liver and other organs, including the spleen.

Colloidal gold may be collected by the reticuloendothelial system and larger amounts are found in the liver. The toxicity of gold and its salts seems to be largely associated with its therapeutic use rather than its industrial use. Dermatitis is the most frequently reported toxic reaction, and stomatitis may accompany this reaction. The skin lesions tend to disappear after the cessation of therapy.

Nephritis with albuminuria, encephalitis, gastrointestinal damage, hepatitis, and blood dyscrasia, including leukopenia, agranulocytosis, thrombopenia, or aplastic anemia have been reported with less frequency. Animal experiments confirm the toxicity of gold salts reported from medical studies. In addition, hypothalamic damage in mice results in obesity after administration of large amounts of gold thioglucose.

HAFNIUM

Hafnium is one of the rarer metals and has limited commercial use. It is found in most minerals containing zirconium. Hafnium has been used in radio tubes, television tubes, incandescent lights, and as a cathode in x-ray tubes. There are no reports of human toxicity. In animals the principal toxic effect is the production of nonhealing ulcers following dermal application (hafnyl chloride) to abraded skin. In a 90-day feeding study, 1 per cent produced slight liver changes.

INDIUM

Occurrence and Use

Indium is produced as a by-product in the manufacture of other metals, chiefly zinc, but also tin and lead. Its industrial uses are in electroplating of nontarnishing silver and copper plate, corrosion-resistant alloys, glass manufacture, in nuclear energy processes, and in the manufacture of containers for foodstuffs.

Absorption, Excretion, Toxicity

Indium is poorly absorbed from the gastrointestinal tract. It is excreted in the urine and feces. Its tissue distribution is relatively uniform. The kidney, liver, bone, and spleen have relatively high concentrations. Intratracheal injections produce similar concentrations, but the concentration in the tracheobronchial lymph nodes is increased. No industrial injury has been reported from the use of indium.

While absorption from indium-plated silver in utensils may occur, this is without known toxic effect. The knowledge of indium toxicity is based largely on results of animal experiments. Sub-cutaneous and intravenous injection is followed by hindleg paralysis, convulsions, and death. The liver and kidneys are affected. Congestion, hem-orrhage, and necrosis are produced in the liver. Hemorrhage and necrosis of the kidneys have been reported. Muscle degeneration has also been reported.

IRON

Occurrence and Use

The principal ores of iron are oxides. The industrial uses of iron are many, mainly in the fabrication of steel. Urban air and water may contain significant amounts of iron industrial or geologic sources. Iron carbonate is a trace mineral and is added to foods. Ferbam, ferric dimethyldithiocarbamate, is an iron-containing fungicide. It has a relatively high threshold limit value of 10 mg/M^3. However, when it is heated, it decomposes and emits highly toxic fumes. Iron is widely distributed in food, animal tissues tending to have considerably more than plants.

Iron is an essential element; its major function in the animal body is the formation of hemoglobin. However, enzymes also contain iron; these include cytochrome and xanthine oxidase. Iron has been used therapeutically since at least 1500 B.C. Its medicinal uses have included treatment of acne, alopecia, hemorrhoids, gout, pulmonary diseases, excessive lacrimation, weakness, edema, and fever, to name a few. The primary therapeutic use of iron salts today is in the treatment of iron-deficiency anemia. Iron is available in nonprescription forms.

Parenteral dosage forms, generally iron carbohydrate complexes, are available for treating iron deficiency that does not respond to oral treatment and for physiologic anemia due to rapid growth, for example, anemia of baby pigs. Every tissue of the body contains iron. Under normal conditions the body burden is about 4g. Hemoglobin is the major iron compound in the body and is highly concentrated in the erythrocytes. Sixty-seven percent of the total iron is contained in hemoglobin. Twenty-seven per cent is stored as ferritin mainly the liver, or as hemosiderin, in cases of excess intake.

Absorption, Excretion, Toxicity

The oral absorption of iron is very complicated and the intestinal mucosa is the principal site for limiting the absorption of iron. In this homeostatic mechanism the divalent form is absorbed into the gastrointestinal mucosa and converted to the trivalent form and attached to ferritin. The ferritin passes into the bloodstream and is then converted to transferring where the iron remains in the trivalent form or is transported to the liver or spleen for storage as ferritin or hemosiderin. The absorption of iron from the gastrointestinal tract may be dependent upon hepatic and pancreatic secretions.

The adequacy of iron stores in the body, however, seems to be the major controlling factor in the absorption of iron by the gastrointestinal tract.Iron has been shown to cross the placenta and concentrate in the fetus. The concentration of the iron in the fetus may serve a valuable physiologic purpose, inasmuch as it prevents anemia caused by rapid growth in the absence of sufficient supplies of iron in the mother's milk. Under normal circumstances the iron contained in food is not well absorbed in nonanemic persons.

With increased in iron beyond the physiologic limits, most is excreted in the feces, but small amounts may accumulate. Some iron may be excreted via the bile. In cases of overload, iron is excreted in the urine, and the presence of high urinary iron concentrations is indicative of excessive iron. Normally, significant quantities of iron are excreted by loss of epithelial cells of the gastrointestinal tract. Women tend to be more anemic than men; menstrual bleeding may account for this tendency. Acute poisoning from accidental ingestion of ferrous sulfate tablets occurs more frequently in children than in adults. About 2,000 such cases occur annually.

Only acute intoxications from aspirin, other unknown medications, and Phenobarbital occur more frequently than acute iron poisoning. Prior to the advent of deferoxamine, the death rate in children was higher from this form of intoxication than from acetylsalicylic acid. Acute toxicity from oral iron preparations is largely due to the irritation of the gastrointestinal tract; vomiting may be the first sign.

There may be some gastrointestinal bleeding, lethargy, restlessness, and gray cyanosis. This may be followed by a short period of recovery, which takes place from several hours to one or two days after the poisoning. This is followed by a third phase of acute iron toxicity in which sings of pneumonitis and convulsions may occur. Gastrointestinal bleeding generally continues throughout this entire period, and some neurologic manifestations, including coma, are predominant during this phase. Sings of hepatic toxicity, such as jaundice, may be observed. Most deaths occur during this time. In those patients surviving three or four days, recovery is generally rapid. Pyloric constriction and gastric fibrosis have been observed, in rare instances, six weeks after the acute phase of iron poisoning.

Marked leukocytosis may also occur. Acute iron poisoning in rabbits produces prolongation of coagulation time and prothrombin time, increased thrombocyte count, and qualitative changes in fibrin formation. Serumglutamic oxalacetic and glutamic pyruvic transaminase are increased. The severity of the clinical course seems to be proportional to the increase in the serum iron concentration. It has been hypothesized that the severe gastrointestinal necrosis facilitates direct access of the iron into bloodstream, circumventing the mucosal block.

Electron microscopic examinations have found livers of rabbits experimentally poisoned with large intravenous doses of iron to have degenerative mitochondria accompanying the marked hepatic degeneration. Fatty degeneration of the myocardium and masses of granular material in the kidneys have been reported. Iron poisoning has been treated by dimercaprol (British antilewissite; BAL), deithylenetri-aminepentaacetate (DTPA), and ethylendediamine-tetraacetate (EDTA). However, these agents have been completely supplanted by deferoxamine. Chronic excessive intake of iron may lead to hemosiderosis or hemochromatosis.

Hemosiderosis refers to a condition in which there is generalized increased iron content in the body tissues, particularly the liver and reticuloendothelial system. Hemochromatosis, on the other hand, indicates demonstrable histologic hemosiderosis and diffused fibrotic changes of the affected organ. Excessive dietary iron intake appears to be the cause of abnormal iron accumulation in the notable condition occurring in South Africa known as "Bantu siderosis." Bantu siderosis is more frequent in men than in women.

The disease probably results form the use of iron pots in food preparation and the brewing of beern in iron containers. This type of disease is marked by iron accumulations in the Kupffer cells of the liver and in the reticuloendothelial cells of the spleen and bone marrow. In addition, a glucose test indicates that 20 percent of the patients with hemochromatosis have abnormal glucose metabolism. Increase in heart disease may also accompany hemochromatosis. It has been reported that high dietary iron intake from other sources, for example red wine, may also play a role in the etiology of hemochromatosis in some areas of the world. In addition, Kaschin-Beck disease, an unusual disorder, has been reported in Asia. This has been ascribed, perhaps in error, to the consumption of drinking water with excessive iron content and result in an arthritic-type disease.Numerous investigators have attempted to produce hemochromatosis in experimental animals by chronic parenteral or oral administration of iron in large doses.

Although it is possible to induce hemosiderosis in the liver and other viscera, fibrosis has not been clearly demonstrated. In experimental animals the production of hemosiderosis, accompanied by cellular injury and fibrosis, apparently requires a cholinedeficient diet. Administration of folic acid to rats receiving a choline-deficient diet has prevented cellular necrosis and fibrosis. Parenteral iron preparations are generally given intramuscularly, although some may be jused intravenously.

Staining at the site of intramuscular injection is a problem with these materials owing to the iron deposition in man and pigs. Iron dextran, administered by subcutaneous injection, has been shown to cause a high incidence of sarcoma in rats. It has been pointed out that the dose used in some of the studies was 200 to 300 times greater than the therapeutic dose in man on a weight basis. There is a high degree of species specificity in the incidence of sarcomas. The incidence is considerably higher in the rat than in the mouse. Iron dextran was removed from the U.S. market in 1960 because of these effects in animals and returned to the market in 1963 Further investigations have revealed that high doses of injectable iron preparations administered intravenously to various species of experimental animals produced teratogenic changes (hydrocephalus, anophthalmia) in the fetuses.

The effect was reduced when deferoxamine was administered. Long-term inhalation exposure to iron, particularly to iron oxide, has resulted in mottling of the lungs, a condition referred to as siderosis. This is considered a benign pneumoconiosis and does not ordinarily cause significant physiologic impairment. However, hematite miners have been reported in certain areas to have from 50 to 70 per cent higher death rates attributable to lung cancer. It has been suggested that at least part of the increased incidence may be due to radioactivity in the mine fields surveyed.

LANTHANONS (RARE EARTHS)

Occurrence and Use

The lanthanons comprise the elements with atomic numbers 57 through 71; yttrium and scandium are included because of similar characteristics. The lanthanons occur in monozite sand, a phosphate mineral, in combination with thorium. Various special chemical methods are available for the separation of cerium. The "heavy" lanthanons, samarium through lutetium, are separated by ion exchange. The "light" elements are separated by conventional crystallization procedures. Before these separation techniques were available, their main use was in mantels for gas lights.

However, they are now used in control rods for atomic reactors, in alloys with nickel and chrome, in microwave devices, lasers and masers, and in television sets. Neodymium and several other rare earths have been tried clinically as anticoagulants. Cerium oxalate has been used to remedy vomiting during pregnancy, and other salts of this element have been used as central nervous system depressants, astringents, and antiseptics.

Absorption, Excretion, Toxicity

Only small amounts of the stable rare earths are absorbed and the gastrointestinal tract. The per cent of oral absorption and retention of cerium is greater in the neonate than in older mice. Subcutaneous injection results in slow excretion, mostly by the gastrointestinal route. Scandium, after intravenous injection, is concentrated in the liver and reticuloendothelial system; the bone concentration is low. Thulium is concentrated in the skeleton. Parenteral injection of rats with the elements lanthanium through samarium revealed an approximate 50 per cent liver and 25 per cent skeletal deposition. Excretion was via the bile with a 15-day half-life. Elements 65 to 71 (terbium to lutetium) have a skeletal retention of 50 per cent or more with a longer biologic half-life. Cerium (^{144}Ce) was taken up in the kidneys, spleen, cartilage, and adrenal cortex, as were ^{180}Tb and ^{169}Yb. Promethium (^{147}Pm) and holmium (^{160}Ho) were taken up in the kidney and cartilage. The oral toxicity of the rare earths is low owing to poor gastrointestinal absorption.

Based on scattered acute parenteral toxicity studies these elements are considered to be only slightly toxic. The signs of acute toxicity observed in rodents consisted of writhing, ataxia, labored respiration, and sedation. The maximum incidence of death occurred at 48 to 96 hours. There was a tendency for the females to be more sensitive than the males. Intravenous injection of various salts of the rare earths, lanthanum through samarium, caused splenic and hepatic degeneration in various rodents. Feeding up to 1 per cent of the rare earths caused liver nuclear vacuolization in the case of gadolinium (Gd), terbium (Tb), thulium (Tm), and ytterbium (Yb). Inhalation exposures in man, while infrequent, have caused sensitivity to heat, itching, and an increased awareness of odor and taste.

Intratracheal administration or inhalation exposure of experimental animals to fluorides or oxides, or a combination thereof, resulted in transient pneumonitis, subacute bronchiolitis, and regional bronchiolar stricturing. The formation of granulomas, while rare, has been reported. Skin damage by the rare earths is apparently not an important factor unless the skin is abraded. Applications to abraded skin cause epilation and scar formation. Intradermal injections produce granulomas in guinea pigs and man. The rare earths irritate the conjunctiva, but not the cornea.

However, in rabbits when the cornea is denuded, scandium, yttrium, lanthanum, cerium, praseodymium, neodymium, samarium, and gadolinium cause permanent corneal opacity. Although intravenous administration of neodymium and other rare earths was useful for their anticoagulant effect, it produced undesirable side effects, primarily hemolysis resulting in hemoglobinuria. Lanthanum oxide fume and dust generated from arc light carbons used to produce intense white illuminations in the lithographic industry may be responsible for workers' frequent complaints of headache and nausea. Cerium is the most widely used of this group, and no records of human injury from industrial or medicinal use are reported.

LITHIUM

Occurrence and Use

The pure metal never occurs in nature, but salts, especially silicates, are common. Lithium is used in alloys, as a catalytic agent, in heat exchangers for air-conditioning systems, as a lubricant. Lithium hydride, which produces hydrogen on contact with water, is used in manufacturing electronic tubes, in ceramics, and in chemical synthesis. Lithium carbonate has recently been used medicinally as an antidepressant. Lithium has been detected in many plant and animal tissues. The daily intake is estimated to be about 2 mg.

Absorption, Excretion, Toxicity

Lithium is readily absorbed from the gastrointestinal tract. Distribution in the human organs is almost uniform. The normal plasma level is about 17 μg/liter. The red cells contain less. Excretion is chiefly through the kidneys, but some is eliminated in the feces. The greater part of lithium is contained in the cells, perhaps at the expense of potassium. In general the body distribution of lithium is quite similar to that of sodium, and it may be competing with sodium at certain sites, for example in renal tubular reabsorption. From the industrial point of viewrexcept for lithium hydride, none of the other salts or the metal itself is hazardous.

Lithium hydride is intensely corrosive. Concentrations of 5 mg/M^3 produced signs of pulmonary irritation. Lithium hydride also has the potential to produce burns on the skin because of the formation of hydroxides. The therapeutic use of lithium carbonate may produce unusual toxic responses. These include neuromuscular changes (tremor, muscle hyperirritability, and ataxia), central nervous system changes (blackout spells, epileptic seizures, slurred speech, coma, psychosomatic retardation, and increased thirst), cardiovascular changes (cardiac arrhythmia, hypertension, and circulatory collapse), gastrointestinal changes (anorexia, nausea, and vomiting) and renal damage (albuminuria and glycosuria).

The latter is believed to be due to temporary hypokalemic nephritis. These changes appear to be more frequent when the serum levels increase above 1.5 mEq/1 suggesting that careful monitoring of this parameter is needed rather than reliance on the amount given. The cardiovascular and nervous system changes may be due to the competitive relationship between lithium and potassium and may thus produce a disturbance in intercellular metabolism. Thyrotoxic reactions, including goiter formation, have also been suggested (Davis and Fann, 1971). While there has been some indication of adverse effects on fetuses following lithium treatment, none was observed in rats (4.05 mEq/kg), rabbits (1.08 mEq/kg), or primates (0.67 mEq/kg). This dose to rats was sufficient to produce maternal toxicity and effects on the pups of treated, lactating dams.

MAGNESIUM

Occurrence and Use

The primary ores of magnesium are magnesite and dolomite. Magnesium may also be obtained from brinewells and salt deposits. It is used in alloys, particularly those requiring light weight. It is used in wire and ribbon for radios as well as for incendiary materials such as flares. Magnesium is an essential nutrient.

Deficiency causes neuromuscular irritability, calcification, and cardiac and renal damage, which can be prevented by supplementation. The deficiency is called "grass staggers" in cattle and "magnesium tetany" in calves. Magnesium is a cofactor of many enzymes and is contained in metallo-enzymes; it is apparently associated with phosphate in these functions. Magnesium citrate, oxide, sulfate, hydroxide, and carbonate are widely taken as antacids or cathartics. The hydroxide, milk of magnesia, is one of the constituents of the universal antidote for poisoning. Topically, the sulfate is also used widely to relieve inflammation.

Magnesium sulfate may be used as a parenterally administered central depressant. Its most frequent use for this purpose is in the treatment of seizures associated with eclampsia of pregnancy and acute nephritis. Nuts, cereals, seafoods, and meats are high dietary sources of magnesium. The average city water contains about 6.5 ppm, but varies considerably, increasing with the hardness of the water.

Absorption, Excretion, Toxicity

Magnesium alts are poorly absorbed from the intestine. In cases of overload this may be due in part to their dehydrating action. Magnesium is absorbed mainly in the small intestine. The colon also absorbs some. Calcium and magnesium are competitive with respect to their absorptive sites and excess calcium may partially inhibit the absorption of magnesium. Magnesium is excreted into the digestive tract by the bile and pancreatic and intestinal juices.

A small amount of radiomagnesium given intravenously appears in the gastrointestinal tract. The serum levels are remarkably constant. There is an apparent obligatory urinary loss of magnesium, which amounts to about 12 mg/day, and the urine is the major route of excretion under normal conditions. Magnesium found in the stool is probably not absorbed. Magnesium is filtered by the glomeruli and reabsorbed by the renal tubules. In the blood plasma about 65 percent is in the ionic form, while the remainder is bound to protein. The former is that which appears in the 'glomerular filtrate.

Mercurial diuretics cause excretion of magnesium as well as potassium, sodium, and calcium. 'Excretion also occurs in the sweat and milk. Endocrine activity, particularly of the adrenocortical hormones, aldosterone, and parathyroid hormone, has an effect on magnesium levels, although these effects may be related to the interaction of calcium and magnesium. Tissue distribution studies indicate that of the 20-g body burden, the majority is intracellular in the bone and muscle.

Bone concentration of magnesium decreases as calcium increases. Most of the remaining tissues have higher concentrations than blood, except for fat and omentum. With age, the aorta tends to accumulate magnesium along with calcium, perhaps as a function of atherosclerotic disease. Freshly generated magnesium oxide can cause metal fume fever if inhaled in sufficient amounts. This is analogous to the effect caused by zinc oxide. Both zinc and magnesium exposure of animals produced similar effects. It is reported that particles of magnesium in the subcutaneous tissue produce lesions that resist healing.

In animals, magnesium subcutaneously or intramuscularly administered produces gas gangrene as a result of interaction with the body fluids and subsequent generation of hydrogen and magnesium hydroxide. The tissue lesion is reversible. Conjunctivitis, nasal catarrh, and coughing up of discolored sputum results from industrial inhalation exposure. With industrial exposures; increases of serum magnesium up to twice the normal levels failed to produce ill effects but were accompanied by calcium increases. Intoxication occurring after oral administration of magnesium salts is rare, but may be present in the face of renal impairment.

The symptoms include a sharp drop in blood pressure and respiratory paralysis due to central nervous system depression.

MANGANESE

Occurrence and Use

The principal ore of manganese is pyrolusite (MnO_2). Manganese and its compounds are used in making steel alloys, dry cell batteries, electrical coils, ceramics, matches, glass, dyes, in fertilizers, welding rods, as oxidizing agents, and as animal food additives. The primary uses in medicine are as antiseptics and germicides. Potassium permanganate ($KMnO_4$) is applied dermally for these effects and for its slight astringent action. This is virtually the only manganese compound of medical use at the present time.

Manganese chloride was administered intravenously for schizophrenia, but was abandoned owing to lack of effect and the danger of flocculation. Manganese is an essential element. It is a cofactor in a number of enzymatic reactions, particularly those involved in phosphorylation, cholesterol, and fatty acids synthesis. Manganese is present in all living organisms. While it is present in urban are (Table 3.3) and in most water supplies (Table 3.2), principal portion of the intake is derived from food. Vegetables, the germinal portions of grains, fruits, nuts, tea, and some spices are rich in manganese.

Absorption, Excretion, Toxicity

The body burden has been estimated at 20 mg (Table 3.1). The liver, kidney, intestine, and pancreas contain the highest concentrations. No significant changes in tissue concentrations occur with age, except that tissues that have low manganese concentrations in adults tend to contain higher amounts in the newborn. The lungs do not accumulate manganese with age despite significant concentrations in urban air. The turnover of manganese is rapid. Injected radiomanganese quickly disappears from the bloodstream; it is concentrated in the mitochondria of the liver and pancreas. Administration of stable manganese in any valence state promotes rapid excretion of radiomanganese.

Serum manganese, normally about. 2.5 μg/1, increases after acute coronary occlusion and is claimed to be a more accurate index of myocardial infarction than glutamic oxalacetic transaminase. Recent studies suggest that the major route of excretion is the gastrointestinal tract via the bile and that the excretion of manganese may be regulated by a homeostatic system maintaining relatively constant tissue levels. This system apparently involves the liver, auxiliary gastro-intestinal mechanisms for excreting excess manganese, and perhaps the adrenal cortex. This regulating mechanism, plus the tendency for extremely large doses of manganese salts to cause gastrointestinal irritation, accounts for the lack of systemic toxicity following oral administration or dermal application.

Industrial toxicity resulting from inhalation exposure, generally to manganese dioxide in mining or manufacturing, is of two types. The first, manganese pneumonitis, is the result of acute exposure. Men working in plants with high concentrations of manganese dust show an incidence of respiratory disease 30 times greater than normal. Pathologic changes include epithelial necrosis followed by mononuclear proliferation. The second and more serious type of disease resulting from chronic inhalation exposure to manganese dioxide, generally over a period of more than two years, involves the central nervous system.

In iron-deficiency anemia the oral absorption of manganese is increased, and it may be that variations in manganese transport related to iron deficiency account for individual susceptibility. Those who develop chronic manganese poisoning (manganism) exhibit a psychiatric disorder characterized by irritability, difficulty in walking, speech disturbances, and compulsive behavior that may include running, fighting, and singing. If the condition persists, a masklike face, retropulsion or propulsion, and a Parkinson-like syndrome develop. The outstanding feature of manganese encephalopathy has been classified as severe selective damage to the sub-thalamic nucleus and pallidum. These symptoms and the pathologic lesions, degenerative changes in the basal ganglia, make the analogy to Parkinson's disease feasible.

In addition to the central nervous system changes, liver cirrhosis is frequently observed. Victims of chronic manganese poisoning tend to recover slowly, even when removed from the excessive exposure. Metal-sequestering agents have not produced remarkable recovery, L-Dopa, which is used in the treatment of Parkinson's disease, has been more consistently effective in the treatment of chronic manganese poisoning than in Parkinson's disease. The syndrome of chronic nervous system effects has not been successfully duplicated in any experimental animals except monkeys and then only by inhalation or intraperitoneal injection.

After intraperitoneal administration of manganese to squirrel monkeys, dopamine and serotonin levels markedly decreased in the caudate nucleus regardless of whether or not behavioral effects were present. Manganese levels were increased in the basal ganglia and cerebellum. Histopathologic examination of animals did not reveal any morphologic changes. Exposure of rats to manganese dioxide at concentrations of 47 mg Mn/M^3 for five hours a day, five days a week for 100 days increased the brain manganese concentration more than four-fold but produced no hematologic, behavioral, or histologic effects.

Experimental chronic oral intake of manganese in rabbits, pigs, and cattle at levels of 1,000 to 5,000 ppm has been reported to reduce the accumulation and utilization of iron. Cereals rich in manganese contain only sightly more than 100 ppm. Thus, it is difficult to see how the effects reported in animals could apply to man.

MOLYBDENUM

Occurrence and Use

The most important mineral source of molybdenum is molybdenite (MoS_2). The United States is the major world producer of molybdenum. The industrial uses of this metal include the manufacture of high-temperature resistant steel alloys for use in gas turbines and jet aircraft engines, production of catalysts, lubricants, and dyes. Molybdenum is widely distributed in nature, being an essential element. It is a cofactor for the enzymes xanthine oxidase and aldehyde oxidase. In plants it is necessary for bacterial fixing of atmospheric nitrogen at the start of protein synthesis. Because of these functions it is ubiquitous in food. Since plankton tend to concentrate molybdenum 25 times that of sea water, shellfish tend to have high concentrations of molybdenum.

Molybdenum is added in trace amounts to fertilizers to stimulate plant growth. The average daily human intake in food is approximately 350 μg. The concentration of molybdenum in urban air is minimal. Fresh water has about 0.35 ppb; however, in certain areas the content may be higher. The use of molybdenized ferrous sulfate has been tried for the treatment of iron-deficiency anemia. However, beneficial effects for the added molybdenum were minimal, and the material appeared to have gastrointestinal side effects equal to or greater than those of ferrous sulfate.

Absorption, Excretion, Toxicity

While molybdenum exists in various valence forms, bio-logic differences with respect to valence are not clear. The soluble hexavalent compounds are well absorbed from the gastrointestinal tract into the liver. Increased molybdenum intake in experimental animals has been shown to increase tissue levels of xanthine oxidase. In man molybdenum is contained principally in the liver, kidney, fat, and blood. Of the approximate total of 9 mg in the body, most is concentrated in the liver, kidney, adrenal, and omentum.

The molybdenum level is relatively low in the newborn and increases until age 20, declining in concentration thereafter. More than half of the molybdenum excreted is in the urine. The blood level, at least in sheep, are in association with the red blood cells. However, molybdenum has been detected in only about 25 percent of the blood samples of human urban population. The excretion of molybdenum is rapid, mainly as molybdate.

Excesses may be excreted also by the bile, partic-ularly the hexavalent forms. Inhalation of molybdenum by guinea pigs has resulted in increased bone levels. Injected radio-molybdenum increased liver and kidney levels but the endocrine glands were also exceptionally high in content. Pastures containing 20 to 100 ppm molybdenum may produce a disease referred to as "teart" in cattle and sheep. It is characterized by anemia, poor growth rate, and diarrhea. Copper or sulfate in the diet prevents the disease, and removal of the animals from pastures containing high levels of molybdenum facilitates their rapid recovery. Prolonged exposure has led to deformities of the joints. Experimental studies have revealed differences in toxicity of molybdenum salts.

Molybdenum sulfide was well tolerated in rats at 500 mg/kg/day and was not injurious to guinea pigs at 28 mg/M^3. Hexavalent com-pounds were more toxic. Molybdenum trioxide at a dose of 100 mg/kg/day caused death in rats and by inhalation was irritating to the eyes and mucous membranes and subsequently lethal. After repeated oral administration at sufficient levels, fatty degeneration of the liver and kidney was induced. In comparison with chromium and tungsten salts, sodium molybdate by intraperitoneal injection was less toxic in mice. Interesting relationships of molybdenum with other metals with respect to toxicity in cattle and sheep have been documented. For example, copper prevents the accumulation of molybdenum in the liver and may antagonize the absorption of molybdenum from food. It is reported that by alternating the intake of copper and molybdenum at weekly intervals, black sheep can be made to grow striped wool.

White wool in black sheep is a sign of copper deficiency. The antagonism of copper is dependent on sulfate in the diet. It has been suggested that sulfate may displace molybdate in the body. It may be that the anemia caused by molybdenum is due to the reduction of sulfide oxidase in the liver, resulting in the formation of copper sulfide, thereby inducing a functional copper deficiency. Feeding of tungstate has also been shown to displace molybdate. In addition, it has been reported that molybdenum may promote fluoride retention and thereby decrease dental caries. There are no data documenting molybdenum toxicity in man due to industrial exposure.

NICKEL

Occurrence and Use

In its principal ores nickel is found in combination with iron or copper. The main uses of nickel are in electronics, coins, steel alloys, batteries, food processing (Ni-Cu Monel), and stainless steel. Nickel is a constituent of urban air (Table 3.3), possibly as a result of fossil fuel combustion.

Incinerators also contribute to the nickel content in the atmosphere. Nickel is hot a normal constituent of water. While some nickel is found as a contaminant from food processing (gelatin and baking powder), relatively large amounts occur naturally in vegetables, legumes, and grains. While *in vitro* nickel will activate enzymes, no functional action has been described in the intact animals.

Absorption, Excretion, Toxicity

The average body burden has been estimated at <10 mg (Table 3.1), but wide geographic variations occur. Nickel is present in the lung, liver, kidney, and intestine of most stillborn infants. The concentration in the lung increases with age. In rats the bones accumulate a major portion of increased intake. Excretion is largely via the feces. Nickel has been found in bile. A mechanism for limiting intestinal absorption has been suggested. Many nickel salts have astringent and irritant properties that limit their absorption.

Normal urine values of about 2.3 μg/100 ml have been reported. Dermatitis (nickel itch) is the most Trequent effect of exposure to nickel. This occurs from direct contact with metals containing nickel such as coins and costume jewelry. It has been estimated that of all eczema, 5 per cent is caused by nickel or nickel compounds. The dermatitis is a sensitization reaction, and contact may, in some cases, produce paroxysmal asthmatic attacks and pulmonary eosinophilia. Nickel carbonyl ($Ni[CO]_4$) is the most toxic of nickel compounds. It has been estimated to be lethal in man at atmospheric exposures of 30 ppm for 20 minutes. This material is formed by nickel or its compounds in the presence of carbon monoxide.

The initial symptoms of toxicity consist of headache and vomiting. These are relieved by fresh air. Delayed symptoms occurring in 12 to 36 hours include dyspnea, cyanosis, leukocytosis, and increased body temperature. Delirium and other central nervous system signs usually appear. Acute chemical pneumonitis results. Death may occur between the fourth and eleventh days. While dietary nickel is excreted largely in the feces, inhalation of nickel carbonyl results in the appearance of significant increases in urinary nickel, both with respect to concentration and relative to that in the stools. The increase of nickel in the urine has been used clinically to confirm exposure, and levels above 0.5 mg/1 are considered serious.

Diethyldithiocarbamate trihydrate (Dithlocarb), a metal binding agent, has been used successfully to treat acute poisoning. Chronic exposure to nickel carbonyl has been implicated epidemiologically in cancer affecting the lungs and nose. These findings have been confirmed by inhalation exposures in experimental animals. In addition, it has been reported that cigarette smoke contains significant amounts of nickel carbonyl. There are no epidemiologic studies that confirm the systemic toxicity of nickel or its com-pounds (except $Ni[CO]_4$).

Doses of nickel sulfate (0.1 to 0.5 mg/kg for 161 days) have induced myocardial and liver damage. High doses of the soluble salts induce giddiness and nausea. Inhalation exposure to nickel and nickel oxide dusts have produced malignant pulmonary neoplasms in guinea pigs and rats. The increase of serum nickel in humans after myocardial infarction is of unknown significance.

NIOBIUM

Occurrence and Use

Niobium (columbium) is found with tantalum in the primary ore of tantalite and columbite. It is used in the manufacture of high-temperature steel alloys, corrosion-resistant chromium-steel alloys, and in electronic equipment. It has a low thermal neutron cross-section and has growing use in nuclear energy and chemistry. Niobium appears to be ubiquitous in nature.

Most grains, meats, and dairy products contain significant amounts of niobium. Fats and oils, of both vegetable and animal origin, tend to have the highest concentrations of niobium.

Absorption, Excretion, Toxicity

Intake from water is about 20 μg. A little less than half the daily intake is absorbed; this is excreted in the urine. In the body niobium is carried mainly in the red blood cells. The total body burden is approximately 100 mg. The red blood cells, liver, kidney, fat, hair, lungs, and pancreas contain the highest concentration, but niobium is found in most other organs. The "normal" concentrations in the red blood dells, serum, and urine are about 5, 0.7, and 0.25 μg/g, respectively.

Niobium is present in the newborn and in milk. Inhalation studies in rats have shown that the largest amounts were retained in the lungs with secondary deposition in the bones. The biologic half-life was 120 days. Incorporation of niobium in the drinking water of mice at 5 ppm plus 1.62 mg/g in the diet caused liver degeneration. *In vitro* studies have suggested that inhibition of adenosine triphosphatase may be involved with the biologic activity of niobium. There are no reports of niobium toxicity in man. Intravenous administration of 30 mg/kg niobium as potassium niobate to dogs and rats produced severe nephrotoxic effects.

Tubular epithelial damage was predominant in the convoluted tubule. Niobium pentachloride produces moderate transient irritation of the eye and severe dermal irritation.

PLATINUM-GROUP METALS

Occurrence and Use

These metals may be grouped together because of their similar chemical properties. Ruthenium (Ru), rhodium (Rh), and palladium (Pd) are the lighter triad, while osmium (Os), iridium (Ir), and platinum (Pt) are the heavier. They are found together in very sparsely distributed deposits. These rare metals are obtained from such deposits or as a by-product of refining other metals, chiefly nickel and copper. The chief use of ruthenium is as a hardener in platinum and palladium alloys used in jewelry and electrical contacts. Rhodium is used in the manufacture of rhodium-platinum alloys, in electroplating, electronic components, movie projectors, and in reflectors for searchlights.

Although rhodium is reported to have slight chemotherapeutic activity against certain mouse viruses, it is not used medicinally at present. The industrial uses of palladium include alloys (especially those used in the communication industry), as a catalyst, for decoration and jewelry, dental alloys, and in cigarette lighters. Palladium in colloidal form has been given medicinally for tubercutosis, gout, and obesity. These uses were without significant benefits. Osmium is used in the-hard points of fountain pens and engraving tools, in staining tissues for electron microscopy, and in finger-printing.

Formerly it was used in the manufacture of electrical lights. Iridium is used to harden platinum jewelry and as an alloy with osmium for hard points of fountain pen and engraving tools. Platinum is widely used as a catalyst in the chemical industry and in electronics. It is alloyed with other metals and used in jewelry.

Absorption, Excretion, Toxicity

Intratracheal injections of radioruthenium chloride are retained in the lungs, suggesting that ruthenium might also be retained after inhalation exposure. Toxicologic information is limited to references in the literature indicating that fumes may be injurious to eyes and lungs. Rhodium trichloride produced death in rats and rabbits within 48 hours after intravenous administration at doses near the LD50 (approximately 200 mg/kg).

Histologic evaluation revealed no changes; however, it was suggested that death was attributable to central nervous system effects. In a single study, incorporation of rhodium (rhodium chloride) or palladium (palladous chloride) into the drinking water of mice at a concentration of 5 ppm over the lifetime of the animals produced a minimally significant increase in malignant tumors. Most of these tumors were classified as of the lymphomaleukemia type? When administered orally, palladium is excreted in the feces. Intravenous administration results in rapid and almost complete excretion in the urine.

Palladium chloride is not readily absorbed from subcutaneous injection. No adverse effects have been reported from industrial exposure. Subcutaneous injection of palladium chloride in rabbits leads to gray-brown discoloration at the site of injection. Intravenous administration in lethal doses causes loss of appetite, hemolysis, renal deposits, and bone marrow injury. Colloid palladium ($Pd[OH]_2$) is reported to increase body temperature, produce discoloration and necrosis at the site of injection, decrease body weight, and cause slight hemolysis. Osmium metals and most of its salts are of little toxicologic significance.

Osmium tetroxide, produced by heating the metal, is toxic. Its action is mainly on the eyes. Lacrimation and halo vision occur, probably owing to effects on the cornea. Irritation of the respiratory tract and headache may occur. One fatal case due to pulmonary irritation has been reported. Some renal toxicity has also been ascribed to osmium tetroxide exposure. No reports implicating iridium as an industrial hazardous agent have appeared. Platinum metal itself is generally harmless, but an allergic dermatitis can be produced in susceptible individuals.

Skin changes are most common between the fingers and in the antecubital fossae. Symptoms of respiratory distress, ranging from irritation to an "asthmatic syndrome" with coughing, wheezing, and shortness of breath, have been reported following exposure to platinum dust. The skin and respiratory changes are termed platinosis. They are mainly confined to persons with a history of industrial exposure to soluble compounds such as sodium chloroplatinate, although cases resulting from wearing platinum jewelry have been reported.

Metallic platinum dust itself is reported not to cause these changes. Toxicity studies of complex platinum bases indicate these are much more toxic than platinum itself. Diamminedichloroplatinum ($Pt[NH_3]_aCl2$), given subcutaneously to experimental animals, produces epileptic-type convulsions and death, preceded by coma. Hyperirritability and cardiac effects, due to action on the vagus center, have also been observed with smaller doses.

RHENIUM

Occurrence and Use

Rhenium is obtained as a by-product of molybdenum and copper processing. It is used in the electrical industry for electronic tubes and in marine engine magnetes, especially because of its resistance to salt corrosion. It is also used as a heater element.

Absorption, Excretion, Toxicity

Rhenium is excreted primarily in the urine. Although rhenium tends to accumulate in the thyroid, it does not bind there as does iodine. No toxic effects have been reported.

RUBIDIUM

Occurrence and Use

Rubidium is a by-product of potassium and molybdenum production. Its chief industrial use is in the manufacture of photoelectric cells. It is widespread in nature at trace levels and it is found in tomatoes, beef, beans, and barley at significant levels. Rubidium is also found in most animal tissues.

Absorption, Excretion, Toxicity

Rubidium seems to be an acceptable substitute for potassium in many physiologic processes. While rubidium will prevent kidney and muscle lesions characteristic of potassium depletion, there is no evidence that rubidium itself is an essential element. Rubidium is fairly well absorbed after oral administration. The highest concentrations are in the heart and skeletal muscles, while bone contains almost none. The muscle acts as the primary storagesite under conditions of excess.

In the blood the largest amount is partitioned in the red blood cells. The urine is the major route of excretion. The feces are of secondary importance in excretion. There is no evidence that rubidium is toxic to man. Rubidium in excess is toxic, especially in animals on potassium-deficient diets. Under this condition hyperirritability, muscle spasms, convulsions, and failure of the young to survive to eanling stage have been demonstrated, development of ventricular extrasysas also been demonstrated.

SELENIUM

Occurrence and Use

Selenium is principally obtained as a by-product of copper refining. Tellurium is also frequently present in these sulfide ores, and some industrial hazards may be due to the combination of these related materials. Selenium is used in the electronics industry for rectifiers, photo cells, and solar batteries, in glass and ceramic manufacturing, as a vulcanizing agent for rubber, in steel manufacturing, and in paints and varnishes. In chemical manufacturing selenium dioxide is used in rare instances as an oxidizing agent and selenium as an oxidant in lubricating and other oils. Selenium has also been used in fungicides, in insecticides, and as an insect repellent.

Selenium is used medicinally as an antidandruff agent. Selenium accumulates in certain plants in sufficient quantities to produce selenium toxicity in livestock. On the other hand, selenium is considered an essential element: Selenium-deficient diets cause liver necrosis in rats and multiple-organ (liver, heart, kidneys, skeletal muscle, and testes) necrosis in mice. In chicks, pancreatic fibrosis, exudative diathesis, and alopecia are responsive to selenium supplementation of deficient diets.

Selenium-responsive diseases in the young turkey include muscle (cardiac, skeletal, and smooth-gizzard) myopathies. Lambs and calves suffer from a muscle disease called stiff lamb disease and white muscle disease when raised in selenium-deficient ranges. In addition, embryo mortality in ewes from selenium-deficient areas is reversed by supplementation. Young pigs also may be supplemented with selenium to good effect if maintained on deficient diets to prevent necrotic liver degeneration and cardiac myopathy (mulberry heart). While the role of selenium as an essential mineral seems well established in animals, the same is not so in man, although that possibility seems likely. The apparent roles of selenium in metabolic chemistry are many.

Either as a replacement for vitamin E, or in conjunction with this vitamin, selenium has a significant role in the biosynthesis of ubiquinone (coenzyme Q). Furthermore, the enzyme glutathione peroxidase, important in maintaining erythrocyte integrity, contains selenium as an essential constituent. These roles for selenium and vitamin E in controlling lipid peroxido formation are perhaps the basis for its appearing essential in the diet. Further cellular roles of selenium involve DNA-RNA, control, at least in part, of ion fluxes across cell membranes, maintenance of the integrity of keratins, and stimulation of antibody synthesis. The reaction of selenium with thiol groups may be the primary source of selenium toxicity. The concentration of selenium in foodstuffs provides another source of exposure.

Seafoods, especially shrimp, meat, milk products, and grains, provide the largest amounts in the diet. River water levels of selenium vary depending on environmental and geologic factors; 0.02 ppm has been reported as a representative estimate. Selenium has also been detected in urban air, presumably from sulfur-containing materials. The calculated human dose from urban air is about 0.02 μg/day.

Absorption, Excretion, Toxicity

Elemental forms of selenium are probably not absorbed from the gastrointestinal tract. The average human body burden is approximately 14.6 mg. The greatest concentration is found in the kidney. The level in the liver is approximately one-half that observed in the kidney. Selenium, like arsenic, is reported to accumulate in the hair. Human blood samples contain selenium with an erythrocyte:plasma ratio of approximately 3:1. Selenium is present in the infant, but selenium levels do not increase with age. Human milk contains selenium. Selenium is excreted in the urine, which normally contains about twice as much as the feces. Increased urinary levels provide a reasonable index of excessive exposure.

Normal selenium urinary output is 0.0 to 15 μg/100ml. Excretory products appear in sweat and expired air. The latter may have a garlicky odor due to dimethyl selenide. Within certain physiologic limits the body appears to have a homeostatic mechanism for retaining trace amounts of selenium and excreting the excess material. Selenium toxicity occurs when the intake exceeds the excretory capacity. Industrial exposure to hydrogen selenide, occurring as a result of a reaction to acid or water with metal selenides, produces "garlic" breath, nausea, dizziness, and lassitude. Eye and nasal irritation may occur.

In experimental animals 10 ppm is fatal. Selenium oxychloride, a vesicant, presents an industrial hazard. In rabbits 0.01 ml applied dermally resulted in death. Percutaneous absorption increased blood and liver selenium concentrations. Acute selenium poisoning produces central nervous system effects, which include nervousness, drowsiness, and sometimes convulsions. Symptoms of chronic inhalation exposure may include pallor, coated tongue, gastrointestinal disorders, nervousness, "garlic" breath, liver and spleen damage, anemia, mucosal irritation, and lumbar pain. It has been suggested that some of these symptoms are due to tellurium impurities. "Blind staggers" caused by excess selenium in livestock consuming 100 to 1,000 ppm is characterized by impairment of vision, weakness of limbs, and respiratory failure.

Clear evidence of chronic selenium toxicity in man occurs only in seleniferous areas when the local foods are processed. These signs of intoxication may include discolored or decayed teeth, skin eruptions, gastrointestinal distress, lassitude, and partial loss of hair and nails. Livestock foraging on plants containing about 25 ppm suffer from "alkali" disease, which is characterized by lack of vitality, loss of hair, sterility, atrophy of hooves, lameness, and—anemia. Fatty necrosis of the liver is frequent. In rats given 3 ppm of the material in drinking water, selenite has been reported to be more toxic than selenate. Selenite produced increased numbers of aortic plaques and was found to be more toxic in female than male mice.

Selenium has produced loss of fertility and congenital defects and is considered embryotoxic and teratogenic on the basis of animal experiments. A recent report has also suggested that selenium causes fetal toxicity and teratogenic effects in humans. Chronic toxicity in experimental rodents consists of hepatic cirrhosis and, in the hands of several investigators, evidence of carcinogenesis, primarily in the liver. There has been considerable discussion as to the significance of these findings, and the lack of human selenium-induced carcinogenesis or increased cancer in livestock in seleniferous areas has led to questioning the significance of the results obtained in rodents. In more recent studies, epidemiologic investigations have indicated a decrease in human cancer death rates (age and sex adjusted) correlated with increasing selenium content of forage crops.

In addition, recent experimental evidence supports the antineoplastic effect of selenium with regard to benzo(*a*)pyrene-and benzanthracene-induced skin tumors in mice, *N*-2-fluorenylacetamide and diethylaminoazobenzene-induced hepatic tumors in rats, and spontaneous mammary tumors in mice. A possible mechanism of the protective effects of selenium has been postulated to involve the inhibition of the formation of malonaldehyde, a product of peroxidative tissue damage, which is carcinogenic. Although the mechanism of toxicity of selenium is not well understood, several related factors are very interesting.

First, selenium may have the ability to replace sulfur in certain tissues, *i.e.,* nails and hooves, and as selenate, it may have an inhibiting effect on many sulf-hydryl enzymes. Second, many of the signs of toxicity can be prevented by high-protein diets, and by methionine in the presence of vitamin E. In addition to the apparent protective effect against some carcinogenic agents, selenium is an antidote to the toxic effect of other metals. At appropriate levels, mutual detoxification of selenium and mercury, selenium and thallium, selenium and copper, selenium and arsenic, and selenium and cadmium has been demonstrated.

The case of silver differs from other selenium-metal interactions in that silver precipitates the symptoms of selenium deficiency in vitamin E-deficient animals, perhaps by the formation of a silver-selenium complex, thereby reducing the effectively available selenium required for normal cellular processes.

SILVER

Occurrence and Use

Silver occurs in manyores. The primary silver ore is argentite (Ag_2S). Silver is also obtained as a by-product of copper, lead, and other metals. Silver is used in electrical applications because of its excellent properties of conduction. Jewelry, coins, and eating utensils are some of the principal uses of the metal. Silver halides are used in photography; silver nitrate is used for making indelible inks and for medicinal purposes. The use of silver nitrate for prophylaxis of ophthalmia neonatorum is a legal requirement in some states. Other medicinal uses of silver salts are as a caustic, germicide, antiseptic, and astringent.

Absorption, Excretion, Toxicity

Silver does not occur regularly in animal or human tissue. The major effect of excessive absorption of silver is local or generalized impregnation of the tissues, a condition referred to as argyria. Silver can be absorbed from the lungs and gastrointestinal tract. Some of the absorbed silver is retained in the cells of the gastrointestinal tract. Intravenous injection produces accumulation in the spleen, liver, bone marrow, lungs, muscle, and skin. The major route of excretion is via the gastrointestinal tract. Urinary excretion has not been reported to occur even after intravenous injection. Industrial argyria, a chronic occupational disease, has two forms, local and generalized.

The local form involves the formation of gray-blue patches on the skin or may manifest itself in the conjunctiva of the eye. In generalized on the skin shows widespread pigmentation, often spreading from the face to most uncovered of the body. In some cases the skin may black with a metallic luster. The eyes may be affected to such a point that the lens and vision are disturbed. The respiratory tract may also be affected in severe cases. Large oral doses of silver nitrate cause severe gastrointestinal irritation due to its caustic action.

Lesions of the kidneys and lungs and the possibility of arteriosclerosis have been attributed to both industrial and medicinal exposures. Large doses of colloidal silver administered intravenously to experimental animals produced death due to pulmonary edema and congestion. Hemolysis and resulting bone marrow hyperplasia have been reported. Chronic bronchitis has also been reported to result from medicinal use of colloidal silver.

STRONTIUM

Occurrence and Use

Strontianite ($SrCO_3$) is the principal mineral form of strontium. The metallurgic uses are limited to the addition of small amounts in alloys of tin and lead. Strontium is also used as a deoxidizer in copper and bronze. Various strontium salts are used in paints and rubber, in the refinement of sugar from beets, and in freezing mixtures and refrigerators. Strontium has also been used as a depilatory and in tooth pastes. Some drinking water supplies naturally contain levels as high as 50 ppm. Some plants tend to concentrate strontium from the soil and levels as high as 26,000 ppm have been reported.

Absorption, Excretion, Toxicity

The biologic action and physiologic function of strontium resemble those of calcium, especially with regard to the bone. There is some evidence that strontium is essential for the growth of animals and especially for the calcification of bones and teeth. Apparently a homeostatic balance between calcium and strontium exists that favours the absorption of calcium and the preferential excretion of strontium. Strontium is absorbed from the gastrointestinal tract in limited amounts, although the rate is somewhat dependent on dietary calcium levels. About one-fourth of the ingested strontium is excreted in the feces, while the remainder is eliminated in the urine.

Milk is also a secondary route of excretion. Excess strontium tends to be stored in the teeth and bones, which are the primary storage sites under normal conditions. The normal concentration in the bone is estimated at about 360 ppm. Human fetuses are reported to contain lower concentrations than adults. The concern over adverse effects of strontium intake is based on the radiation damage, since radiostrontium is a nuclear fallout contaminant.

Chemically, toxicity from strontium is almost nil. No adverse effects from industrial use have been reported.

Electrocardiographic changes have been produced after intravenous injection of massive doses. Increased salivation, nausea, diarrhea, and death due to respiratory paralysis have been produced experimentally.

TANTALUM

Occurrence and Use

Tantalum is found with niobium in tantalite or columbite. It is used in electronics, in tungsten cutting alloys, in chemical manufacturing as a catalyst, and for acid-resistant materials. Tantalum is used as a supporting gauze in the repair of hernias, as a dressing for burns, in prosthetic appliances, and in local radiation for bladder cancer after neutron activation.

Absorption, Excretion, Toxicity

Oral salts of tantalum are poorly absorbed. After intramuscular injection the liver, bone, and kidney contain significant amounts. A few animal experiments have suggested that after inhalation, tantalum may produce some pulmonary effects, benign and nonfibrotic in nature. No adverse effects have been reported as a result of industrial exposure. Implantation of tantalum has not shown any adverse tissue reaction in either man or experimental animals.

TELLURIUM

Occurrence and Use

Tellurium is found in various sulfide ores along with selenium and is produced as a by-product of metal refineries. Its industrial uses include applications in the refining of copper and in the manufacture of rubber. Tellurium vapor is used in "daylight" lamps. It is used in various alloys as a catalyst and is a semiconductor. Condiments, dairy products, nuts, and fish have high concentrations of tellurium.

Food packaging contains some tellurium; higher concentrations are found in aluminum cans than tin cans. Some plants, such as garlic, accumulate tellurium from the soil. Potassium tellurate has been used to reduce sweating.

Absorption, Excretion, Toxicity

Of the 600 mg of average body burden in man (Table 3.1), the majority is in the bone. The kidney is the highest in content among the soft tissues. Some data suggest that tellurium also accumulates in the liver. Soluble tetravalent tellurites, absorbed into the body after oral administration, are reduced to tellurides, partly methylated, and then exhaled as dimethyl telluride. The latter is responsible for the garlic odor in persons exposed to tellurium compounds, but does not account for a high percent of the excreted tellurium. Tellurium in the food is probably in the form of teilurates.

The urine and bile are the principal routes of excretion. Sweat and milk are secondary routes of excretion. Teilurates and tellurium are of low toxicity, but tellurites are generally more toxic. Acute inhalation exposure results in decreased sweating, nausea, a metallic taste, and sleeplessness. The typical garlic breath is a reasonable indi-cator of exposure to tellurium by the dermal, inhalation, or oral route.

Serious cases of tellurium intoxication from industrial exposure have not been reported. In rats, chronic exposure to high doses of tellurium dioxide has produced decreased growth and necrosis of the liver and kidney. Sodium tellurite at 2 ppm in drinking water or potassium tellurate at 2 ppm of tellurium plus 0.16 μg/g in the diet of mice for their lifetime produced no effects in the tellurate group. The females of the tellurite (tetravalent) group did not live as long. In rats, 500 ppm in the diet of pregnant females induced hydrocephalus in the offspring.

Abnormalities of and reduction in numbers of mitochondria were thought to be possible cellular causes of the transplacental effect. One of the few serious recorded cases of tellurium toxicity resulted from accidental poisoning by injection of tellurium into the ureters during retrograde pyelography. Two of the three victims died. Stupor, cyanosis, vomiting, garlic breath, and loss of consciousness were observed in this unlikely incident. Dimercaprol treatment for tellurium increases the renal damage. While ascorbic acid decreases the characteristic garlic odor, it may also adversely affect the kidneys in the presence of increased amounts of tellurium.

THALLIUM

Occurrence and Use

Thallium is obtained as a by-product of iron, cadmium, and zinc. It is used as a catalyst, in certain alloys, optical lenses, jewelry, low-temperature thermometers, dyes and pigments, and scintillation counters. It has been used medicinally as a depilatory. Thallium compounds, chiefly thallous sulfate, have been used as rodenticides and insecticides.

Absorption, Excretion, Toxicity

Thallium is not a normal constituent of animal tissues. It is absorbed through the skin and gastrointestinal tract. After parenteral administration it can be identified in the urine within a few hours. The highest concentrations after poisoning are in the kidney and urine. The intestines, thyroids, testes, pancreas, skin, bone, and spleen have lesser amounts. The brain and liver concentrations are still lower. Thallium is excreted slowly. Following the initial exposure, large amounts are excreted in the urine during the first 24 hours.

After that period the feces may be an important route of excretion. There have been many cases of poisoning from both the medicinal and the rodenticide use of thallium. Acute poisoning is characterized by gastrointestinal irritation, acute ascending paralysis, and psychic disturbances. Acute toxicity studies in rats have indicated that thallium is quite toxic. It has an oral LD50 of approximately 30 mg/kg. The estimated lethal dose in humans, however, is 8 to 12 mg/kg. Rat studies also indicate that thallium oxide, while relatively insoluble, is more toxic orally than by the intravenous or intraperitoneal route.

The acute cardiovascular effects of thallium ions probably result from the exchange against cellular potassium. The signs of subacute or chronic thallium poisoning in rats were hair loss, cataracts, and hindleg paralysis occurring with some delay after the initiation of dosing. Renal lesions were observed at gross necropsy. Histologic changes revealed damage of the proximal and distal renal tubules. The central nervous system changes were most severe in the mesencephalon where necrosis was observed. Perivascular cuffing was also reported in several other brain areas.

Electron microscope examination indicated that the mitochondria in the kidney may have been the first organelles affected. Liver mitochondria also revealed degenerative changes. The livers of newborn rats whose dams had been treated throughout pregnancy showed these changes. Similar mitochondria changes were observed in the intestine, brain, seminal vesicle, and pancreas. It has been suggested that thallium may combine with the sulfhydryl groups in the mitochondria and there interfere with oxidative phosphorylation. A teratogenic response to thallium salts characterized as achondroplasia (dwarfism) has been described in rats. In man fatty infiltration and necrosis of the liver, nephritis, gastroenteritis, pulmonary edema, degenerative changes in the adrenals, degeneration of peripheral and central nervous system, alopecia, and in some cases death have been reported as result of long-term systemic thallium intake.

These cases usually are caused by the contamination of food or the use of hi as a depilatory. Industrial poisoning is a special risk in the manufacture of fused halides Sroduction of lenses and windows. Loss of vision plus the other signs of thallium poison-ing have been related to industrial exposures.

TIN

Occurrence and Use

Cassiterite (SnO_2) is the only important ore of tin. It is used in the manufacture of tinplate, in food packaging, and in solder, bronze, and brass. Stannous and stannic chlorides are used in dyeing textiles. Organic tin compounds have been used as fungicides, bactericides, and slimicides, as well as in plastics as stabilizers.

Absorption, Excretion, Toxicity

Effective absorption after oral administration of even soluble tin salts such as sodium stannous tartrate is limited. Ninety percent of the tin administered in this manner is recovered in the feces. The small amounts absorbed are reflected by increases in the liver and kidneys. Injected tin is excreted by the kidneys, with smaller amounts in the bile. The normal urine level is about 14μg/100ml. The majority of inhaled tin or its salts remains in the lungs, most extracellularly, with some in the macrophages, in the form of SnO_2. The organic tins, particularly triethyltin, may be somewhat better absorbed.

The tissue distribution of tin from this material shows highest concentrations in the blood and liver, with smaller amounts in the muscle, spleen, heart, or brain. Tetraethyltin is converted to triethyltin *in vivo.* Chronic inhalation of tin in the form of dust or fumes leads to benign pneumoconiosis. Tin hydride (SnH_4) is more toxic to mice and guinea pigs than is arsine; however, its effects appear mainly in the central nervous system and no hemolysis is produced. Orally, tin or its inorganic compounds require relatively large doses (500 mg/kg for 14 months) to produce toxicity. The use of tin in food processing seems to demonstrate little hazard.

The average U.S. daily intake, mostly from foods as a result of processing, is estimated at 17 mg. Inorganic tin salts given by injection produce diarrhea, muscle paralysis, and twitching. Organic tin compounds tend to be considerably more toxic. An outbreak of almost epidemic nature took place in France due to the oral ingestion of organic tin compounds used for skin disorders. One hundred deaths resulted from this incident.

Excessive industrial exposure to triethyltin has been reported to produce headaches, visual defects, and EEG changes that were very slowly reversed. Experimentally, triethyltin produces depression and cerebral edema. The resulting hyperglycemia may be related to the centrally mediated depletion of catecholamines from the adrenals. Acute burns or subacute dermal irritation has been reported among workers as a result of tributyltin. Triphenyltin has been shown to be a potent immunosuppressant. Inhibition in the hydrolysis of adenosine triphosphate and uncoupling of oxidative phosphorylation taking place in the mitochondria have been suggested as the cellular mechanisms of tin toxicity.

TITANIUM

Occurrence and Use

Rutile (TiO_2) and ilmenite ($FeTiO_3$) are the primary ores of titanium. Its industrial uses are as a deoxidizer, in permanent magnets, in corrosion-resistant alloys, in pigments in welding rods, in electrodes and lamp filaments, and in surgical appliances. Titanium dioxide salve has been used in the treatment of burns. Titanium has been detected in some foods: butter, corn oil, shrimp, lettuce, pepper, and othercondiments. Titanium is found in North American rivers at levels of 2 to 107 μg/1. The mean concentration in municipal U.S. drinking water is 2.1 μg/1. Titanium is also a contaminant of urban air (Table 3.3).

Absorption, Excretion, Toxicity

Approximately 3 per cent of an oral dose of titanium is absorbed. The majority of that absorbed is excreted in the urine. The normal urine concentration has been estimated at 10 μg/1. The estimated body burden of titanium is about 15 mg. Most of it is in the lungs, probably as a result of inhalation exposure. Inhaled titanium tends to remain in the lungs for long periods. It has been estimated that about one-third of the inhaled titanium is retained in the lungs.

The geographic variation in lung burden is to some extent dependent on air concentration. For example, concentrations of 430, 1,300, and 91 ppm in ashed lung tissue have been reported for the U.S., Delhi, and Hong Kong, respectively. The mean concentrations of 8 and 6 ppm for the liver and kidney, respectively, were reported in the United States. Newborns have little titanium. Lung burdens tend to increase with age. Slight fibrosis of lung tissue has been reported following inhalation exposure to titanium dioxide pigment, but the injury was not disabling. Otherwise, titanium dioxide has been considered physiologically inert by all routes (ingestion, inhalation, dermal, and subcutaneous). The metal and other salts are also relatively nontoxic except for titanic acid, which, as might be expected, will produce irritation.

TUNGSTEN

Occurrence and Use

Wolframite([Fe,Mn] Wo_4) and scheelite are the chief ores of tungsten. It is used in making high-speed tool steel and in other alloys. In addition, tungsten is used in filaments for x-ray tubes, radio tubes, and light bulbs as well as in pigments and waterproofing textiles.

Absorption, Excretion, Toxicity

Tungsten is not a normal complement of animal tissues. It is absorbed to some extent from the gastrointestinal tract and retained largely in the bone, although smaller amounts have been assayed in the spleen, liver, and kidney. The oral toxicity of tungsten compounds varies depending on the salt. The signs of toxicity by oral and parenteral administration are nervous prostration, diarrhea, coma, and death due to respiratory paralysis.

Oral toxicity is apparently not a significant problem in man. Some controversy exists over the effects of tungsten by inhalation. It is difficult to ascertain whether tungsten or cobalt is the causative agent in the pneumoconiosis of the tungsten carbide tool industry. Animal experiments suggest that the effects on the lungs are due not so much to tungsten itself as to other components, especially cobalt. Tungsten has been shown experimentally to interact with two other elements. The addition of soluble tungsten to the diet reduced the mortality and liver lesions characteristic of high levels of selenium intake. Increased tungsten intake also decreased molybdenum deposition in the liver of rats and reduced intestinal xanthine oxidase. Sufficient amounts of tungsten caused molybdenum deficiency in chicks.

URANIUM

Occurrence and Use

The chief raw material of uranium is pitchblende or carnotite ore. This element is largely limited to use as a nuclear fuel.

Absorption, Excretion, Toxicity

The uranyl ion is rapidly absorbed from the gastrointestinal tract. About 60 per cent is carried as a soluble bicarbonate complex, while the remainder is bound to plasma protein.

Sixty percent is excreted in the urine within 24 hours. About 25 per cent may be fixed in the bone. Following inhalation of the insoluble salts, retention by the lungs is prolonged. Uranium tetrafluoride and uranyl fluoride can produce a typical toxicity because of hydrolysis to HF. Skin contact (burned skin) with uranyl nitrate has resulted in nephritis.

The soluble uranium compound (uranyl ion) and those that solubilize in the body by the formation of a bicarbonate complex produce systemic toxicity in the form of acute renal damage.

The classic impairment of renal function by uranium may result in death. However, if exposure is not severe enough, the renal tubular epithelium is regenerated and recovery occurs. Renal toxicity with the classic signs of impairment, including albuminuria, elevated blood urea nitrogen, and loss of weight, is brought about by filtration of the bicarbonate complex through the glomerulus, resorption by the proximal tubule, liberation of uranyl ion, and subsequent damage to the proximal tubular cells. Inhalation exposure of rats, dogs, and monkeys to uranium dioxide dust at a concentration of 5 mg U/M^3 for up to five years produced accumulation in the lungs and tracheobronchial lymph nodes that accounted for 90 percent of the body burden. No evidence of toxicity was observed despite the unusually long duration of the experimental investigation.

VANADIUM

Occurrence and Use

Vanadium occurs in several ores. Carnolite is one of commercial importance from which the metal is usually obtained. Vanadium can also be obtained as a by-product of petroleum refinement. Vanadium pentoxide is used as a catalyst in the production of various materials, of which sulfuric acid may be the most important. It is used in the hardening of steel, the manufacture of pigments, in photography, and in insecticides. Various salts of vanadium have been used medicinally as an antiseptic, as a spirochetocide, as antituberculosis and antianemia agents, and as a general tonic. These uses are without proven efficacy.

Vanadium is a ubiquitous element. It is common in many foods; significant amounts are found in milk, seafoods, cereals, and vegetables. Vanadium has a natural affinity for fats and oils; food oils have high concentrations. Municipal water supplies may contain on the average about 1 to 6 ppb. Urban air contains some vanadium, perhaps due to the use of petroleum products or from refineries. There has been some suggestion that vanadium is useful, if not essential, in various biologic systems. A hematopoietic effect has been postulated. Iron-deficient rats respond more rapidly to added iron in the diet when vanadium is also present. Vanadium also decreases cholesterol and phospholipid content in livers of experimental animals. An anticaries activity of vanadium has also been demonstrated under experimental conditions.

Absorption, Excretion, Toxicity

The average body burden of vanadium has been estimated at about 30 mg. The largest single compartment is the fat. Bone and teeth stores contribute to the body burden. It has been postulated that some homeostatic mechanism maintains the normal levels of vanadium in the face of excessive intake, since the element, in most forms, is moderately absorbed. The principal route of excretion of vanadium is the urine. The normal serum level is 35 to 48 μg/100 ml.

When excess amounts of vanadium are in the diet, the concentration in the red cells tends to increase.

Parenteral administration increases levels in the liver and kidney, but these increased amounts may only be transient. The lung tissue may contain some vanadium, depending on the exposure by that route, but normally the other organs contain negligible amounts. The toxic action of vanadium is largely confined to the respiratory tract. Bronchitis and bronchopneumonia are more frequent in workers exposed to vanadium compounds. In industrial exposures to vanadium pentoxide dust a greenish-black discoloration of the tongue is characteristic. Irritant activity with respect to the skin and eyes has also been ascribed to industrial exposure.

Gastrointestinal distress, nausea, vomiting, abdominal pain, cardiac palpitation, tremor, nervous depression, and kidney damage, too, have been linked with industrial vanadium exposure. Medicinal use of vanadium compounds (V_2O_5) produced gastrointestinal disturbances, slight abnormalities of clinical chemistry related to renal function, and nervous system effects. Acute vanadium poisoning in animals is characterized by marked effects on the nervous system, hemorrhage, paralysis, convulsions, and respiratory depression.

Short-term inhalation exposure of experimental animals tends to confirm the effects on the lungs as well as the effect on the kidney. In addition, experimental investigations have suggested that the liver, adrenals, and bone marrow may be adversely affected by subacute exposure at high levels. It has been postulated that heart disease is related to vanadium air pollution and it may act with cadmium to produce these adverse effects.

ZINC

Occurrence and Use

The principal ore of zinc is sphalerite (largely ZnS). The principal uses of zinc are in the manufacture of galvanized iron, bronze, white paint, rubber, glazes, enamel, glass, and paper and as a wood preservative, $ZnCI_2$, for its fungicidal action. Zinc is ubiquitous and is considered an essential trace element. Its necessary roles involve enzymes and enzymatic functions, protein synthesis, and carbohydrate metabolism. It is necessary for normal growth and development in mammals and birds.

Human dwarfism and lack of sexual development have been related to zinc deficiency. Zinc is present in a number of metalloenzymes, including carbonic anhydrase, carboxypeptidase, alcohol dehydrogenase, glutamic dehydrogenase, lactic dehydrogenase, and alkaline phosphatase. Therapeutically, zinc compounds are used as topical astringents, dermal products, antiseptics, and emetics. The total exposure to zinc is increased through the widespread use of zinc undecylenate preparations for athlete's foot and zinc pyridinethione in antidandruff shampoos.

Zinc is also contained in medicinal preparations of insulin and in zinc bacitracin. Veterinary uses of zinc are similar. In addition, zinc is used as a nutritional supplement to prevent or treat parakeratosis, a disease of pigs caused by zinc deficiency. Zinc is omnipresent in the environment, being found in water, in air, and in all living organisms. In both natural and contaminated states it is almost always accompanied by cadmium. The zinc: cadmium ratio plays a vital role in the effect zinc has on living organisms. Zinc is found in natural water supplies, but the content may be increased if the water flows through galvanized, copper, or plastic pipes.

Seafoods, meats, whole grains, dairy products, nuts, and legumes are high in zinc content. Vegetables are lower. Zinc applied to the soil is taken up in growing vegetables. Zinc atmospheric levels are increased over industrial areas. The average American daily intake is approximately 12.6 mg, mostly from food. Geographic variations in zinc tissue levels may be due to the reduction of zinc by the refining of grains.

Absorption, Excretion, Toxicity

Under normal conditions, not all of the zinc in the diet is absorbed. Phytate (inositol hexaphosphate) present in cereal grains markedly impairs the absorption of zinc, probably by forming an insoluble calcium-zinc-phytate complex in the upper small intestine. Normally, the muscle, liver, kidney, and pancreas contain large amounts. Relatively high concentrations of zinc are also found in the male reproductive system and the epididymis, prostate, and testes of various species. The eyes also contain high concentrations. The zinc in the blood is largely contained in the red blood cells.

Data indicate that there are high levels of zinc in the newborn, but that the level decreases with age. Injection of ^{65}Zn has shown that the liver stores large amounts initially, but the red blood cells and the bone also tend to accumulate zinc. There is a correlation between zinc plasma and bone concentrations, but the zinc concentrations of the kidney and liver may not be decreased in zinc deficiency. Zinc is eliminated principally by the gastrointestinal tract. The pancreatic fluid contains significant amounts, while additional quantities are found in the bile. The urine contains significantly less than the feces (about 20 per cent of fecal amounts).

Milk also contains significant concentrations of zinc. The antagonistic action between cadmium and zinc in the rat with respect to increased growth, prevention of dermal lesions, interaction with copper metabolism, testicular damage, teratogenic changes, and maintenance of circulation and body temperature by adequate intake of zinc has been demonstrated. Therefore it was suggested that, in a sense, cadmium could be considered as an antimetabolite of zinc. Similar interactions between iron, copper, or molybdenum versus zinc have been suggested, but are not well documented. Accidental oral poisoning has been reported in humans as a result of consuming acidic food or beverages from galvanized containers.

The symptoms of such intoxications consist of fever, vomiting, stomach cramps, and diarrhea. These signs are typical of ingestion of large doses of soluble zinc compounds. However, because these signs, especially the ones related to gastroenteritis, are similar to those resulting from ingestion of cadmium and the zinc in galvanized pans containing 1.0 per cent cadmium, it was postulated that some of these cases may be due to cadmium. With regard to industrial exposure, the metal fume fever resulting from inhalation of freshly formed fumes of zinc oxide presents the most significant effect. Only the freshly formed material is potent, presumably because of flocculation in the air, thereby preventing deep penetration into the lungs. After the initial response resulting in "chills," repeated exposures often cause no reaction.

Workers have noted that this effect appears most frequently on Mondays or after holidays. The primary symptom, fever, has been reproduced in rabbits, and it was postulated that the increased body temperature is due to the action of the fumes on endogenous pyrogen in the leukocytes. Even in the most severe cases recovery is usually complete in 24 to 48 hours. There is no evidence that chronic effects have resulted from oxide inhalation. While zinc oxide fumes are the most common cause of metal fume fever, inhalation of other metal oxides may induce this reaction. Dermal toxicity following exposure to $ZnCl_2$ has resulted from consistently handling these salts, and inhalation of mists or fumes may give rise to irritation of the gastrointestinal or respiratory tract.

A gray cyanosis, dermatosis, and ulceration of the nasal passages have resulted from the inhalation of this caustic material. The ocular hazard of zinc salts varies with the salt. Zinc chloride, while used as an astringent in eye drops at 0.2 to 0.5 per cent, may cause damage at higher concentrations. Zinc sulfate in concentrations as high as 20 per cent has been applied to the cornea for therapeatic purposes. Attempts to produce zinc toxicity by incorporation of as much as 0.25 per cent in the diet of rats have not been successful.

At levels above this the homeostatic mechanism breaks down; growth retardation, hypochromic anemia, and defective mineralizations of bone occur. Displacement of copper and altered phosphatase activity are perhaps the mechanisms of action. Zinc 2-pyridinethiol-oxide, an antidandruff agent, provides an example of a unique species-specific toxicity. This material in dogs causes blindness, resulting from retinal detachment, following nine oral daily doses of 25 mg/kg. Further studies did not reveal such effects at similar doses in monkeys or rodents.

This effect in dogs is judged to be related to the chelation of zinc, which could be accomplished with sodium pyridinethione and hydroxypyridinethione in the tapetum lucidim of the dogs, a structure not present in man or monkeys. However, during the course of these investigations it was revealed that this zinc salt caused a chelinergic-like effect. Antidotal treatment of pyridoxine and nicotinic acid was lifesaving in the dog, but not in primates, suggesting different routes of metabolism. Testicular tumors have been produced by direct intratesticular injection in rats and chickens. This effect is probably related to the concentration of zinc normally in the gonads and may be hormonally dependent. Other routes have not produced carcinogenic effects by zinc salts.

ZIRCONIUM

Occurrence and Use

Zircon ($ZrSiO_2$) is the primary ore of zirconium. It is used in the nuclear industry as a shielding material, in metal alloys, as a catalyst in organic reactions, in the manufacture of water-repellent textiles, in dyes, in pigments on ceramics, in abrasives, and cigarette lighter flints. The metallurgic difficulties probably prevent even wider application. Zirconium oxychloride has been used as an antiperspirant.

Zirconium carbonate and oxide are used for dermatitis. Intravenous injection of zirconium has been advocated for prophylactic use to prevent skeletal deposition of certain radioelements, especially plutonium. The daily oral intake in man has been estimated at 3.5 mg. Lamb, pork, eggs, dairy products, and grains contain the highest concentrations. Plant uptake of zirconium from soil and fertilizer has been demonstrated. Zirconium has been detected in rivers at 0.1 ppb. Because the common salts are insoluble, the water concentrations are of small and doubtful significance in urban water supplies.

Absorption, Excretion, Toxicity

The average body burden is 250 mg. Fat, gallbladder, aorta, liver, red blood cells, diapharm, lung, kidney, muscle, brain, pancreas, stomach, spleen, and testes have concentrations ranging from 18.7 in fat to 1.88 in testes in terms of micrograms per gram of tissue (wet weight). Zirconium is excreted by the intestine, probably in the bile. Zirconium levels are negligible in the urine.

Milk is a secondary route of excretion. Significant amounts of zirconium are found in fetuses. While metabolism studies are lacking, tissue concentrations indicate that significant amounts of zirconium may be absorbed orally. It has been suggested that some homeostatic mechanism exists with respect to zirconium. Inhalation exposure to water-soluble $ZrOCl_2$ indicated that the highest concentrations of zirconium occurred in the lungs and pulmonary lymph nodes. Deposition and retention in the bone (femur) were greater than in the liver. Granulomatous lesions, probably of allergic epithelioid origin, have been observed following the use of deodorant sticks and poison ivy lotions containing zirconium.

Rabbits developed pulmonary granulomata following zirconium lactate exposure. Inhalation exposure of $ZnCU_4$ (6 mg Zr/M^3) for 60 days produced slight decrease in hemoglobin and red blood count in dogs and increased mortality in rats and guinea pigs. Zirconium oxide at 75 mg/M^3 caused no effect. The oral toxicity of zirconium compounds is low. No evidence of industrial disease related to zirconium exposure has been documented.

4 Metals in Catalysis and Electron Transport

Although the amounts present within living cells are very small, the ions of the transition metals Fe, Co, Ni, Cu, and Mn are extremely active centers for catalysis, especially of reactions that take advantage of the ability of these metals to exist in more than one oxidation state. Iron, copper, and nickel are also components of the electron carrier proteins that function as oxidants or reductants in many biochemical processes. These metals are all nutritionally essential, as are chromium and vanadium. Among the heavier transition elements molybdenum is a constituent of an important group of enzymes that includes the sulfite oxidase of human liver and nitrogenase of nitrogen-fixing bacteria. Tungsten occasionally substitutes for molybdenum.

A. IRON

Iron is one of the most abundant elements in the earth's crust, being present to the extent of ~4% in a typical soil. Its functions in living cells are numerous and diverse. The average overall iron content of both bacteria and fungi is ~ 1 mmol/kg, but that of animal tissues is usually less. Seventy per cent of the 3-5 g of iron present in the human body is located in the blood's erythrocytes, whose overall iron content is ~ 20 mM. In other tissues the total iron averages closer to 0.3 mM and consists principally of storage forms.

The total concentration of iron in all of the iron-containing *enzymes* of tissues amounts to only about 0.01 mM. Although these concentrations are low, the iron is concentrated in oxidative enzymes of membranes and may attain much higher concentrations locally. Only a few parasitic or anaerobic bacteria, *e.g.*, the lactic acid bacteria, possess no oxygen-requiring enzymes and are almost devoid of both iron and copper. All other organisms appear to require iron for life.

Uptake by Living Cells

A major problem for cells is posed by the relative insolubility of ferric hydroxide and other compounds from which iron must be extracted by the organism. A consequence is that iron is often taken up in a chelated form and is transferred from one organic ligand, often a protein, to another with little or no existence as free Fe^{3+} or Fe^{2+}. As can be calculated from the estimated solubility product of $Fe(OH)_3$ (Eq. 4.1) the equilibrium concentration of Fe^{3+} at pH7 is only 10^{-17} M.

$$K_{sp} = [Fe^{3+}][OH^-]^3 < 10^{-38}\ M^4 \qquad ...(4.1a)$$

or

$$[Fe^{3+}]/[H^+]^3 < 10^4\ M^4 \text{ at } 25°\ C \qquad ...(4.1b)$$

For a 2 μm^3 bacterial cell this amounts to just one free Fe^{3+} ion in almost 100 million cells at any single moment. The importance of chelated forms of iron becomes obvious. It is also evident from Eq. 4.1 that , in addition to chelation, a low external pH can also facilitate uptake of Fe^{3+} by organisms. The values of the formation constants for chelates of Fe^{2+} typically lie between those of Mn^{2+} and Co^{2+}. For example, $K_1 = 10^{1.3}\ M^{-1}$ for formation of the Fe^{2+} chelate of EDTA. The smaller arid more highly charged Fe^{3+} is bonded more strongly ($K_1 = 10^{14.3}\ M^{-1}$), These binding constants are independent of pH.

However, the binding of any metal ion is affected by pH. A fact of considerable biochemical significance is the stronger binding of Fe^{3+} to oxygen-containing ligands than to nitrogen atoms, while Fe^{2+} tends to bind preferentially to nitrogen. It is also significant that Fe^{3+} bound to oxygen ligands tends to exchange readily with other ferric ions in the medium, whereas Fe^{3+} bound to nitrogen-containing ligands such as heme exchanges slowly. This fact is important for both iron-transport compounds and enzymes.

Siderophores

If a suitably high content of iron (*e.g.,* 50 μM or more for *E. coli*) is maintained in the external medium, bacteria and other microorganisms have little problem with uptake of iron. However, when the external iron concentration is low, special compounds called siderophores are utilized to render the iron more soluble. For example, at iron concentrations below 2 μM, *E. coli* and other enterobacteria secrete large amounts of *enterobactin* (Fig. 4.1). The stable Fe^{3+}-enterobactin complex is taken up by a transport system that involves receptors on the outer bacterial membrane. Siderophores from many bacteria have in common with enterobactin the presence of *catechol* (*ortho*-dihydroxybenzene) groups that chelate the iron. The three catechol groups of enterobactin are carried on a cyclic serine triester structure.

A variety of both cyclic and linear structures are found among other catechol siderophores. For example, *parabactin* and *agrobactin* (Fig. 4.1) contain a backbone of spermidine[20]. After the Fe^{3+}-enterobactin complex enters a bacterial cell the ester linkages of a siderophore are cleaved by an esterase. Because of the extremely high formation constant of $\sim 10^{52}\ M^{-1}$ for the complex[11] the only way for a cell to release the Fe^{3+} is through this irreversible destruction of the iron carrier. Reduction to Fe^{2+} may be involved in release of iron from some siderophores. The first known siderophore, isolated in 1952 by Neilands,[22] is *ferrichrome* (Fig. 4.1), a cyclic hexapeptide containing *hydroxamate* groups at the iron-binding centers. Oxygen atoms

O
C
R N O⁻
R′

Hydroxamate group

form the bonds to iron in this compound also. Ferrichrome binds Fe^{3+} with a formation constant of $\sim 10^{29}\ M^{-1}$. The binding is not as tight as with enterobactin and the iron can be released by enzymatic reduction to Fe^{2+} which is much less tightly bound than is Fe^{3+}.

The released ferrichrome can be secreted and used repeatedly to bring in more iron. Ferrichrome is produced by various fungi and bacilli and is only one of a series of known hydroxamate siderophores. Since iron is essential to virtually all parasitic organisms the ability to obtain iron is often the limiting factor in establishing an infection. In *E. coli* there are seven different outer *membrane receptors* for siderophores. One of these, the gated porin *FepA*, is specific for ferric enterochelin. With the assistance of another protein, *TonB*, it allows the ferric siderophore to penetrate the outer membrane. A different receptor, *FhuA*, binds ferrichrome. Both FepA and FhaA are large 22-strand porins resembling the 16-strand porin. However, they are nearly 7 nm long with internal diameters three times those of the 16-strand porins. In addition, loops of polypeptide chain on the outer edges can close while an N-terminal domain forms a "cork" that remains in place until the Fe^{3+}-siderophore complex enter the

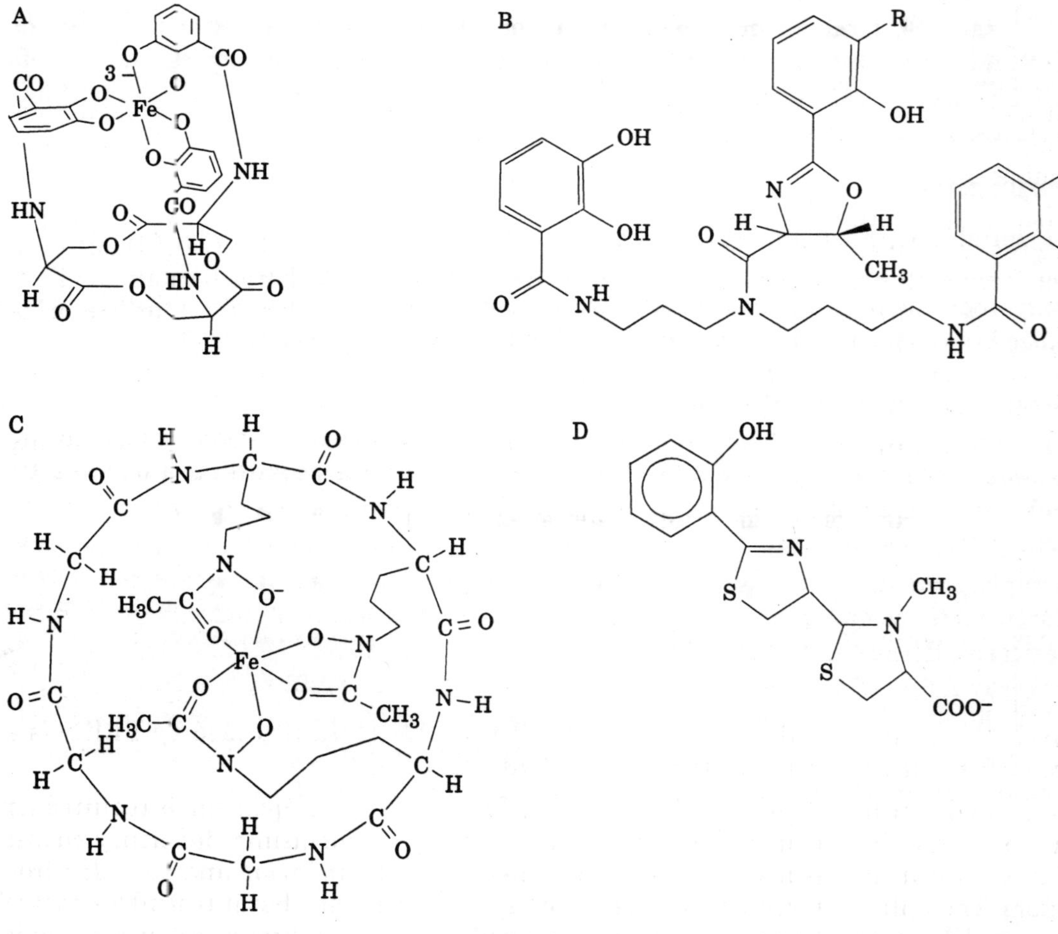

Fig. 4.1. Structures of several siderophores and of their metal complexes. (A) Enterobactin of *E. coli* and other enteric bacteria (B) parabactin (R = H) from *Paracoccus denitrificans* and agrobactin (R = OH) from *Agrobacterium tumefaciens* (C) ferrichrome (D) pyochelin from *Pseudamonas aeruginosa*.

channel and binds. Like the hatches in an air lock on a spacecraft, the outer loops then close, after which the inner cork unwinds to allow the siderophore complex to enter the periplasmic space. The apparatus requires an energy supply, which apparently is provided by an additional complex consisting of proteins TonB, ExbA, and ExbD. They evidently couple the electrochemical gradient across the cell membrane with the operation of the channel gates. Some bacteria use a different strategy for passage across the outer membrane. The channel of a receptor protein contains a molecule of an iron-free siderophore. When a molecule of Fe^{3+}-siderophore binds in the outside part of the channel the Fe^{3+} jumps to the inner siderophore, which then dissociates from the receptor, carrying the Fe^{3+}-siderophore complex into the periplasmic space.

This mechanism also seems to be available in *E. coli.* The *ferric uptake regulation* (Fur) protein binds excess free Fe^{2+}, the resulting complex acting as a represser of all of the iron uptake genes in *E. coli*. Additional proteins are required for passage through the membrane. These are ABC transporters, which utilize hydrolysis of ATP as an energy source. For uptake of the Fe^{3+}-ferrichrome complex protein *FhuD* is the periplasmic binding protein, *FhuC* is an integral membrane component, and the cytosolic *FhuC* contains the ATPase center. Another ABC transporter, found in many bacteria, carries unchelated Fe^{3+} across the inner membrane. The binding protein component for *Hemophihis influenzae,* designated Hit, resembles one lobe of mammalian transferrin.

The siderophore receptors of bacteria have been "parasitized" by various bacteriophages and toxic proteins. For example, FepA is also a receptor for the toxic colicins B and D and tonB is a receptor for bacteriophage T1. Some bacteria do not form siderophores but take up Fe^{2+}. *Even E. coli,* when grown anaerobically, synthesizes an uptake system for Fe^{2+}. It utilizes a 75-residue peptide encoded by gene *feoA* and a 773-residue protein encoded by *feoB*.

Uptake of iron by eukaryotic cells

The yeast *Saccharomyces cerevisiae* utilizes two systems for uptake of iron. A low-affinity system transports Fe^{2+} with an apparent K_m of ~ 30 μM, while a high-affinity system has a K_m of –0.15 μM. Study of these systems has been greatly assisted by the use of genetic methods developed for both bacteria and yeast. The low-affinity iron uptake depends upon a protein transporter encoded by gene FET4 and on a reductase encoded by genes FRE1 and FRE2, proteins that are embedded in the cytoplasmic membrane. It might seem reasonable that the FET3 copper oxidoreductase should keep Fe^{2+} *reduced* while it is transported. However, it appears to *oxidize* Fe^{2+} to Fe^{3+}. The high-affinity uptake system is more puzzling. It requires a permease encoded by *FTRI* and an additional protein encoded by FET3. The Fet3 protein is a copper oxidoreductase related to *ceruloplasmin*. The protein *Frelp* (encoded by FRE1) is a metalloreductase that reduces Cu^{2+} to Cu^{+}, as well as Fe^{3+} to Fe^{2+}.

It is essential for copper uptake. It has long been known that ceruloplasmin is required for mobilization of iron from mammalian tissues. Hereditary ceruloplasmin deficiency causes accumulation of iron in tissues. Yeast also contains both *mitochondrial* and *vacuolar* iron transporters. The uptake of iron by animals is not as well understood but it resembles that of yeast. A general divalent cation transporter that is coupled to the membrane proton gradient is involved in intestinal iron uptake. Ascorbic acid promotes the uptake of iron, presumably by reducing it to Fe(II), which is more readily absorbed than Fe(III), and also by promoting ferritin synthesis. Uptake is also promoted by meat in the diet. Within the body iron is probably transferred from one protein to another with only a transient existence as free Fe^{2+}.

An average daily human diet contains ~15 mg of iron, of which ~ 1 mg is absorbed. This is usually enough to compensate for the small losses of the metal from the body, principally through the bile. Once it enters the body, iron is carefully conserved. The 9 billion red blood cells destroyed daily yield 20-25 mg of iron which is almost all reused or, stored. The body apparently has no mechanism for excretion of large amounts of iron; a person's iron content is regulated almost entirely by the rate of uptake. This rate is increased during pregnancy and, in young women, to compensate for iron lost in menstrual bleeding.

Nevertheless, control of iron uptake is imperfect and perhaps 500 million people around the world suffer from iron deficiency. For others, an excessive intake of iron or a genetic defect lead to accumulation of iron to toxic levels, a condition called hemochromatosis. This condition may also arise in any disease that leads to excessive destruction of hemoglobin or accumulation of damaged erythrocytes. Examples are β-thalassemia and cerebral malaria. Treatment with chelating agents designed to remove iron is often employed.

Transferrins

Within the body iron is moved from one location to another while bound as Fe^{3+} to transferrins, a family of related 680- to 700-residue 80-kDa proteins. Each transferrin molecule contains two Fe^{+3} binding sites, one located in each of two similar domains of the folded peptide chain. A dianion, usually CO_3^{2-}, is bound together with each Fe^{3+}. Milk transferrin (*Iactoferrin* also found in leukocytes), hen egg transferrin (ovotransferrin), and rabbit and human serum transferrin all have similar structures. Each Fe^{3+} is bonded to oxygen anions from two tyrosine side chains, an aspartate carboxylate, an imidazole group, and the bound carbonate ion.

Transferrin of blood plasma is encoded by a separate gene but has a similar structure. Transferrin of chickens appears to be identical to conalbumin of egg whites. The iron-binding proteins of body fluids are sometimes given the group name siderophilins. Transferrins may function not only in transport of iron throughout the body but also as iron buffers that provide a relatively constant iron concentration within tissues. The entrance of iron into the body through the intestinal mucosal cells may involve the transferrin present in those cells and the influx of iron may also be regulated by blood plasma transferrin.

There is also a nontransferrin pathway. Transferrins bind Fe^{2+} weakly and it is likely that a transferrin- Fe^{2+}-HCO_3^- complex formed initially undergoes oxidation to the Fe^{3+}-CO_3^{2-} complex within cells and within the bloodstream. A conformational change closes the protein around the iron ions.[56] In yeast the previously mentioned copper oxidoreductase encoded by the FET3 gene appears to not only oxidize Fe^{2+} but also transfer the resulting Fe^{3+} to transferrin. Ceruloplasmin may play a similar role in mammals. Iron is transferred from the plasma transferrin into cells of the body following binding of the Fe^{3+} –transferrin complex to specific receptors. The surface of an immature red blood cell (reticulocyte) may contain 300,000 transferrin receptors, each capable of catalyzing the entry of ~ 36 iron ions per hour.

The receptor is a 180-kDa dimeric glycoprotein. When the Fe^{3+}-transferrin complex is bound, the receptors aggregate in coated pits and are internalized. The mechanism of release of the Fe^{3+} may occur by different mechanisms in the two lobes. The pH of the endocytic vesicles containing the receptor complex is probably lowered to ~ 5.6 as in lysosomes. This may protonate the bound CO_3^{2-} in the complex and assists in the release of the Fe^{3+}, possibly after reduction to Fe^{2+}. Both the apotransferrin and its receptor are returned to the cell surface for reuse, the apotransferrin being released into the blood. Chelating agents such as pyrophosphate, ATP, and citrate as well as simple anions may also assist in removal of iron from transferrin. The same transmembrane transporter that is involved in intestinal iron uptake is needed to remove iron from the endosome after release.

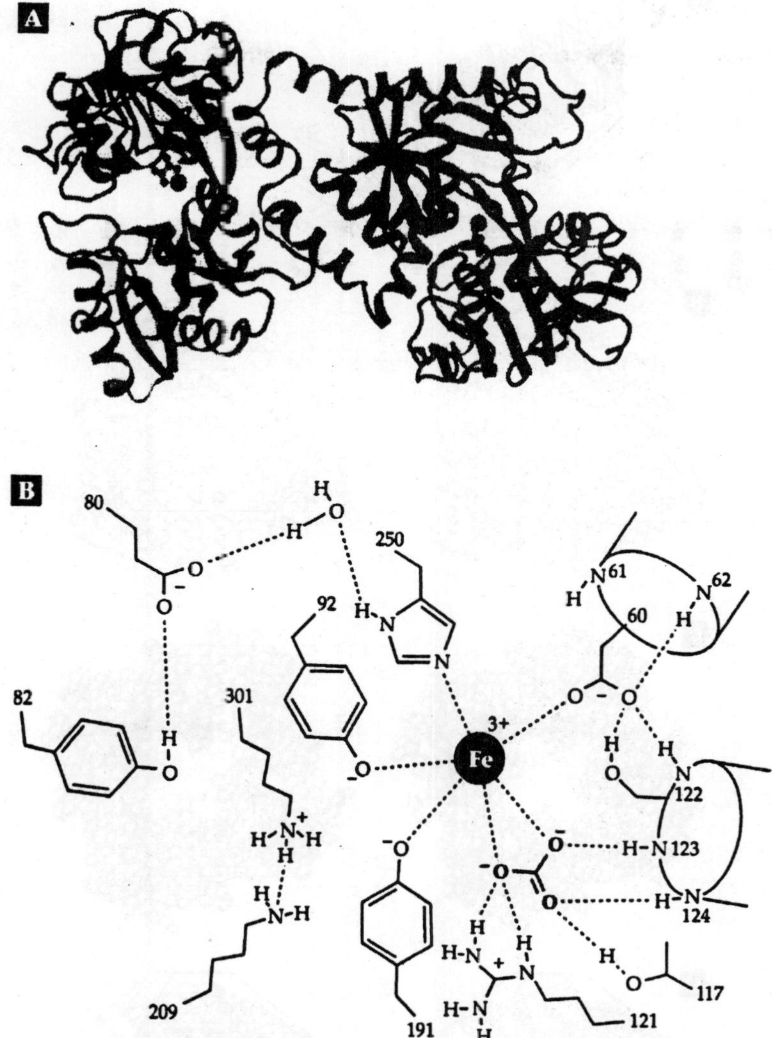

Fig . 4.2. (A) Ribbon drawing of the polypeptide chain of a transferrin, human lactoferrin. The N lobe is to the left and the C lobe to the right. Each active site contains bound Fe^{3+} and a molecule of oxalate dianion which replaces the physiological CO_3^{2-}. From Baker *et al.* Courtesy of Edward Baker. (B) Schematic diagram showing part of the hydrogen-bond network involved in binding the Fe^{3+} in the N lobe of hen ovotransferrin. Some side chain groups and water molecules have been omitted. The positions of hydrogen atoms and the charge state of acid-base groups are uncertain. Most of the hydrogen-bond distances (O - - O, O - - N, N - - N) indicated by dashed lines are between 0.27 and 0.3 nm. Release of the bound Fe^{3+} may be accomplished in part by protonation of the bound CO_3^{2-} to form HCO_3^-.

Storage of Iron

Within tissues of animals, plants, and fungi much of the iron is packaged into the red-brown water-soluble protein *ferritin*, which stores Fe(III) in a soluble, nontoxic, and readily

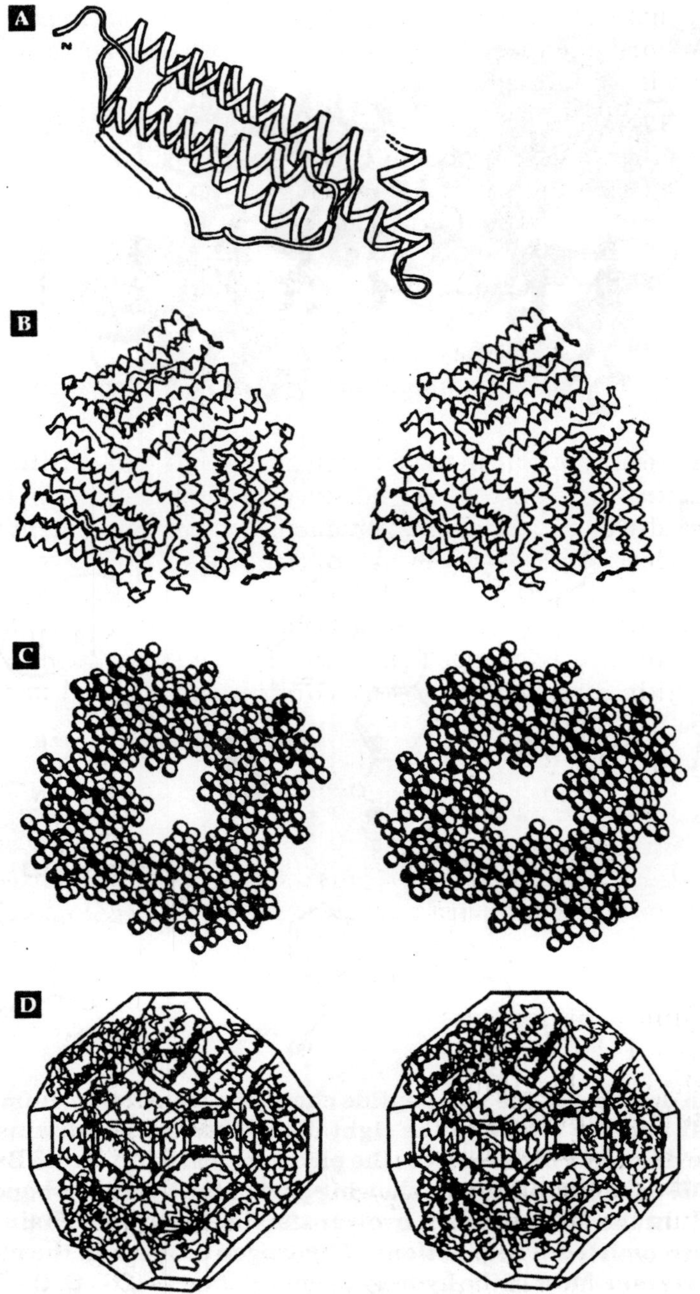

Fig. 4.3. Structure of the protein shell of ferritin (apoferritin). (A) Ribbon drawing of the 163-residue monomer. (B) Stereo drawing of a hexamer composed of three dimers. (C) A tetrad of four subunits drawn as a space-filling diagram and viewed down the four-fold axis from the exterior of the molecule. (D) A half molecule composed of 12 subunits inscribed within a truncated rhombic dodecahedron.

available form. Although bacteria store very little iron, some of them also contain a type of ferritin. On the other hand, the yeast S. *cerevisieae* stores iron in polyphosphate-rich granules, even though a ferritin is also present.

Ferritin contains 17-23% iron as a dense core of hydrated ferric oxide ~7 nm in diameter surrounded by a protein coat made up of twenty-four subunits of molecular mass 17- to 21-kDa. Each subunit is folded as a four-helix bundle (Fig. 4.3). Mammalian ferritins consist of combinations of subunits of two or more types. For example, human ferritins contain similarly folded 19-kDa L (light) and 21-kDa H (heavy) sub-units. The twenty-four subunits are arranged in a cubic array (Fig. 4.3). The outer diameter of the 444-kDa apoferritin is ~12 nm. The completely filled ferritin molecule contains 23% Fe and over 2000 atoms of iron in a crystalline lattice. Larger ferritins may contain as many as 4500 atoms of iron with the approximate composition $[Fe(O)OH]_8FeOPO_3H_2$. Phosphate ions are sometimes bound into surface layers of the ferritin cores.

Ferritin cores are readily visible in the electron microscope, and ferritin is often used as a labeling reagent in microscopy. Another storage form of iron, *hemosiderin,* seems to consist of ferritin partially degraded by lysosomes and containing a higher iron content than does ferritin. Depositions of hemosiderin in the liver can rise to toxic levels if excessive amounts of iron are absorbed. Iron can be deposited in ferritin by allowing apoferritin to stand with an Fe(II) salt and a suitable oxidant, which may be O_2. Physiological transfer of Fe(III) from transferrin to ferritin is thought to require prior reduction to Fe(II). The reoxidation by O_2 to Fe(III) for deposition in the ferritin core catalyzed by *ferrooxidase sites* located in the centers of the helical bundles of the H-chains.

$$2\,Fe^{2+} + O_2 + 4\,H_2O \rightarrow 2Fe(O)OH_{core} + H_2O_2 + 4H^+ \quad ...(4.2)$$

The ferrooxidase site is a dinuclear iron center in which two iron ions (probably Fe^{2+}) are bound as in Fig. 4.4. They are then converted to Fe^{3+} ions by O_2, which may bind initially to the Fe^{2+}, forming a transient blue intermediate that is thought to have a peroxodiferric structure, perhaps of the following type. Reaction of this intermediate with H_2O

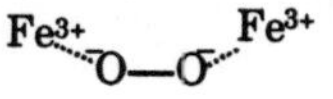

$$Fe^{3+}\cdots O{-}O\cdots Fe^{3+}$$

may yield H_2O_2 plus a biomineral precursor $Fe^{2+} - O - Fe^{2+}$, which is incorporated into the core. Ferritin H subunits predominate in tissues with high oxygen levels, *e.g.,* heart and blood cells, while the L subunits predominate in tissues with slower turnover of iron, *e.g.,* liver. The L subunits lack ferrooxidase activity but, in the centers of their helical bundles, contain polar side chains that may help to initiate growth of the mineral core. Removal of Fe(III) from storage in ferritin cores may require reduction to Fe(II) again, possibly by ascorbic acid or glutathione.

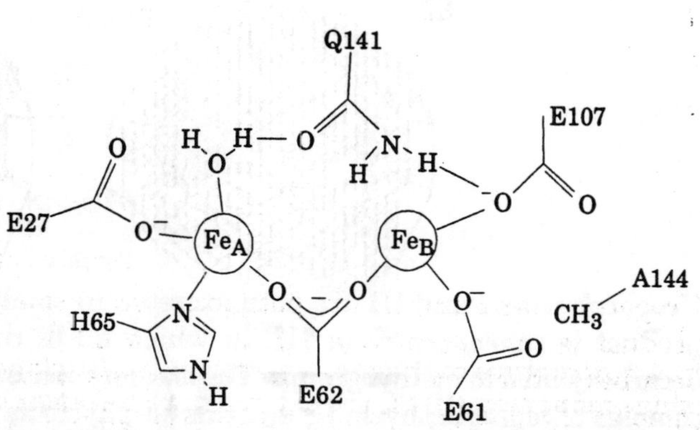

Q141, E107, E27, Fe_A, Fe_B, A144, CH_3, H65, E62, E61

Fig. 4.4. The dinuclear iron center or ferroxidase center of human ferritin based on the structure of a terbium(III) derivative.

Some bacterial ferritins contain a bound cytochrome *b* which may assist in reduction. Released iron in the Fe^{2+} state can be incorporated into iron-containing proteins or into heme. The enzyme *ferrochelatase* catalyzes the transfer of free Fe^{2+} into protoporphyrin IX to form protoheme. Iron in the Fe(II) state may also be oxidized to Fe^{3+} through action of the copper-containing ceruloplasmin be incorporated into heme by direct transfer from ferritin.

Heme Proteins

In 1879, German physiological chemist Hoppe Seyler showed that two of the most striking pigments of nature are related. The red iron-containing *heme* from blood and the green magnesium complex *chlorophyll a* of leaves have similar ring structures. Later, H. Fischer proved their structures and provided them with the names and numbering systems that are used today. This information is summarized in the following section.

Some Names to Remember

Porphins are planar molecules which contain large rings made by joining four pyrrole rings with methine bridges. In the *chlorins*, found in the chlorophylls, one of the rings reduced. The specific class of porphins known as *porphyrins* have eight substituents around the periphery of the large ring. Like the chlorins and the *corrins* of vitamin B_{12}, the porphyrins are all formed biosynthetically from *porphobilinogen*. This compound is polymerized in two ways to give porphyrins of types I and III. In type I porphyrins, polymerization of porphobilinogen has taken place in a regular way so that the sequences of the carboxymethyl and carboxyethyl side chains (often referred to as acetic acid and propionic acid side chains, respectively) are the same all the way around the outside of the molecule.

However, most biologically important porphyrins belong to type III, in which the first three rings A, B, and C have the same sequence of carboxymethyl and carboxyethyl side chains, but in which ring D has been incorporated in a reverse fashion. Thus, the carboxyethyl side chains of rings C and *D* are adjacent to each other. Porphyrins containing all four carboxymethyl and four carboxy-ethyl side chains intact are known as *uroporphyrins*.

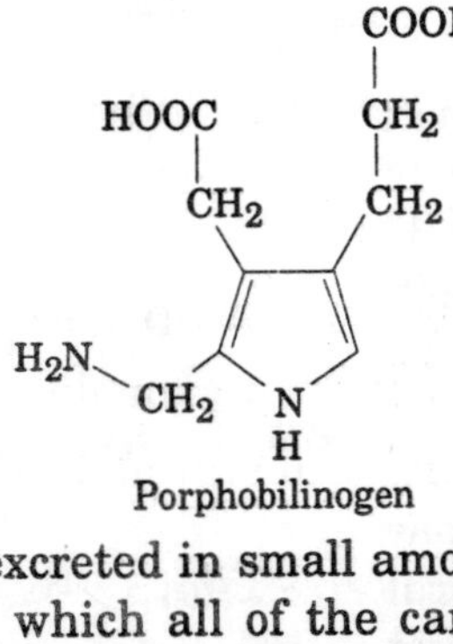

COOH
HOOC CH_2
CH_2 CH_2
H_2N CH_2 N
H

Porphobilinogen

Uroporphyrins I and III are both excreted in small amounts in the urine. Another excretion product is *coproporphyrin* III, in which all of the carboxymethyl side chains have been decarboxylated to methyl groups. The feathers of the tropical touraco are coloured with copper(II) complex of coproporphyrin III and this porphyrin as well as others are commonly found in birds' eggs. The heme proteins are all derived from protoporphyrin IX, which is formed by decarboxylation and dehydrogenation of two of the carboxyethyl side chains of uroporphyrin III to vinyl groups.

Hemes and Heme Proteins

Protoporphyrin IX contains a completely conjugated system of double bonds. In the center two hydrogen atoms are attached, one each to two of the nitrogens; they are free to move to other nitrogens in the center with rearrangement of the double bonds. Thus, there is tautomerism as well as resonance within the heme ring. The two central hydrogens can be replaced by many metal ions to form stable chelates. The complexes with Fe^{2+} are known as *hemes* and the Fe^{2+} complex with protoporphyrin IX as *protoheme*. Heme complexes of Fe^{2+} may be designated as *ferrohemes* and the Fe^{3+} compounds as *ferrihemes*. The Fe^{3+} protoporphyrin IX is also called *hemin*, and may be crystallized as a chloride salt.

Iron tends to have a coordination number of six, and other ligands can attach to the iron from the two axial positions on opposite sides of the planar heme. If these are nitrogen ligands, such as pyridine or imidazole, the resulting compounds, called *hemochromes*, have characteristic absorption spectra. An example is cytochrome b_5, which contains two axial imidazole groups. Several modifications of protoheme are indicated (Fig. 4.5). To determine which type of heme exists in a particular protein, it is customary to split off the heme by treatment with acetone and hydrochloric acid and to convert it by addition of pyridine to the pyridine hemochrome for spectral analysis.

By this means, protoheme was shown to occur in hemoglobin, myoglobin, cytochromes of the *b* and P450 types, and catalases and many peroxidases. Cytochromes *a* and a_3 contain *heme a,* while one of the terminal oxidase systems of enteric bacteria contains the closely related heme *o* (Fig. 4.5). A second terminal oxidase of those same bacteria contains *heme d* (formerly a_2). *Heme c* (present in cytochromes *c* and/) is a variation in which two SH groups of the protein have added to the vinyl groups of protoheme to form two thioether linkages. A few cytochromes c have only one such linkage. In myeloperoxidase three covalent linkages, different than those in cytochrome *c,* join the heme to the protein. There is a possibility that heme *a* may sometimes form a Schiff base with a lysyl amino group through its formyl group. Heme *d* is a chlorin, as is *acrylochlorin heme* from certain bacterial nitrite reductases.

Siroheme, which is found in both nitrite and sulfite reductases of bacteria, an isobacteriochlorin in which both the A and B rings are reduced. It apparently occurs as an amide *siroamide*. *Desulfovibrio.* Heme d_1 of nitrite reductases of denitrifying bacteria is a dioxo-bacteriochlorin derivative. As in myoglobin, hemoglobin, and cytochrome *c*, one axial coordination position on the iron of most heme proteins (customarily called the *proximal* position) is occupied by an imidazole group of a histidine side chain. However, in cytochrome P450 and chloroperoxidase a thiolate ($-S^-$) group from a cysteinyl side Chain, and in catalase a phenolate anion from a tyrosyl side chain, occupies the proximal position.

The sixth or *distal* coordination position is occupied by the sulfur atom of methionine in cytochrome *c* and most other cytochromes with low-spin iron but cytochromes b_5 and c_3 have histidine. The high-spin heme proteins, such as cytochromes *c'*, globins, peroxidase, and catalase, usually have no ligand other than weakly bound H_2O in the distal position. Hemes are found in all organisms except the anaerobic clostridia and lactic acid bacteria. Heme proteins of blood carry oxygen reversibly, whereas those of *terminal oxidase systems, hydroxylases, and oxygenases* "activate" oxygen, catalyzing reactions with hydrogen ions and electrons or with carbon compounds.

The heme-containing *peroxidases and catalases* catalyze reactions not with O_2 but with H_2O_2. Another group of heme proteins includes most of the cytochromes, which are purely electron-transferring compounds.

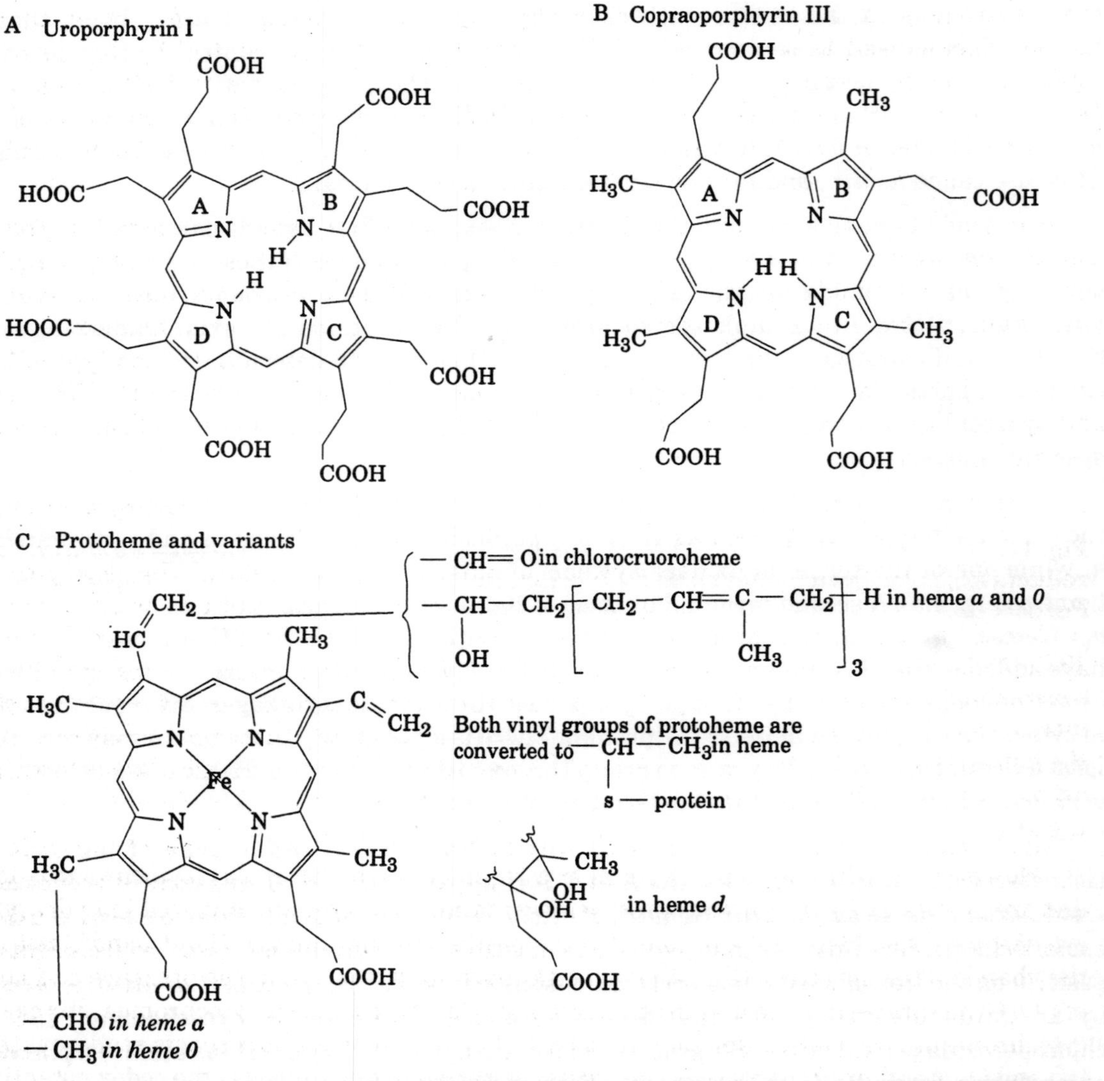

Fig. 4.5. Structures of some biologically important porphyrins. (A) Uroporphyrin I. (B) Coproporphyrin III. Note that a different tautomeric form is pictured in B than in A. Tautomerism of this kind occurs within all of the porphyrins. (C) Protoheme, the Fe^{2+} complex of protoprophyrin IX, present in hemoglobin, cytochromes *b*, and other proteins.

The Cytochromes

The iron in the small proteins known as cytochromes acts as an electron carrier, the α bond, undergoing alternate reduction to the +2 state and oxidation to the +3 state. The cytochromes, discovered in 1884 by McMunn, were first studied systematically in the

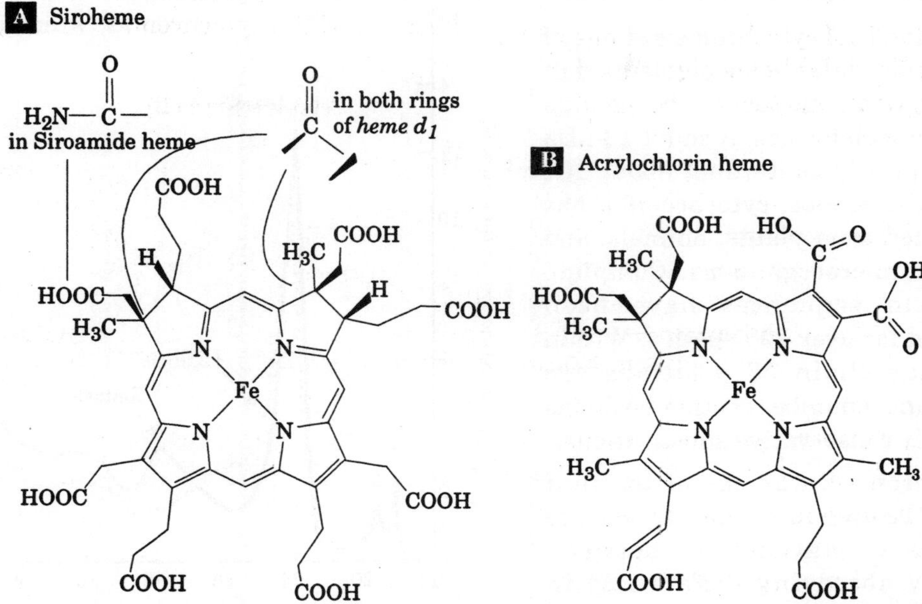

Fig. 4.6. Structures of isobacteriochlorin prosthetic groups. (A) Siroheme from nitrite and sulfite reductases; (B) acrylochlorin heme from dissimilatory nitrite reductases of *Pseudomonas* and *Paracoccus*.

1920s by Keilin have been isolated from many sources. The classification into groups *a, b,* and *c* according to the position of the longest wavelength light absorption band follows a practice introduced by Keilin. However, it is now customary to designate a new cytochrome by giving the heme type *(a, b, c* or *d)* together with the wavelength of the ∝ band, *e.g.*, cytochrome C_{552} or cyt b_{557}. Cytochromes of the *b* type including bacterial cytochrome o contain protoheme. Because the sixth position is ligated, most cytochromes *b* do not react with O_2.

However, cytochromes *o* and *d* serve as terminal electron acceptors (cytochrome oxidases) and are oxidizable by O_2. Another protohemecontaining cytochrome, involved in hydroxylation *cytochrome* P450. Here the 450 refers to the position of the intense "Soret band" (also called the yband) of the spectrum (Fig. 4.7) in a difference spectrum run in the presence and absence of CO. Other properties are also used in arriving at designations for cytochromes. For example, cytochrome a_3 has a spectrum similar to that of cytoc hrome *a* but it reacts readily with both CO and O_2. Another property that distinguishes various cytochromes is the redox potential $E^{\circ\prime}$, which in this discussion is given for pH 7.0.

Cytochromes carry electrons between other oxidoreductase proteins of widely varying values of E^0. Because of the various heme environments cytochromes have greatly differing values of E^0, allowing them to function in many different biochemical systems. For mitochondrial cytochrome *c* the value of $E^{\circ\prime}$ is ~+ 0.265 V but for the closely related cytochrome *f* of chloroplasts it is ~+0.365 V and for cytochrome c_3 of *Desulfovibrio* about –0.330 V. There is more than an 0.6-volt difference between $E^{\circ\prime}$ of cytochromes *f* and c_3. Cytochromes *b* tend to have lower $E^{\circ\prime}$ values, close to zero, than most cytochromes *c*, while cytochrome a_3 has $E^{\circ\prime}$ ~ +0.385 V.

The c-type Cytochromes

Mitochondrial cytochrome *c* is one of the few intracellular heme pigments that is soluble in water and that can be removed easily from membranes. A small 13-kDa protein typically containing about 104 amino acid residues, cytochrome *c* has been isolated from plants, animals, and eukaryotic microorganisms. Complete amino acid sequences have been determined for over 100 species. Within the peptide chain 28 positions are invariant and a number of other positions contain only conservative substitutions.

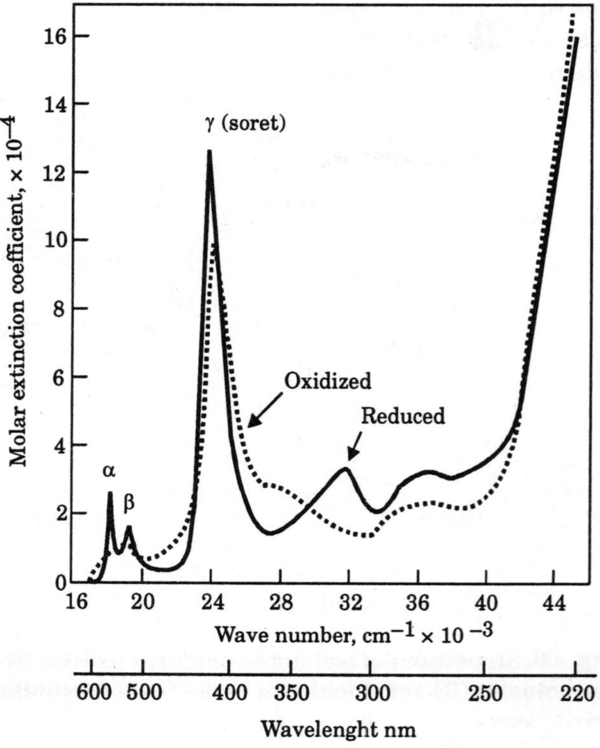

Fig. 4.7. Absorption spectra of oxidized and reduced horse heart cytochrome *c* atpH 6.8. From data of Margoliash and Frohwirt.

Cytochrome c was one of the first proteins to be used in attempting to trace evolutionary relationships between species by observing differences in sequence. Humans and chimpanzees have identical cytochrome *c*, but 12 differences in amino acid sequence occur between humans and the horses and 44 between human and *Neurospora*. The related cytochrome c_2 of the photosynthetic bacterium *Rhodospillum rubrum* is thought to have diverged in evolution 2×10^9 years ago from the precursor of mammalian cytochrome *c*. Even so, 15 residues remain invariant. Structural studies on cytochromes of the *c* and c_2 types show that the heme group provides a core around which the peptide chain is wound. The 104 residues of mitochondrial cytochrome *c* are enough to do little more than envelope the heme.

In both the oxidized and reduced forms of the protein, methionine 80 (to the left in Fig. 4.8A) and histidine 18 (to the right) fill the axial coordination positions of the iron. The heme is nearly "buried" and inaccessible to the surrounding solvent. The shorter chains of the 82- to 86-residue cytochromes c_{550} (from *Pseudomonas*), c_{553}, and c_{555} (from *Chlorobium*) as well as the longer 112-residue polypeptide of cytochrome c_2 from *Rhodospirillurn rubrum* have nearly the same folding pattern as that in mitochondria] cytochrome *c*. However, the 128-residue chain of the dimeric cytochrome *c′* from *Rhodospirillum inolischianum* forms an antiparallel four-helix bundle. This is the same folding pattern present in the ferritin monomer, hemerythrin, and many other proteins including cytochrome b_{562} of *E. coli*.

Cytochrome *f*, which functions in photosynthetic electron transport, is also a *c*-type cytochrome but with a unique protein fold. Most cytochromes have only one heme group per polypeptide chain, but the 115-residue cytochrome c_3 from the sulfate-reducing bacterium *Desulfovibrio binds* four hemes (Fig. 4.8*c*). Each one seems to have a different redox potential in the -0.20 to -0.38 V range. Another *c*-type cytochrome, also from *Desulfovibrio,* contains six hemes in a much larger 66-kDa protein and functions as a nitrite reductase. Triheme and octaheme proteins are also known.

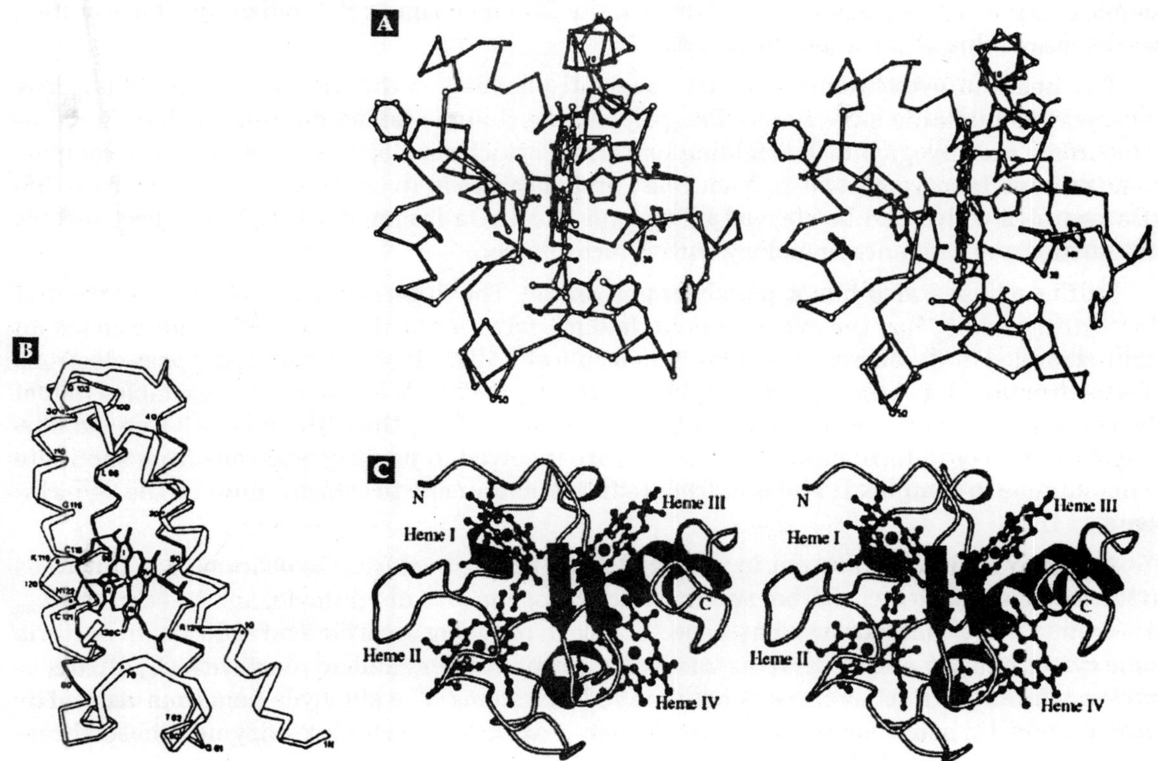

Fig. 4.8. Structures of three cytochromes of the *c* type. (A) Horse mitochondrial cytochrome *c*; (B) a subunit of the dimeric cytochrome *c'* from *Rhodospirillittn molischianum;* (C) cytochrome C_3 from *Desulfovibrio desulfuricans.* (A) and (B) courtesy of Salemme (C).

Many cytochromes *c* are soluble but others are bound to membranes or to other proteins. A well-studied tetraheme protein binds to the reaction centers of many purple and green bacteria and transfers electrons to those photosynthetic centers. Cytochrome c_2 plays a similar role in *Rhodobacter,* forming a complex of known three-dimensional structure. Additional cytochromes participate in both cyclic and noncyclic electron transport in photosynthetic bacteria and algae. Some bacterial membranes as well as those of mitochondria contain a *cytochrome* bc_1 *complex* whose structure.

Cytochromes b, a, and o

Protoheme-containing cytochromes *b* are widely distributed. There are at least five of them in *E. coli.* Whether in bacteria, mitochondria, or chloroplasts, the cytochromes *b* function within electron transport chains, often gathering electrons from dehydrogenases and passing them on to *c*-type cytochromes or to iron-sulfur proteins. Most cytochromes *b* are bound to or embedded within membranes of bacteria, mitochondria, chloroplasts, or endoplasmic reticulum (ER). For example, cytochrome b_5 delivers electrons to a fatty acid desaturating system located in the ER of liver cells and to many other reductive biosynthetic enzymes. The protein contains 132 amino acid residues plus another 85 largely hydrophobic N-terminal residues that provide a nonpolar tail which is thought to be buried in the ER membranes. Solubilization of the protein causes loss of this N-terminal sequence.

The heme in cytochrome b_5 is not covalently bonded to the protein but is held tightly between two histidine side chains. The polypeptide chain is folded differently than in either cytochrome *c* or myoglobin. The folding pattern of cytochrome b_5 is also found in the complex heme protein *flavocytochrome* b_2 from yeast probably also in liver *sulfite oxidase*. Both are 58-kDa peptides which can be cleaved by trypsin to 11-kDa fragments that have spectroscopic similarities and sequence homology with cytochrome b_5.

Sulfite oxidase also has a molybdenum center. The 100-residue N-terminal portion of flavocytochrome b_2 has the cytochrome b_5 folding pattern but the next 386 residues form an eight-stranded $(\alpha/\beta)_8$ barrel that binds a molecule of FMN. All of these proteins pass electrons to cytochrome *c* (Fig. 4.8). In contrast, the folding of *cytochrome* b_{562} of E. *coli* resembles that of cytochrome *c'*. However, it has methionine side chains as both the fifth and sixth iron ligands. Cytochromes *b of* mitochondrial membranes are involved in passing electrons from succinate to ubiquinone in complex II and also from reduced ubiquinone to cytochrome *c*, in the 248-kDa complex III.

A similar complex is present in photosynthetic purple bacteria. Cytochrome b_{560} functions in the transport of electrons from succinate dehydrogenase to ubiquinone, and cytochrome b_{560} of secretory vesicle membranes has a specific role in reducing ascorbic acid radicals. In bacteria some cytochromes *b* and d_1 serve as terminal electron carriers able to react with O_2, nitrite, or nitrate, while others act as carriers between redox systems. The aldehyde heme *a* is utilized by animals and by some bacteria in *cytochrome c oxidase*, a complex enzyme whose three-dimensional structure is known.

Mechanisms of Biological Electron Transfer

The heme groups of the cytochromes as well as many other transition metal centers act as carriers of electrons. For example, cytochrome *c* may accept an electron from reduced cytochrome c_1 and pass it to cytochrome oxidase or cytochrome *c* peroxidase. The electron moves from one heme group to another over distances as great as 2 nm.

Similar electron-transfer reactions between defined redox sites are met in photosynthetic reaction centers, in metalloflavoproteins, and in mitochondrial membranes. What are the factors that determine the probability of an electron transfer reaction and the rate at which it may occur? They include: (1) The distance from the electron donor to the acceptor. (2) The thermodynamic driving force $\Delta G°$ for the reaction. This can be approximated using the difference in standard electrode potentials between donor and acceptor. $\Delta G° = -96.5\ AE°$ kJ/ mol at 25° C. (3) The chemical makeup of the material through which the electron transfer takes place. (4) Any changes in the geometry or charge state of the donor or acceptor that accompany the transfer. (5) The orientation of the acceptor and donor groups. It is usually assumed that the Franck-Condon principle is obeyed, *i.e.*, that the electron jump occurs so rapidly ($<10^{-12}$ s) that there is no change in the positions of atomic nuclei. Subsequent rearrangement of nuclear positions may occur at rates that allow a rapid overall reaction.

The various factors that affect the rate of electron transfer were incorporated by Marcus into a quantitative theory: Electron transfer is often discussed in terms of this classic Marcus theory together with effects of quantum mechanical tunneling. According to Marcus the electron transfer rate from a donor to an acceptor at a fixed separation depends upon $\Delta G°$, a nuclear reorganization parameter (λ), and the electronic coupling strength $|H_{AB}|$ between reactant and product in the transition state (Eq. 4.3).

$$k_{ET} = (4\pi^3/h^2\lambda k_BT)\,|H_{AB}|^2 \exp[-\Delta G° + \lambda)^2/4\lambda k_BT] \qquad ...(4.3)$$

Here $|H_{AB}|$ is a quantum mechanical matrix whose strength decreases exponentially with the distance of separation R as $e^{-\beta R}$ where β is a coefficient of the order of 9-14 nm^{-1}. At the closest contact (R = 0) the rate k_{KT}, by extrapolation from experimental data on small synthetic compounds, is close to the molecular vibration frequency of 10^{13} s^{-1}.

At distances greater than 2 nm the rate would be negligible were it not for other factors. Using mutant proteins as well as a variety of redox pairs and electron-transfer distances the validity of the Marcus equation with respect to the thermodynamic driving force and distance dependence has been verified. This is even true for cytochrome *c* mutants functioning in living yeast cells. A huge amount of experimental work with proteins has been done to test and refine the theories of electron transport. For example, electron donor groups with various reduction potentials have been attached to various sites on the surface of a protein containing a heme or other electron accepting group.

Ruthenium complexes such as Ru(III) $(NH_3)_5^{3+}$ form tight covalent linkages to imidazole nitrogens such as that of His 33 of horse heart cytochrome *c*. This metal can be reduced rapidly to Ru(II) by an external reagent, after which the transfer of an electron from the Ru(II) across a distance of 1.2 nm to the heme Fe(III) can be followed spectroscopically. The reduction potentials $E°$ for the Ru(III)/Ru(II) and Fe(III)/Fe(II) couples at pH 7 in these compounds are 0.16 and 0.27 V, respectively. Thus, an electron will jump spontaneously from the Ru(II) to the Fe(III) with $\Delta G°$ = -15.4 kJ/mol.

A rate constant of ~5 s^1, which was nearly independent of temperature, was observed. Since the structures of Fe(II) and Fe(III) forms of cytochrome *c* differ only slightly, the electron transfer apparently occurs with only a small amount of geometric rearrangement. The distribution of charges and dipoles within the protein may be such that the Fe^{2+} and Fe^3+ complexes have almost equal thermodynamic stability.

Electron-transfer Pathways?

In spite of the success of the Marcus theory, rates of electron-transfer from the iron of cytochrome *c* have been found to vary for different pathways. For example, transfer of an electron from Fe(II) in reduced cytochrome c to an Ru(III) complex on His 33 was fast (–440 s^{-1}) but the rate of transfer to an equidistant Fe(III) ion on Met 65 was at most 0.6 s^{-1}. These results suggested that distinct electron-transfer pathways exist. One suggestion was that the sulfur atom of Met 80 donates an electron to Fe^{3+} leaving an electron-deficient radical. The "hole" so created could be filled by an electron jumping in from the –OH group of the adjacent Tyr 67, which might then accept an electron from an external acceptor via Tyr 74 at the protein surface.

Do electrons flow singly or as pairs from the surface through hydrogen-bonded paths? Use of both semisynthesis and directed mutation of cytochromes *c* is permitting a detailed study of these effects. A striking result is that substitution of the conserved residue phenylalanine 82 in a yeast cytochrome *c* with leucine or isoleucine retards electron transfer by a factor of ~ 10^4.

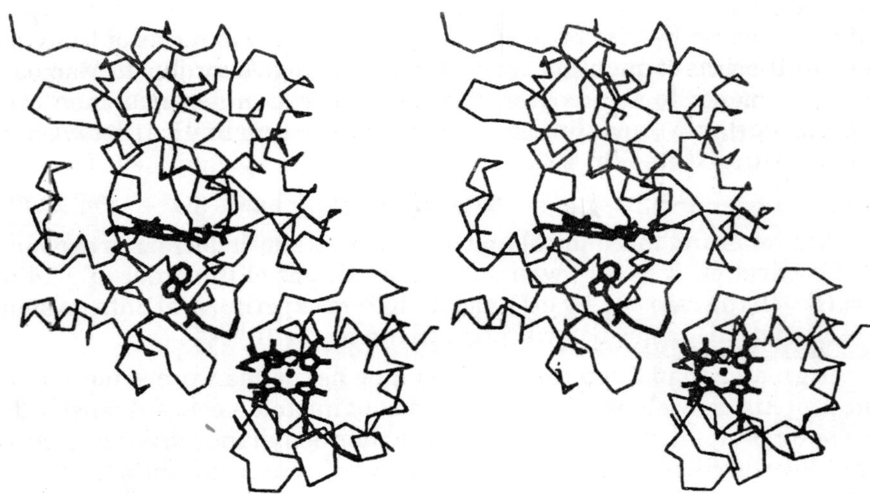

Fig. 4.9. Stereoscopic α-carbon plot of yeast cytochrome *c* peroxidase (top) and yeast cytochrome *c* (below) as determined from a cocrystal by Pelletier and Kraut. The heme rings of the two proteins appear in bold lines, as does the ring of tryptophan 191 and the backbone of residues 191 -193 of the cytochrome *c* peroxidase.

"Docking."

It is now recognized that there are distinct "docking sites" on the surface of electron-transport proteins. For rapid electron transfer to occur the two electron carriers must be properly oriented and docked by formation of correct polar and nonpolar interactions. Early indications of the importance of docking came from study of modified cytochromes *c*. Each one of the 19 lysine side chains was individually altered by acylation or alkylation to remove the positive charge or to replace it with a negative charge. The rate of electron transfer into cytochrome *c* from hexacyano-ferrate was decreased by a factor of 1.3-2.0 when any one lysine at positions 8,13, 27, 72, or 79, which are clustered around the heme edge, was modified. Modification of Lys 22, 55, 99, or 100, distant from this edge, had no effect.

Electron transfer *into* cytochrome *c* from its natural electron donor ubiquinol: cytochrome *c* reductase was also strongly inhibited by modification of lysines that surround the heme edge. Modification of these lysines also inhibited electron transfer out of cytochrome *c* into its natural acceptors, cytochrome *c* oxidase and cytochrome *c* peroxidas. A major factor in these effects is probably the large dipole moment in the cytochrome *c* that arises from the unequal distribution of surface charges. This charge distribution must assist in the proper docking of the cytochrome with its natural electron donors and acceptors.

The positive surface charges presumably also facilitate the reaction with hexacyanoferrate (II) or ascorbate, both of which are negatively charged reductants that react rapidly with cytochrome *c*. Measurements of many kinds have been made between natural donoracceptor pairs such as cytochrome *c*-cytochrome b_5, cytochrome *c*-cytochrome *c* peroxidase, trimethylamine dehydrogenase-FMN to Fe_4S_4 center, and methylamine dehydrogenase (TTQ radical)-amicyanin (Cu^{2-}). Designed metalloproteins are being studied as well. Femtosecond laser spectroscopy is providing a new approach.

Coupling and Gating of Electron Transfer

Electrons are thought to be transferred into or out of cytochrome *c* through the exposed edge of the heme. The rate depends upon effective coupling, which in turn may depend upon orientation as well as the structure and dynamics of the protein. Proteins with much β structure appear to provide stronger coupling than those that are largely composed of a helices. The high nuclear reorganization energy λ of α helices may block electron transfer along some pathways. A conformational change, transfer of a proton, or binding of some other specific ion before electron transfer occurs can be the "gating" process that determines the rate of electron transfer. Electron transfer can also be "coupled" to an unfavourable, but fast, equilibrium.

Effects of Ionic Equilibria on Electron Transfer

The charge on an ion of Fe^{2+} in a heme is exactly balanced by two negative charges on the porphyrin ring. However, when the Fe^{2+} loses an electron to become Fe^{3+} an extra positive charge is suddenly present in the center of the protein. This change in charge will have a powerful electrostatic effect on charged groups in the immediate vicinity of the iron and even at the outer surface of the molecule. For example, an anion from the medium or from a neighbouring protein molecule might become bound to the heme protein (Eq. 4.4).

$$X^- + [\text{Fe(II)}\,(\text{heme})] \rightarrow [\text{Fe(III)}\,(\text{heme})] + X^- + e^- \quad \text{...(4.4)}$$

In this case the presence of a high concentration of X^- in the medium would favour the oxidation of Fe(II) to Fe(III). The reduced heme would be a better reducing agent and the oxidized form a weaker oxidant than in the absence of X^-. If –YH were a group in the protein the loss of an electron could cause –YH to dissociate so that Y^- and the Fe(III) would interact more tightly.

$$\text{HY} - [\,\text{Fe (II)}\,(\text{heme})\,] \rightarrow Y^- - [\,\text{Fe (III)}\,(\text{heme})\,]^+ + H^+ + e^- \quad \text{...(4.5)}$$

We see that electron transfer can be accompanied by loss of a proton and that $E^{\circ\prime}$ may become pH dependent (see also Eq. 4.18). Even with cytochrome *c,* although there is little structural change upon electron transfer, there is an increased structural mobility in the oxidized form. This may be important for coupling and could also facilitate associated proton-transfer reactions. For example, it is possible that in some cytochromes the imidazole ring in the fifth coordination position may become deprotonated upon oxidation. This possibility is of special interest because cytochromes are components of proton pumps in mitochondrial membranes.

Reactions of Heme Proteins with Oxygen or Hydrogen Peroxide

As Ingraham remarked, "Living in a bath of 20% oxygen, we tend to forget how reactive it is." From a thermodynamic viewpoint, all living matter is extremely unstable with respect to combustion by oxygen. Ordinarily, a high temperature is required and if we are careful with fire, we can expect to escape a catastrophe. However, one mole of properly chelated copper could catalyze consumption of all of the air in an average room within one second. Biochemists are interested in both the fact that O_2 is kinetically stable and unreactive and also that oxidative enzymes such as cytochrome *c* oxidase are able to promote rapid reactions. Two oxygen atoms, each with six valence electrons, might reasonably be expected to form dioxygen, O_2, as a double-bonded structure with one a and one κ bond as follows (left):

:Ö= Ö: :Ọ̈— Ọ̈:

Dioxygen

However, O_2 is paramagnetic and contains two un-paired electrons. From this evidence O_2 might be assigned the structure on the right. The oxygen molecule is very stable, and it is relatively difficult to add an electron to form the reactive *superoxide anion radical* O_2^-.

:Ö Ö:⁻

Superoxide anion radical

For this reason, oxidative attack by O_2 tends to be slow. However, once an electron has been acquired, it is easy for additional electrons to be added to the structure and further reduction occurs more easily. The biochemical question suggested is, "How can some heme proteins carry O_2 reversibly without any oxidation of the iron contained in them while others *activate* oxygen toward reaction with substrates?" Among this latter group, cytochrome *c*, oxidase transfers electrons to both oxygen atoms so that only H_2O is a product, whereas the hydroxylases and oxygenases, incorporate either one or two of the atoms of O_2. respectively, into an organic substrate. Before examining these reactions let us reconsider the heme oxygen carriers.

Oxygen-carrying Proteins

In examined the behaviour of hemoglobin in the cooperative binding of four molecules of O_2 and studied its structural relationship to the monomeric muscle protein myoglobin. The iron in functional hemoglobin and myoglobin is always Fe(II) and is only very slowly converted by O_2 into the Fe(III) forms methemoglobin or metmyoglobin. Erythrocytes contain an enzyme system for immediately reducing methemoglobin back to the Fe(II) state. Binding of O_2 to the iron in the heme is usually considered not to cause a change in the oxidation state of the metal. However, oxygenated heme has some of the electronic characteristics of an Fe^{3+}–OO^- peroxide anion.

Bonding of the heme iron to oxygen is thought to occur by donation of a pair of electrons by the oxygen to the metal. In deoxyhemoglobin the Fe(II) ion is in the "high-spin" state; four of the five *3d* orbitals in the valence shell of the iron contain one unpaired electron and the fifth orbital contains two paired electrons. The binding of oxygen causes the iron to revert to the "low-spin" state in which all of the electrons are paired and the paramagnetism of hemoglobin is lost. The stability of hemeoxygen complexes is thought to be enhanced by "back-bonding," *i.e.*, the donation of an electron pair from one of the filled *d* orbitals of the iron atom to form a bond with the adjacent oxygen. This can be indicated symbolically as follows:

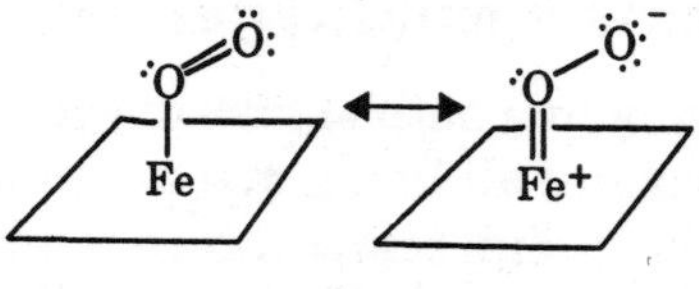

[Fe–O=Ö: ⟷ Fe⁺=O–Ö:⁻]

These structures, which have been formulated by assuming that one of the unshared electron pairs on O_2 forms the initial bond to the metal, are expected to lead to an angular geometry which has been observed in X-ray structures of model compounds, in oxymyoglobin, and in oxyhemoglobin. Neutron diffraction studies have shown that the outer-most oxygen atom of the bound O_2 is hydrogen bonded to the H atom on the N^{ε} atom of the distal imidazole ring of His E7 (Fig. 4.10). Carbon monoxide binds with the C ≡ O axis perpendicular to the heme plane and unable to form a corresponding hydrogen bond. This decreases the affinity for CO and helps to protect us from carbon monoxide poisoning.

All oxygen-carrying heme proteins have another imidazole group that binds to iron on the side opposite the oxygen site. Without this proximal imidazole group, heme does not combine with oxygen. Coordination with heterocyclic nitrogen compounds favors formation of low-spin iron complexes and simple synthetic compounds that closely mimic the behaviour of myoglobin have been prepared by attaching an imidazole group by a chain of appropriate length to the edge of a heme ring.

Similar compounds bearing a pyridine ring in the fifth coordination position have a low affinity for oxygen. Thus, the polarizable imidazole ring itself seems to play a role in promoting oxygen binding. The π electrons of the imidazole ring may also participate in bonding to the iron as is indicated in the following structures. The TI bonding to the iron would allow the iron to back-bond more strongly to an O_2 atom entering the sixth coordination position. These diagrams illustrate another feature found frequently in heme proteins: The N-H group of the imidazole is hydrogen bonded to a peptide backbone carbonyl group.

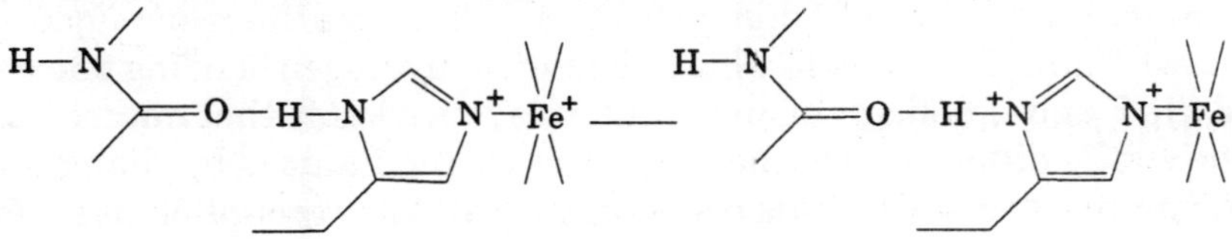

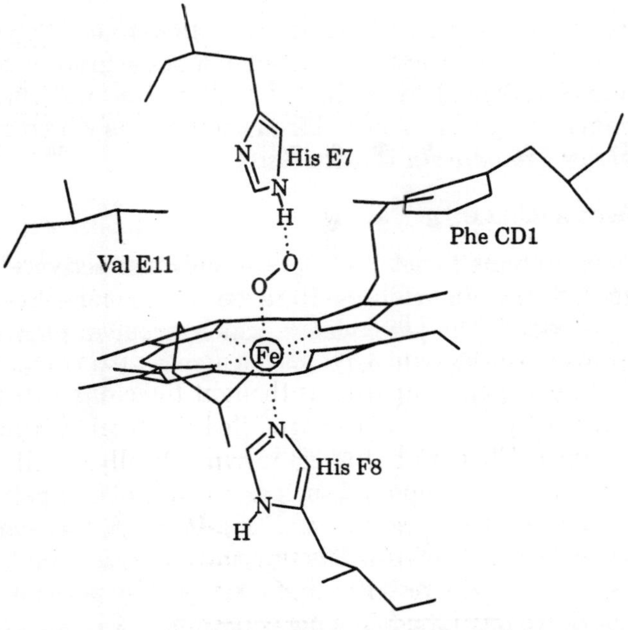

Fig. 4.10. Geometry of bonding of O_2 to myoglobin and position of hydrogen bond to N^{ε} of the distal histidine E7 side chain.

The coordination of the heme iron to histidine also appears to provide the basis for the cooperativity in binding of oxygen by hemoglobin. The radius of high-spin iron, whether Fe(II) or Fe(III), is so large that the iron cannot fit into the center of the porphyrin ring but is displaced toward the coordinated imidazole group by a distance of –0.04 nm for Fe(II). Thus, in deoxyhemoglobin both iron and the imidazole group lie further from the center of the ring than they do in oxyhemoglobin. In the latter, the iron lies in the center of the porphyrin ring because the change to the low-spin state is accompanied by a decrease in ionic radius.

The change in protein conformation induced by this small shift in the position of the iron ion was described. However, the exact nature of the linkage between the Fe position and the conformational changes is not clear. The mechanical response to the movement of the iron and proximal histidine, described, may explain this linkage. However, oxygenation may also induce a change in the charge distribution within the hydrogen-bond network of the protein. The carbonyl group shown in the foregoing structure is attached to the F helix hydrogen bonded to other amide groups.

Electron withdrawal into the heme-oxygen complex would tend to strengthen the hydrogen bond as indicated by the resonance forms shown and also to weaken competing hydrogen bonds. This could affect the charge distribution in the upper end of the F helix and could conceivably induce a momentary conformational change that could facilitate the rearrangement of structure that was discussed. The $\alpha_1\beta_2$ contact in which a change of hydrogen bonding takes place is located nearby behind the F and G helices. In any event, it is remarkable that nature has so effectively made use of the subtle differences in the properties of iron induced by changes in the electron distribution within the *d* orbitals of this transition metal. A few groups of invertebrates, *e.g.*, the sipunculid worms, use a nonheme iron-containing protein, *hemerythrin*, as an oxygen carrier.

Its 113-residue subunits are often associated as octamers of C_4 symmetry, each peptide chain having a four-helix bundle structure. Instead of a heme group, each monomer contains two atoms of high-spin Fe(II) held by a cluster of histidine and carboxylate side chains. Hemerythrin is a member of a group of such diiron oxoproteins which are considered further. The copper oxygen carrier *hemocyanin* is discussed.

Catalases and Peroxidases

Many iron and copper proteins do not bind O_2 reversibly but "activate" it for further reaction. We will look at such metalloprotein oxidases. Here we will consider heme enzymes that react not with O_2 but with peroxides. The peroxidases, which, occur in plants, animals, and fungi, catalyze the following reactions (Eq. 4.6, 4.7):

$$H_2O_2 + AH_2 \rightarrow 2\,H_2O + A \qquad (4.6)$$

$$ROOH + AH_2 \rightarrow R\text{—}OH + H_2O + A \qquad (4.7)$$

Here AH_2 is an oxidizable organic compound such as an alcohol or a pair of one-electron donor molecules. Catalases, which are found in almost all aerobic cells, may sometimes account for as much as 1% of the dry weight of bacteria. The enzyme catalyzes the break-down of H_2O_2 to water and oxygen by a mechanism similar to that employed by peroxidases. If rewritten with H_2O_2 for AH_2 and O_2 for A, we have the following equation:

$$2H_2O \xrightarrow{\text{Catalase}} 2H_2O + O_2 \qquad ...(4.8)$$

The action of catalase is very fast, almost 10^4 times faster than that of peroxidases. The molecular activity per catalytic center is about $2 \times 10^5\ s^{-1}$.

Catalase exerts a protective function by preventing the accumulation of H_2O_2 which might be harmful to cell constituents. The complete intolerance of obligate anaerobes to oxygen may result from their lack of this enzyme. Support for this protective function comes from the existence of the human hereditary condition *acatalasemia*. Persons with extremely low catalase activity are found worldwide but are especially numerous in Korea. In Japan it is estimated that there are 1800 persons lacking catalase. Because about half of them have no symptoms, catalase might be judged unessential. However, many of the individuals affected develop ulcers around their teeth. Apparently, hydrogen peroxide produced by bacteria accumulates and oxidizes hemoglobin to methemoglobin depriving the tissues of oxygen. Catalase from most eukaryotic species is tetrameric. The protein from beef liver consists of 506-residue subunits.

Human catalase is similar. The proximal ligand to the heme Fe^{3+} is a tyrosinate anion (Tyr 358), while side chains of His 75 and Asn 148 lie close to the heme on the distal H_2O_2-binding side. Larger ~650-residue fungal and bacterial Catalases have a similar folding pattern but an extra C-terminal domain with a flavodoxin-like structure. Catalase is gradually inactivated by its very reactive substrate. As isolated, beef liver catalase usually contains about two subunits in which the heme ring has been oxidatively cleaved to *biliverdin* and various other alterations have been found. Each subunit of mammalian catalases normally contains a bound molecule of NADPH which helps to protect against inactivation by H_2O_2. Catalases from *Neurospora* and from *E. coli* contain heme *d* rather than protoporphyrin.

Some lacto-bacilli, lacking heme altogether, form a manganese-containing pseudocatalase. Of the plant peroxidases, which are found in abundance in the peroxisomes, the 40-kDa monomeric *horseradish peroxidase* has been studied the most. It occurs in over 30 isoforms and has an extracellular role in generating free radical intermediates for polymerization and crosslinking of plant cell wall components. Secreted fungal peroxidases, *e.g.,* such as those from *Coprinus* and *Arthromyces,* form a second class of peroxidases with related structues. A third class is represented by *ascorbate peroxidase* from the cytosol of the pea and by the small 34-kDa *cytochrome c peroxidase* from yeast mitochondria. The latter has a strong preference for reduced cytochrome *c* as a substrate (Eq. 4.9).

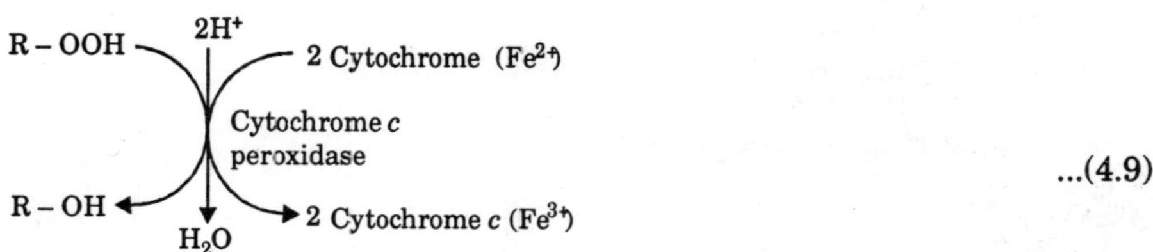

...(4.9)

Because the three-dimensional structures of the peroxidase, its reductant cytochrome *c,* and the complex of the two are known, cytochrome c peroxidase is the subject of much experimental study. Other fungal peroxidases, some of which contain manganese rather than iron, act to degrade lignin. A lignin peroxidase from the white wood-rot fungus *Phanerochaete chrysosporium* has a surface tryptophan with a specifically hydroxylated Cβ carbon atom which may have a functional role in catalysis. The human body contains *lactoperoxidase*, a product of exocrine secretion into milk, saliva, tears, etc., and peroxidases with specialized functions in *saliva*, the *thyroid*, *eosinophils*, and *neutrophils*.

The functions are largely protective but the enzymes also participate in biosynthesis. Mammalian peroxidases have heme covalently linked to the proteins. The active site structure of peroxidases quite highly conserved. As in myoglobin, an imidazole group is the proximal heme ligand, but it is usually hydrogen bonded to an aspartate carboxylate as a catalytic diad.

In cytochrome *c* peroxidase there is also a buried tryptophan, which has already been highlighted in Fig. 4.9. A conserved and essential feature on the distal side is another histidine, which is hydrogen bonded to an asparagine and which can also hydrogen bond to the substrate H_2O_2. Fungal peroxidases also have a conserved arginine on the distal side. However, even an octapeptide with a bound heme cut from cytochrome *c* acts as a "microperoxidase" with properties similar to those of natural peroxidases.

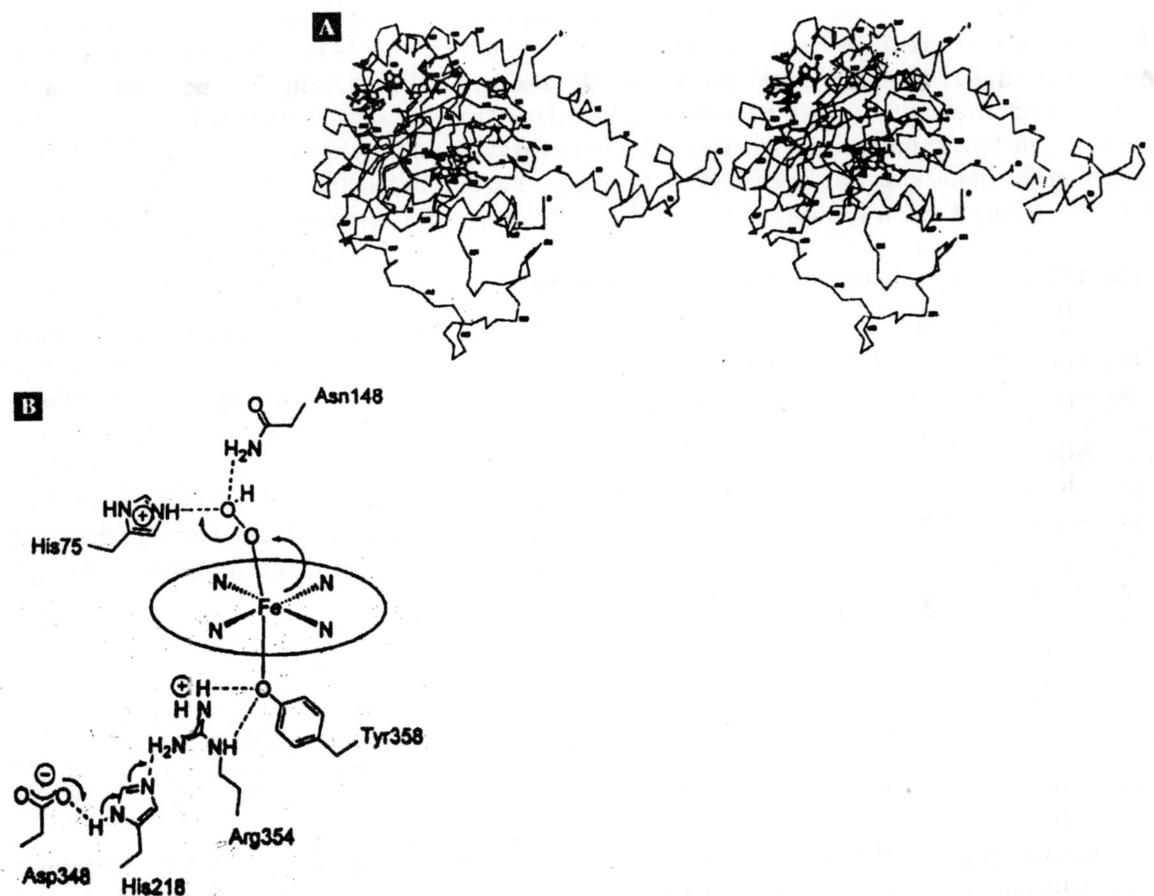

Fig. 4.11. (A) Stereo drawing showing folding pattern for beef liver catalase and the positions of the NADPH (upper left) and heme (center). From Fita and Rossmann. (B) Diagram of proposed structure of an Fe(III)-OOH ferric peroxide complex of human catalase (see also Fig. 4.14). A possible mechanism by which the peroxide is cleaved indicated by the arrows. His 75 and Asn 148 are directly involved, and a charge relay system below the ring may also participate.

Peroxidases and catalases contain high-spin Fe(III) and resemble metmyoglobin in properties. The enzymes are reducible to the Fe(II) state in which form they are able to combine (irreversibly) with O_2. We see that the same active center found in myoglobin and hemoglobin is present but its chemistry has been modified by the proteins. The affinity for O_2 has been altered drastically and a new group of catalytic activities for ferriheme-containing proteins has emerged.

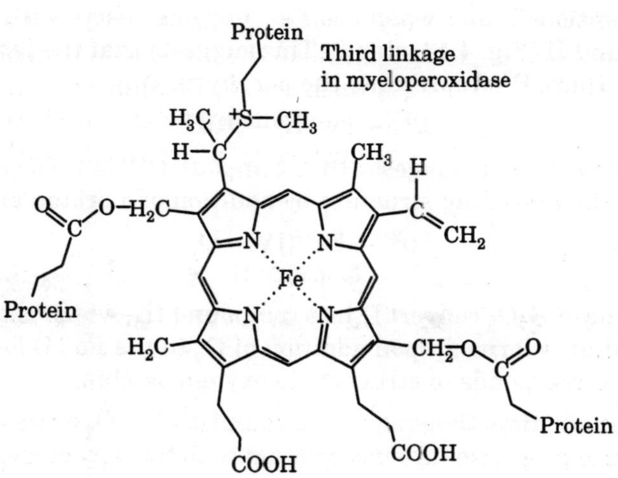

Fig. 4.12. Linkage of heme to mammalian peroxidases. There are two ester linkages to carboxylate side chains from the protein. Myeloperoxidase contains a third linkage.

Mechanisms of Catalase and Peroxidase Catalysis

Attention has been focused on a series of strikingly coloured intermediates formed in the presence of substrates. When a slight excess of H_2O_2 is added to a solution of horseradish peroxidase, the dark brown enzyme first turns olive green as *compound I* is formed, and then pale red as it turns into *compound II*. The latter reacts slowly with substrate AH_2 or with another H_2O_2 molecule to regenerate the original enzyme. This sequence of reactions is indicated by the coloured arrows in Fig. 4.14 step *a* – *d*. Titrations with such reducing agents as ferrocyanide or K_2IrBr_6 have established that compound I is converted into compound II by a one-electron reduction and compound II to free peroxidase by another one-electron reduction.

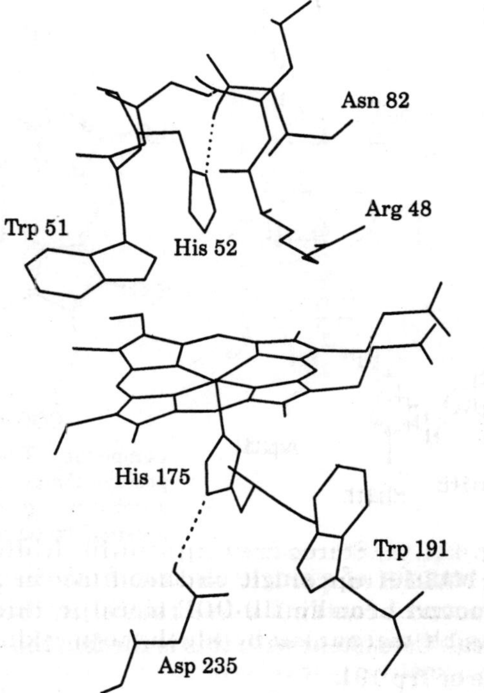

Fig. 4.13. The active site of yeast cytochrome *c* peroxidase. Access for substrates is through a channel above the front edge of the heme ring as viewed by the reader. A pathway for entrance of electrons may be via Trp 191 and His 175.

Thus, the iron in compound I may formally.be designated Fe(V) and that in II as Fe(IV). However, this does not tell us whether or not the oxygen atoms of H_2O_2 are present in compounds I and II. The enzyme in the Fe^{3+} form can be reduced to

Fe^{2+}, as previously mentioned, and when the Fe^{2+} enzyme reacts with H_2O_2 it is apparently converted into compound II (Fig. 4.14, step *f*). This suggests that the latter is an Fe^{2+} complex of the peroxide anion. Here, P^{2-} represents the porphyrin ring:

$$P^{2-} — Fe^{+}(II) — OOH$$

However, spectroscopic evidence suggests that compound II is a *ferryl iron* complex which could be derived from the preceding structure by addition of a proton and loss of water.

$$p^{2-} — Fe^{2+}(IV) = O$$

Compound II

High concentrations of H_2O_2 convert II into compound III, which is thought to be the same as the *oxyperoxidase* that is formed upon addition of O_2 to the Fe(II) form of the free enzyme (Fig. 4.14, step *g*) and corresponds in structure to oxyhemoglobin.

Compound I was at one time thought to be a complex of H_2O_2 or its anion with Fe(III), but its magnetic and spectral properties are inconsistent with this structure. Rather, it too appears to contain ferryl iron bound to an electron-deficient porphyrin π-cation radical. The reaction with peroxide probably involves initial formation of a peroxide anion complex (Fig. 4.14 step *a*)which is cleaved with release of water (step *b*). The resulting $Fe(V)^{+} = O$ compound is converted to compound

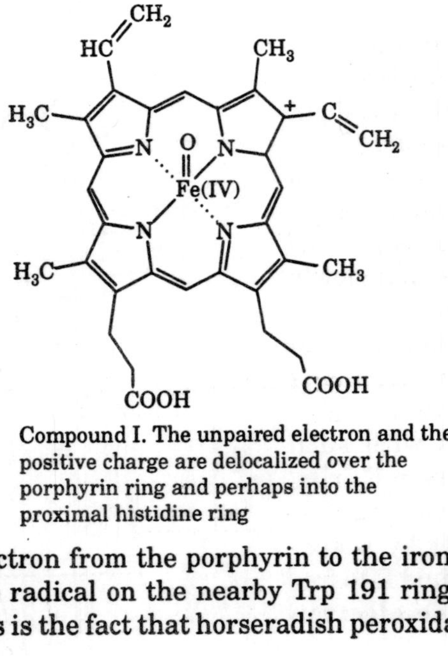

Compound I. The unpaired electron and the positive charge are delocalized over the porphyrin ring and perhaps into the proximal histidine ring

I by transfer of a single electron from the porphyrin to the iron. In cytochrome *c* peroxidase compound I contains a free radical on the nearby Trp 191 ring instead of on the porphyrin radical. Consistent with this is the fact that horseradish peroxidase contains phenylalanine in place of Trp 191.

If we consider the fate of substrate AH_2 during the action of a peroxidase, we see that donation of an electron to compound I to convert it into II (Fig. 4.14 step *c*) will generate a free radical •AH as well as a proton. The radical may then donate a second electron to II to form the free enzyme. Alternatively, a second molecule of AH_2 may react (Fig. 4.14 step *c*) to form a second radical •AH. The two •AH radicals may then disproportionate to form A and AH_2 or they may leave the enzyme and react with other molecules in their environment. Compound II of horse radish peroxidase is able to exchange the oxygen atom of its Fe(IV) = O center with water rapidly at pH 7, presumably by donation of a proton from the nearby histidine side chain (corresponding to it is 52 of Fig. 4.13). This histidine presumably also functions in proton transfer during reactions with substrates.

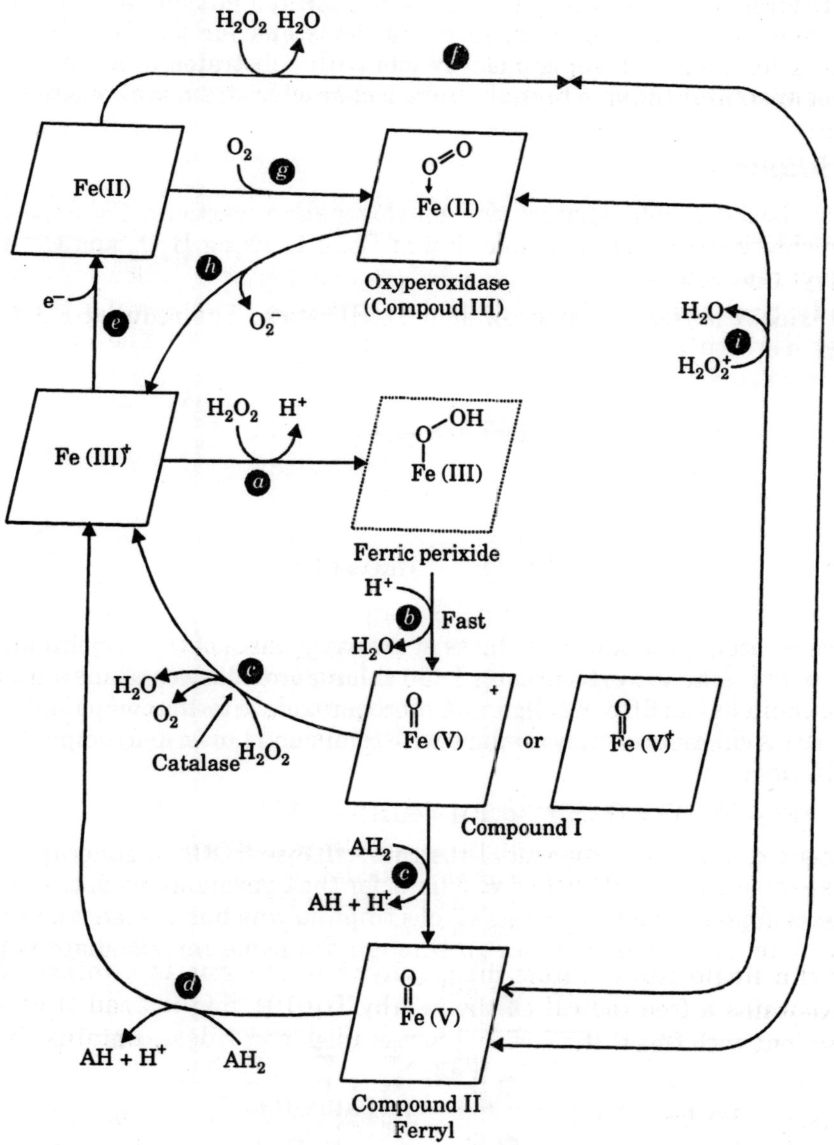

Fig. 4.14. The catalytic cycles and other reactions of peroxidases and catalases. The principal cycle for peroxidases is given by the coloured arrows. That of catalases is smaller, making use of step *a*, *b*, and *c'*, which is marked by a light green line.

The catalase compound I appears to be converted in a two-electron reduction by H_2O_2 directly to free ferricatalase without intervention of compound II (Fig. 4.14 step *c*). The catalytic histidine probably donates a proton to help form water from one of the oxygen

$$\bullet P^{+}\text{—}Fe(IV) = O + H_2O_2 \rightarrow H_2O + O_2 + P\text{—}Fe(III)^{+} \text{ Compound I} \qquad ...(4.10)$$

atoms of the H_2O_2. Nevertheless, compound II does form slowly, especially if a slow substrate such as ethanol is present. The previously mentioned bound NADPH apparently reduces compound II formed in this way, converting the inactivated enzyme back to active catalase. This may involve unusual one-electron oxidation steps for the NADPH. Under some circumstances compound II of peroxidases reacts with substrates in a two-electron process with transfer of an oxygen atom to the substrate, a characteristic also of reactions catalyzed by cytochromes.

Haloperoxidases

Many specialized peroxidases are active in halogenation reactions. *Chloroperoxidases* from fungi catalyze chlorination reactions like that of Eq. 4.11 using H_2O_2 and Cl^- as well as the usual peroxidase reaction.

Chloroperoxidase is isolated in a low-spin Fe(III) state. The reduced Fe(II) enzyme is a high-spin form

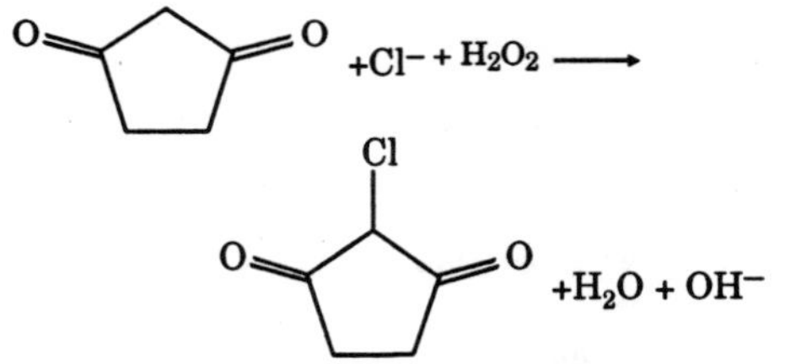

O O $+Cl^- + H_2O_2 \longrightarrow$

Cl

O O $+H_2O + OH^-$ ...(4.11)

with spectroscopic properties similar to those of the oxygenases of the cytochrome P450 family, which are discussed. Like the cytochromes P450 chloroperoxidase contains a thiolate group of a cysteine side chain as the fifth iron ligand. Chloroperoxidase forms compounds I and II, as do other peroxidases. A chloride ion may combine with compound I to form a complex of hypochlorite with the Fe(III) heme.

$$\bullet P^{+}\text{—} Fe(IV) = O + Cl^- \rightarrow P\text{—}Fe(III)^{+}\text{—}OC1 \qquad (4.12)$$

This intermediate could then halogenate substrate AH, lose HOCl, or generate Cl_2 by reaction with Cl^-. These are all well-established reactions for the enzyme. In each case the chlorine in the peroxidase complex can be viewed as an electrophile which is transferred to an attacking nucleophile. A fourth reaction that can go through the same intermediate is conversion of

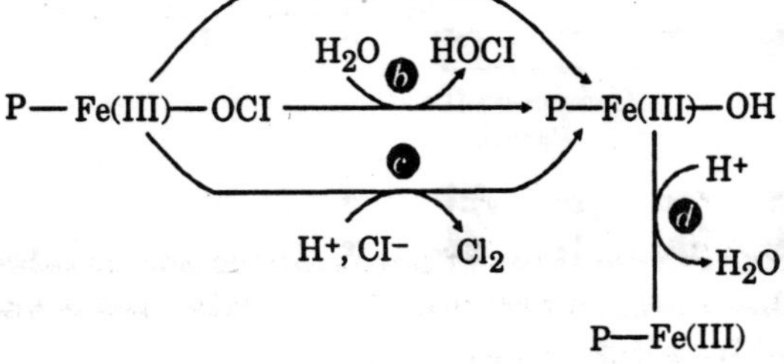

AH (a) ACI

H_2O (b) HOCI

P—Fe(III)—OCI ⟶ P—Fe(III)—OH

(c) H^+

H^+, CI^- Cl_2 (d) H_2O ...(4.13)

P—Fe(III)

alkenes to α,β-halohydrins. *Lactoperoxidase* of milk, reacting with I^- + H_2O_2, promotes an analogous iodination of tyrosine and histidine residues of proteins. With radioactive $^{125}I^-$ or $^{131}I^-$ it provides a convenient and much used method for labeling of proteins in exposed surfaces

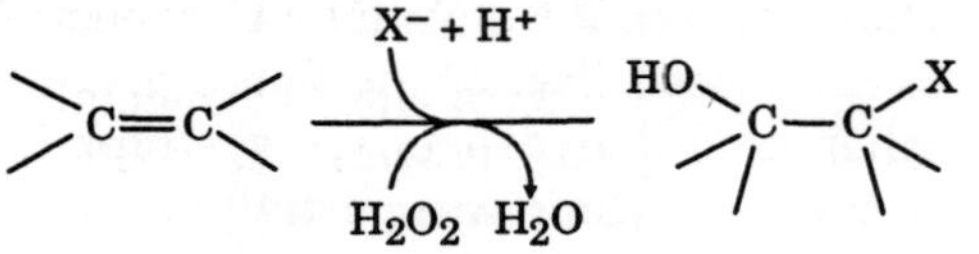

$$>C=C< \xrightarrow[H_2O_2 \;\; H_2O]{X^- + H^+} HO-C-C-X \quad ...(4.14)$$

I⁻, H_2O_2 peroxidase

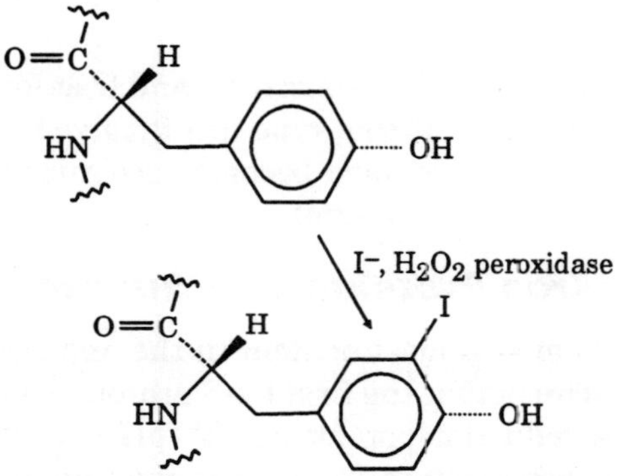

...(4.15)

of membranes (Eq. 4.15). lodinated tyrosine derivatives are formed in the thyroid gland by a similar reaction catalyzed by *thyroid peroxidase*. Even horseradish peroxidase can oxidize iodide ions but neither it nor lactoperoxidase will carry out chlorination or bromination reactions. *Myeloperoxidase*, present in specialized lysosomes of polymorphonuclear leukocytes (neutrophils), utilizes H_2O_2 and a halide ion to kill ingested bacteria.

Phagocytosis induces increased respiration by the leukocyte and generation of H_2O_2, partly by the membrane-bound NADPH oxidase described. Some of the H_2O_2 is used by myeloperoxidase to attack the bacteria, apparently through generation of HOCl by peroxidation of Cl^-. Human myeloperoxidase is a tetramer of two 466-residue chains and two 108-residue chains, which carry the covalently linked heme. Another oxygen-dependent killing mechanism that may also be used by neutrophils is the generation of the reactive *singlet oxygen*. This can occur by reaction of hypochlorite with H_2O_2 (Eq. 4.16) from an enzyme-bound hypochlorite intermediate such as the that shown in Eq. 4.13. Hereditary deficiency of myeloperoxidase is relatively common.

$$OCl^- + H_2O_2 \rightarrow \underset{\text{singlet } O_2}{O_2(^1\Delta g)} + H_2O + Cl^- \quad ...(4.16)$$

Lactoperoxidase and chloroperoxidase also generate singlet oxygen. The possible biological significance is discussed. Eosinophil peroxidase appears to promote formation of *hydroxyl radicals*. *Bromoperoxidases* are found in many red and green marine algae.

Many of them contain *vanadium* and function by a mechanism different than that used by heme peroxidases. Another related nonheme enzyme is the selenoprotein glutathione peroxidase. It reacts by a mechanism very different those discussed here, as does *NADPH peroxidase*, a flavoprotein with a cysteine sulfinate side chain in the active site. A lignin-degrading peroxidase from the white wood rot fungus *Phanewchaete chrysosporium* is a simple heme protein, while other peroxidases secreted by this organism contain Mn.

The Iron-Sulfur Proteins

Not all of the iron within cells is chelated by porphyrin groups. Hemerythrin has been known for many years, but the general significance of nonheme iron proteins was not appreciated until large-scale preparation of mitochondria was developed by Crane in about 1945. The iron content of mitochondria was found to far exceed that of the heme proteins present. In 1960, Beinert, who was studying the mitochondrial dehydrogenase systems for succinate and for NADH, observed that when the electron transport chain was partially reduced by these substrates and the solutions were frozen at low temperature and examined, a strong EPR signal was observed at $g = 1.94$.

The signal was obtained only upon reduction by substrate, and fractionation pointed to the nonheme iron proteins. Six or more proteins of this type are involved in the mitochondrial electron transport chain, and numerous others have become recognized as members of the same large family of *iron-sulfur proteins* (Fe-S proteins).

Ferredoxins, High-potential Iron Proteins, and Rubredoxins

The presence of nonheme iron proteins is most evident in the anaerobic clostridia, which contain no heme. It was from these bacteria that the first Fe-S protein was isolated and named *ferredoxin*. This protein has a very low reduction potential of $E°'$(pH 7) = –0.41 V. It participates in the pyruvate-ferredoxin oxidoreductase reaction, in nitrogen fixation in some species, and in formation of H_2. A small green-brown protein, the ferredoxin of *Peptococcus aerogenes* contains only 54 amino acids but complexes eight atoms of iron. If the pH is lowered to ~1, eight molecules of H_2S are released. Thus, the protein contains eight "labile sulfur" atoms in an iron sulfide linkage. There are also eight iron atoms.

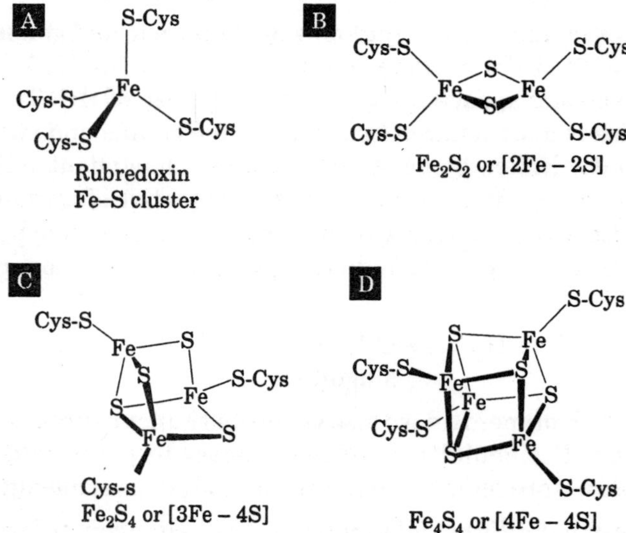

Fig. 4.15. Four different iron-sulfur clusters of a type found in many proteins.

Another group of related electron carriers, the *high-potential iron proteins* (HIPIP) contain four labile sulfur and four iron atoms per peptide chain. X-ray studies showed that the 86-residue polypeptide chain of the HIPIP of *Chromatium* is wrapped around a single *iron-sulfur cluster* which contains the side chains of four cysteine residues plus the four iron and four sulfur atoms (Fig. 4.15D). This kind of cluster is referred to as [4Fe-4S], or as Fe_4S_4. Each cysteine sulfur is attached to one atom of Fe, with the four iron atoms forming an irregular tetrahedron with an Fe-Fe distance of ~0.28 nm. The four labile sulfur atoms (S^{2-}) form an interpenetrating tetrahedron 0.35 nm on a side with each of the sulfur atoms bonded to three iron atoms.

The cluster is ordinarily able to accept only a single electron. The iron-sulfur cluster structure was a surprise, but after its discovery it was found that ions such as $[Fe_4S_4(S\text{-}CH_2CH_2COO\text{-})_4]^{6-}$ assemble spontaneously from their components and have a similar cluster structure. Thus, living things have simply improved upon a natural bonding arrangement. The bacterial ferredoxins from *Peptococcus* (Fig. 4.16B), *Clostridium*, *Desnlfovibrio,* and other anaerobes each contain two Fe_4S_4 clusters with essentially the same structure as that of the *Chroniatium* HIPIP. Each cluster can accept one electron. Much of the amino acid sequence in the first half of the ferredoxin chain is repeated in the second half, suggesting that the chain may have originated as a result of gene duplication.

Many invariant positions are present in the sequence, including those of the cysteine residues forming the Fe-S cluster. Ferredoxins with single Fe_4S_4 clusters are also known. The ferredoxins have reduction potentials $E^{\circ\prime}$ (pH 7) from about –0.4 V to as low as –0.6 V. However, the corresponding values for HIPIP proteins range from +0.05 to +0.50 V at pH 7. This wide range of potentials initially seemed strange because the structures of the active centers of both the clostridial ferredoxins and the *Chrainatinm* HIPIP appear virtually identical. Part of the explanation lies in the fact that Fe_4S_4 clusters can exist in three oxidation states (Eq. 4.17) that differ, one from another, by a single electron.

$$[Fe_4S_4]^{+} \xrightarrow{-e^-} [Fe_4S_4]^{2+} \xrightarrow{-e^-} [Fe_4S_4]^{3+} \qquad \ldots(4.17)$$

Reduced Fd; Oxidized Fd; reduced high potential iron protein; Oxidized high potential iron protein; Super-oxidized Fd

Here the charges shown are those on the cluster. The cysteine ligands from the protein each add an additional negative charge. The *Chromatium* HIPIP and the clostridial ferredoxins have the middle oxidation state in common. The cluster is a little smaller in the more oxidized states; in the *Chromatium* HIPIP the Fe-Fe distance changes from 0.281 to 0.272 nm upon oxidation.

Rubredoxins

The simplest of the Fe-S proteins are the rubredoxins. These proteins contain iron but no labile sulfur. The rubredoxin of *Clostridium pasteur-ianum* is a 54-residue peptide containing four cysteines whose side chains form a distorted tetrahedron about a single iron atom (Fig. 4.16A). Not shown for any of the structures in Fig. 4.16 are NH—S hydrogen bonds that connect backbone NH groups of the peptide chain to the sulfur atoms of the cysteine groups, forming the clusters. These bonds may have important effects on properties of the cluster. Rubredoxins also participate in electron transport and can substitute for ferredoxins in some reactions. Larger 14-kDa and 18-kDa rubredoxins able to bind two iron ions participate in electron transport in a hydroxylase system of *Pseudomoms.* A smaller 7.9-kDa 2-Fe *desulforedoxin* functions in sulfate-reducing bacteria.

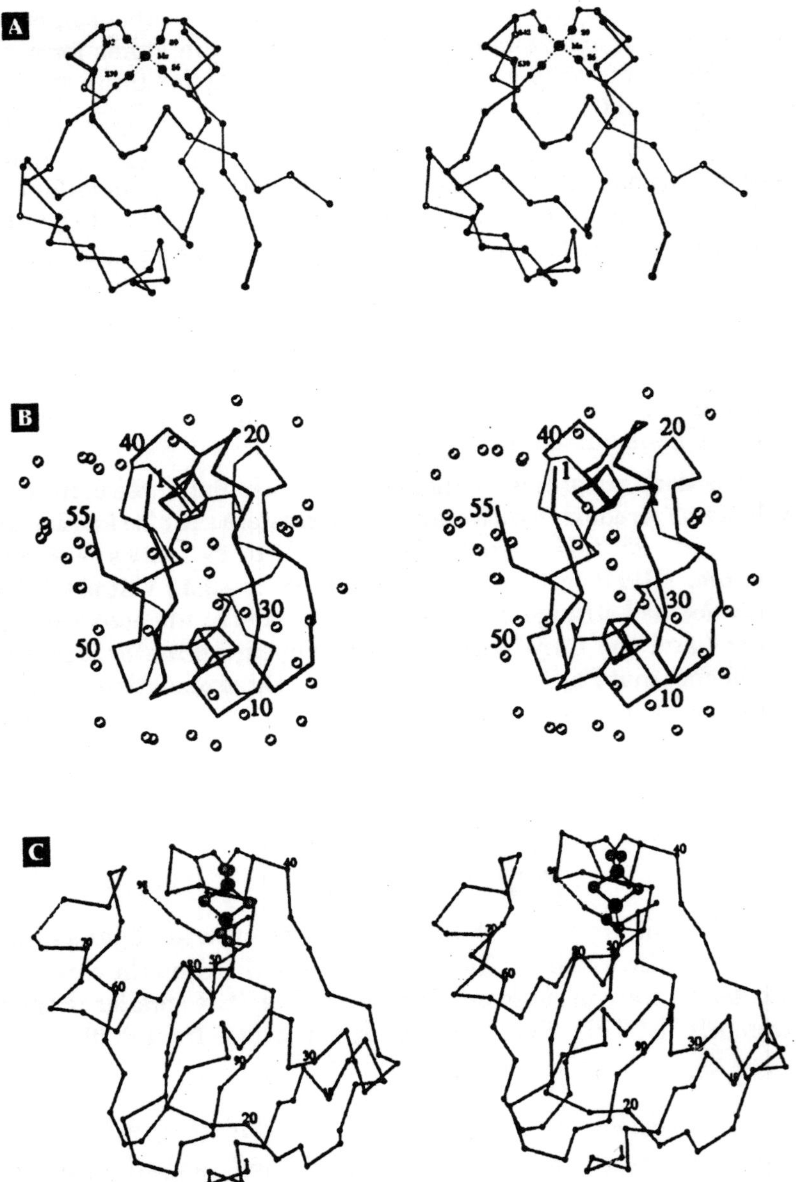

Fig. 4.16. (A) Superimposed stereoscopic α-carbon traces of the peptide chain of rubredoxin from *Clostridium pasteurianum* with either Fe^{3+} (solid circles) or Zn^{2+} (open circles) bound by four cysteine side chains. From Dauter *et al.* (B) Alpha-carbon trace for ferredoxin from *Clostridium acidurici*. The two Fe_4S_4 clusters attached to eight cysteine side chains are also shown. The open circles are water molecules. Based on a high-resolution X-ray structure by Duee *et al.* Courtesy of E. D. Duee. (C) Polypeptide chain of a chloroplast-type ferredoxin from the cyanobacterium *Spirulina platensis*. The Fe_2S_2 cluster is visible at the top of the molecule.

Chloroplast-type Ferredoxins

Members of a large class of [2Fe-2S] or Fe_2S_2 ferredoxins each contain two iron atoms and two labile sulfur atoms with the linear structure of 4.15B. Best known are the chloroplast ferredoxins, which transfer electrons from photosynthetic centers of chloroplasts to the flavoprotein reductase that reduces $NADP^+$ to NADPH. The structure of a cyanobacterial protein of this type in shown in Fig. 4.16C. A second group of Fe_2S_2 ferredoxins are found in bacteria including *E. coli* and in human mitochondria.

For example, in steroid hormone-forming tissues the ferredoxin *adrenodoxin* carries electrons to cytochromes P450. Its Fe_2S_2 center receives electrons from adrenodoxin reductase. The nitrogenfixing *Clostridhim pasteuricmum* also contains a ferredoxin of this class. In *Pseudomonas putida* (Chapter 18) the related 106-residue *putidaredoxin* transfers electrons to cytochromes P450.

The 3Fe-4S clusters

A 106-residue ferredoxin from *Azotobacter vinelandii* contains seven iron atoms in two Fe-S clusters that operate at very different redox potentials of-0.42 and +0.32 V. Other similar seven-iron proteins are known. From EPR measurements it appeared that both clusters function between the 2^+ and 3^+ (oxidized and superoxidized) states of Eq. 4.17, despite the widely differing potentials. A super-reduced all-Fe(H) form with $E^{\circ\prime}$ (pH 7) = –0.70 V can also be formed. X-ray crystallographic studies have revealed that the protein contains one Fe_4S_4 cluster (Fig. 4.15C). The structures and environments of the Fe_3S_4 clusters are similar to those of Fe_4S_4 clusters but they lack one iron and one Fe_3S_4 cluster.

The structures and environments of the Fe_3S_4 clusters are similar to those of Fe_4S_4 clusters but they lack one iron and one cysteine side chain. An Fe_4S_4 cluster may sometimes lose S^{2-} to form an Fe_3S_4 cluster such as the one in the *A. vinelandii* ferredoxin. A less likely possibility is isomerization to a linear Fe_3S_4 structure.

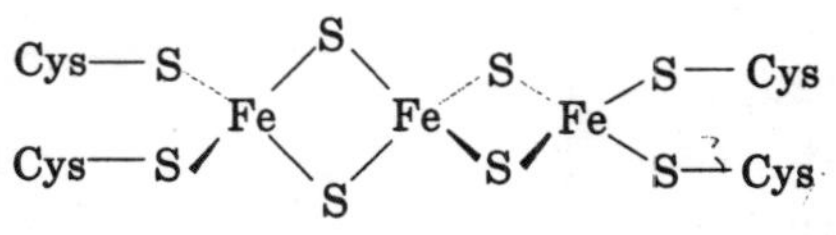

Cys—S　　S　　S　　S—Cys
Fe　　Fe　　Fe
Cys—S　　S　　S　　S—Cys
S

Aconitase isolated under aerobic conditions contains an Fe_3S_4 cluster and is catalytically inactive. Incubation with Fe^{2+} activates the enzyme and recon-verts the Fe_3S_4 to an Fe_4S_4 cluster.

Properties of Iron-sulfur Clusters

These clusters were viewed for many years as unstable and unable to exist outside of a protein. However, if protected from oxygen and manipulated in the presence of soluble organic thiols they are stable and "cofactor-like." Intact clusters can be "extruded" from proteins by treatment with thiols in nonaqueous media. Both Fe_4S_4 and Fe_2S_2 clusters as well as more complex forms have been synthesized and nonenzymatic cluster interconversions have been demonstrated.

Binding to proteins stabilizes the clusters further, but some (Fe-S) proteins are labile and difficult to study. This is evidently because of partial exposure of the cluster to the surrounding solvent. Not only can O_2 cause oxidation of exposed clusters but also superoxide, nitric oxide, and peroxynitrite can react with the iron. Aconitase has only three cysteine side chains available for coördination with Fe and the protein is unstable. Apparently, a superoxidized $[Fe_4S_4]^{3+}$ cluster is formed in the presence of O, but loses Fe^{2+} to give an $[Fe_3S_4]^+$ cluster.

Another interesting cluster conversion is the joining of two Fe_2S_2 clusters in a protein to form a single Fe_4S_4 cluster at the interface between a dimeric protein. Such a cluster is present in the nitrogenase iron protein probably also in biotin synthase.

The clusters in such proteins can also be split to release the monomers. Synthetic iron-sulfur clusters have weakly basic properties and accept protons with a pK_a of from 3.9 to 7.4. Similarly, one clostridial ferredoxin, in the oxidized form, has a pK_a of 7.4; it is shifted to 8.9 in the reduced form. If we designate the low-pH oxidized form of such a protein as HOx^+ and the reduced form as HRed, we can depict the reduction of each Fe_4S_4 cluster as follows.

$$HOx^+ + e^- \rightarrow Hred \qquad (4.18)$$

Comparing with using the Michaelis pH functions for HOx^+ and HRed, it is easy to show that the value of $E^{\circ\prime}$ ($E_{1/2}$) at which equal amounts of oxidized protein (HOx^+ + Ox) and reduced protein (HRed + Red^-) are present is given, in which K_{ox} and K_{red} are the K_a values for dissociation of the protonated oxidized and reduced forms, respectively.

$$E_{1/2} = E^0(\text{low pH}) + 0.0592 \log \left[\frac{(1+K_{red}/[H^+])}{(1+K_{ox}/[H^+])}\right] V \qquad ...(4.19)$$

At the high pH limit this becomes $E_{1/2} = E^\circ$ (low pH) + 0.0592 ($pK_{ox} - pK_{red}$) and V = −0.371 + 0.0592 (7.4-8.9) V = −0.431 V. Thus, the value of $E_{1/2}$ changes from -0.371 to −0.460 V as the pH is increased. In the pH range between the pK_a values of 7.4 and 8.9 reduction of the protein will lead to binding of a proton from the medium and oxidation to loss of a proton. Human and other vertebrate ferredoxins also show pH-dependent redox potentials. This suggests, as with the cytochromes, a possible role of Fe-S centers in the operation of proton pumps in membranes.

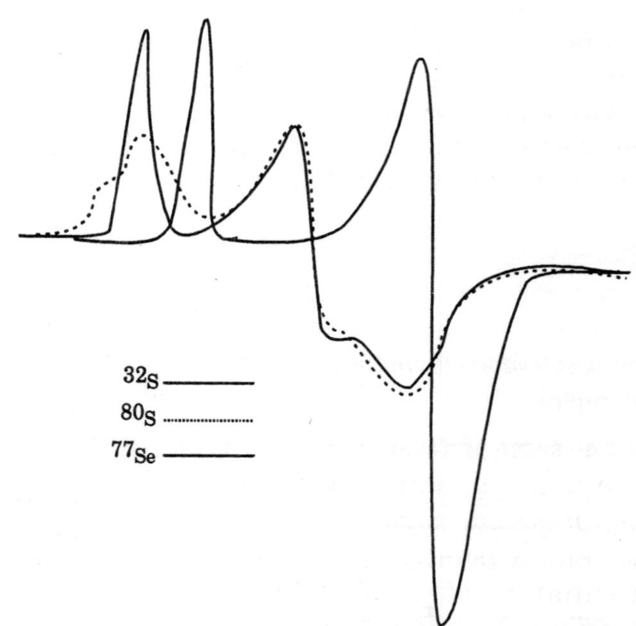

Fig. 4.17. Electron paramagnetic resonance spectrum of the Fe-S protein putidaredoxin in the natural form (^{32}S) and with labile sulfur replaced by selenium isotopes. Well-developed shoulders are seen in the low-field end of the spectrum of the ^{77}Se (spin = 1/2)-containing protein.

Nevertheless, many ferredoxins, such as that of *C. pasteurianiim,* show a constant value of $E^{\circ\prime}$ from pH 6.3 to 10 and appear to be purely electron carriers. Both the iron and labile sulfur can be removed from Fe-S proteins and the active proteins can often be reconstituted by adding sulfide and Fe^{2+} ions.

Using this approach, the natural isotope ^{56}Fe (nuclear spin zero) has been exchanged with ^{57}Fe, which has a magnetic nucleus and ^{32}S has been replaced by ^{77}Se. The resulting proteins appear to function naturally and give EPR spectra containing hyperfine lines that result from interaction of these nuclei with unpaired electrons in the Fe_4S_4 clusters (Fig. 4.17). These observations suggest that electrons accepted by Fe_4S_4 clusters are not localized on a single type of atom but interact with nuclei of both Fe and S. The native proteins as well as many mutant forms are being studied by NMR and other spectroscopic techniques, by theoretical computations, and by protein engineering and "rational design."

Functions of Iron-sulfur Enzymes

Numerous iron-sulfur clusters are present within the membrane-bound electron transport chains discussed. Of special interest is the Fe_2S_2 cluster present in a protein isolated from the cytochrome be complex (complex III) of mitochondria. First purified by Rieske *et al.*, this protein is often called the *Rieske iron-sulfur protein*. Similar proteins are found in cytochrome be complexes of chloroplasts.

In some bacteria Rieske-type proteins deliver electrons to oxygenases. The 196-residue mitochondrial protein has an unusually high midpoint potential, E_m of ~ 0.30 V. The Fe_2S_2 cluster, which is visible in the atomic structure of complex III, is coordinated by two cysteine thiolates and two histidine side chains. In a *Rhodobacter* protein they occur in the following conserved sequences: C133-T-H-L-G-C138 and C153-P-C-H-G-S158. One iron is bound by C133 and C153 and the other by H135 and H156 as follows:

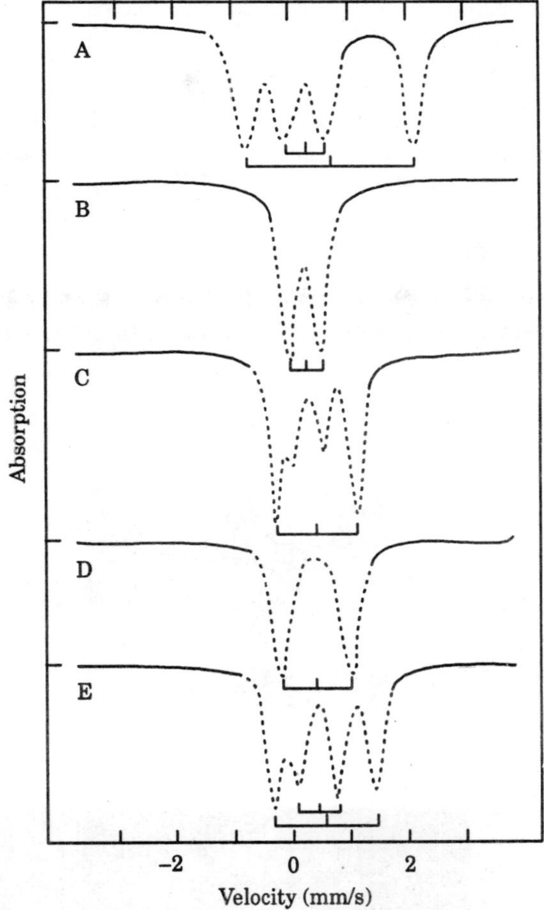

Fig. 4.18. Mossbauer X-ray absorption spectra of ironsulfur clusters. Quadrupole doublets are indicated by brackets and isomer shifts are marked by triangles. (A) $[Fe_2S_2]^{1+}$ cluster of the Rieske protein from *Pseudonionns mendocina*, at temperature T = 200 K. (B) $[Fe_3S_4]^{1+}$ state of D. *gigas* ferredoxin II, *T* = 90 K. (C) $[Fe_3S_4$f state of *D. gigas* ferredoxin II, T = 15K. (D) $[Fe_4S_4]^{2+}$ cluster of *E. coli* FNR protein, *T* = 4.2 K. (E) $[Fe_4S_4]^{1+}$ cluster of *E. coli* sulfite reductase, *T* = 110 K. From Beinert et *al.*

H135

C133 — S, S, Fe(III), Fe(II), N, N—H

C133 — S, S, N, N—H

H156

These proteins may also have an ionizable group with a pK_a of ~ 8.0, perhaps from one of the histidines that is linked to the oxidation-reduction reaction.

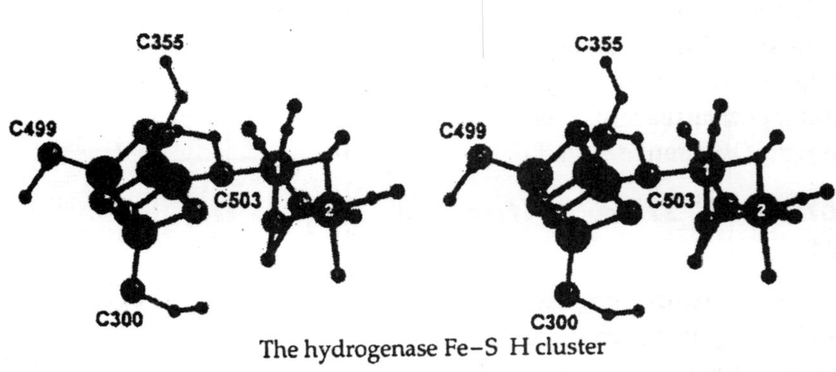

The hydrogenase Fe–S H cluster

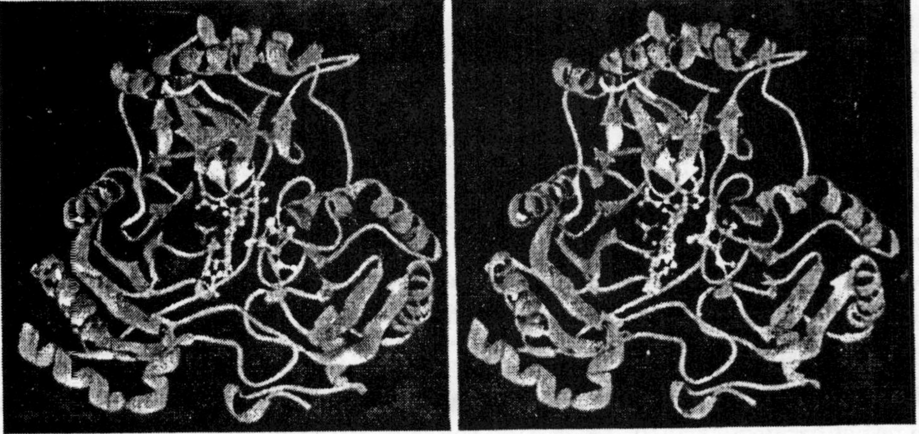

Fig. 4.19. Stereoscopic view of *E. coli* assimilatory sulfite reductase. The siroheme the center with one edge toward the viewer and the Fe_4S_4 cluster is visible on its right side is in. A single S atom from a cysteine side chain bridges between the Fe of the siroheme and the Fe_4S_4 cl

Iron-sulfur clusters are found in flavoproteins such as NADH dehydrogenase and trimethylamine dehydrogenase the siroheme-containing *sulfite reductases* and *nitrite reductases*. These two reductases are found both in bacteria and in green plants. Spinach nitrite reductase, which is considered further, utilizes reduced ferredoxin to carry out a six-electron reduction of NO_2^- to NH_3 or of SO_3^{2-} to S^{2-}. The 61-kDa monomeric enzyme contains one siroheme and one Fe_4S_4 cluster.

A sulfite reductase from *E. coli* utilizes NADPH as the reductant. It is a large $\beta_a\alpha_4$ oligomer. The 66-kDa a chains contain bound flavin (4 FAD + 4 FMN), while the 64-kDa (3 subunits contain both siroheme and a neighboring Fe_4S_4 cluster (Fig. 4.19). The iron of the siroheme and the closest iron atom of the cluster are bridged by a single sulfur atom of a cysteine side chain. A somewhat similar double cluster is present in *all-Fe hydrogenases* from *Clostridium pasterurianum.* These enzymes also contain two or three Fe_4S_4 clusters and, in one case, an Fe_2S_2 cluster. At the presumed active site a special *H cluster* consists of an Fe_3S_4 cluster with one cysteine sulfur atom shared by an adjoining Fe_2S_2 cluster. The role of the iron-sulfur clusters in many of the proteins that we have just considered is primarily one of single-electron transfer. The Fe-S cluster is a place for an electron to rest while waiting for a chance to react. There may sometimes be an associated proton pumping action. In a second group of enzymes, exemplified by aconitase, an iron atom of a cluster functions as a *Lewis acid* in facilitating removal of an –OH group in an α,β dehydration of a carboxylic acid.

A substantial number of other bacterial dehydratases as well as an important plant dihydroxyacid dehydratase also apparently use Fe-S clusters in a catalytic fashion. Fumarases A and B from *E. coli,* L-serine dehydratase of a *Peptostreptococcus* species, and the dihydroxyacid dehydratase may all use their Fe_4S_4 clusters in a manner similar to that of aconitase. However, the Fe-S enzymes that dehydrate R-lactyl-CoA to crotonyl-CoA, 4-hydroxybutyryl-CoA to crotonyl-CoA, and R-2-hydroxyglutaryl-CoA to E-glutaconyl-CoA must act by quite different mechanisms, perhaps similar to those utilized in vitamin B_{12}-dependent reactions. In these enzymes, as in pyruvate formate lyase, the Fe-S center may act as a radical generator. The molybdenum-containing enzymes considered in Section F also contain Fe-S clusters.

Nitrogenases contain a more complex Fe-S-Mo cluster. Carbon monoxide dehydrogenase contains 2 Ni, ~11 Fe, and 14 S^{2-} as well as Zn in a dimeric structure. In these enzymes the Fe-S clusters appear to participate in catalysis by undergoing alternate reduction and oxidation. For a few enzymes such as aconitase and amido-transferases, an Fe-S cluster plays a *regulatory role* in addition to or instead of a catalytic function. Cytosolic aconitase is identical to *iron regulatory proteins* 1 (IRP1), which binds to iron responsive elements in RNA to inhibit translation of genes associated with iron uptake.

A high iron concentration promotes assembly of the Fe_4S_4 cluster. Another example is provided by the *E. coli* transcription factor SOXR. This protein, which controls a cellular defense system against oxygen-derived superoxide radicals, contains two Fe_2S_2 centers. Oxidation of these centers by superoxide radicals appears to induce the transcription of genes encoding superoxide dismutase and other proteins involved in protecting cells against oxidative injury.

The (μ-oxo) Diiron Proteins

Both the ferroxidase center of ferritin the oxygen-carrying hemerythrin are members of a family of diiron proteins with similar active site structures. The pair of iron atoms, with the bridging (μ) ligands such as O_2, HO^-, HOO^-, O_2^{2-}, is often held between four ∝ helices, (as can be seen in (Fig. 4.20*e*). In most cases each iron is ligated by at least one histidine, one glutamate side chain, and frequently a tyrosinate side chain. As many as three side chains may bind to each iron.

In addition, one or two carboxylate groups from glutamate or aspartate side chains bridge to both irons, as does the μ-oxo group, which is typically H_2O or ^-OH. Examples of this structure are illustrated in Figs. 4.4 and 4.20. Although the active sites of all of the diiron proteins appear similar, the chemistry of the catalyzed reactions is varied.

Hemerythrin

When both iron atoms are in the Fe(II) state, hemerythrin, like hemoglobin, functions as a carrier of O_2. In the oxidized Fe(III) *methemerythrin* form the iron atoms are only 0.32 nm apart. Three bridging (μ) groups lie between them: two carboxylate groups and a single oxygen atom which may be either O^{2-} or OH^-. One coordination position on one of the hexacoordinate iron atoms is open and appears to be the site of binding of oxygen. The O_2 is thought to accept two electrons, oxidizing the two iron atoms to Fe(HI) and itself becoming a peroxide dianion O_2^{2-}. The process is completely reversible.

The conversion of the oxygen to a bound peroxide ion is supported by studies of resonance Raman spectra suggest that the peroxide group is protonated. In the diferrous protein the μ-oxo bridge is thought to be an ^-OH group. Upon oxygenation the proton could be shared with or donated to the peroxo group (Eq. 4.20). A similar binding of O_2 as a peroxide dianion appears to occur in the copper-containing hemocyanins.

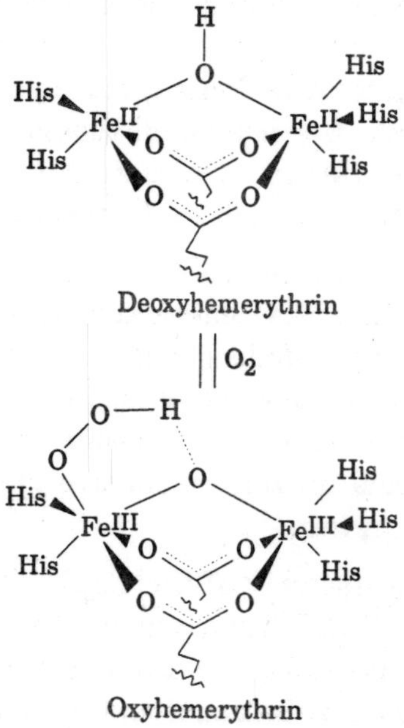

Deoxyhemerythrin

O_2

Oxyhemerythrin

...(4.20)

Purple Acid Phosphatases

Diiron-tyrosinate proteins with acid phosphatase activity occur in mammals, plants, and bacteria. Most are basic glycoproteins with an intense 510- to 550-nm light absorption band. Well-studied members come from beef spleen, from the uterine fluid of pregnant sows (*uteroferrin*), and from human macrophages and osteoclasts. One of the two iron atoms is usually in the Fe(III) oxidation state, but the second can be reduced to Fe(II) by mild reductants such as ascorbate. This half-reduced form is enzymatically active and has a pink colour and a characteristic EPR signal. Treatment with oxidants such as H_2O_2 or hexacyannoferrate (III)

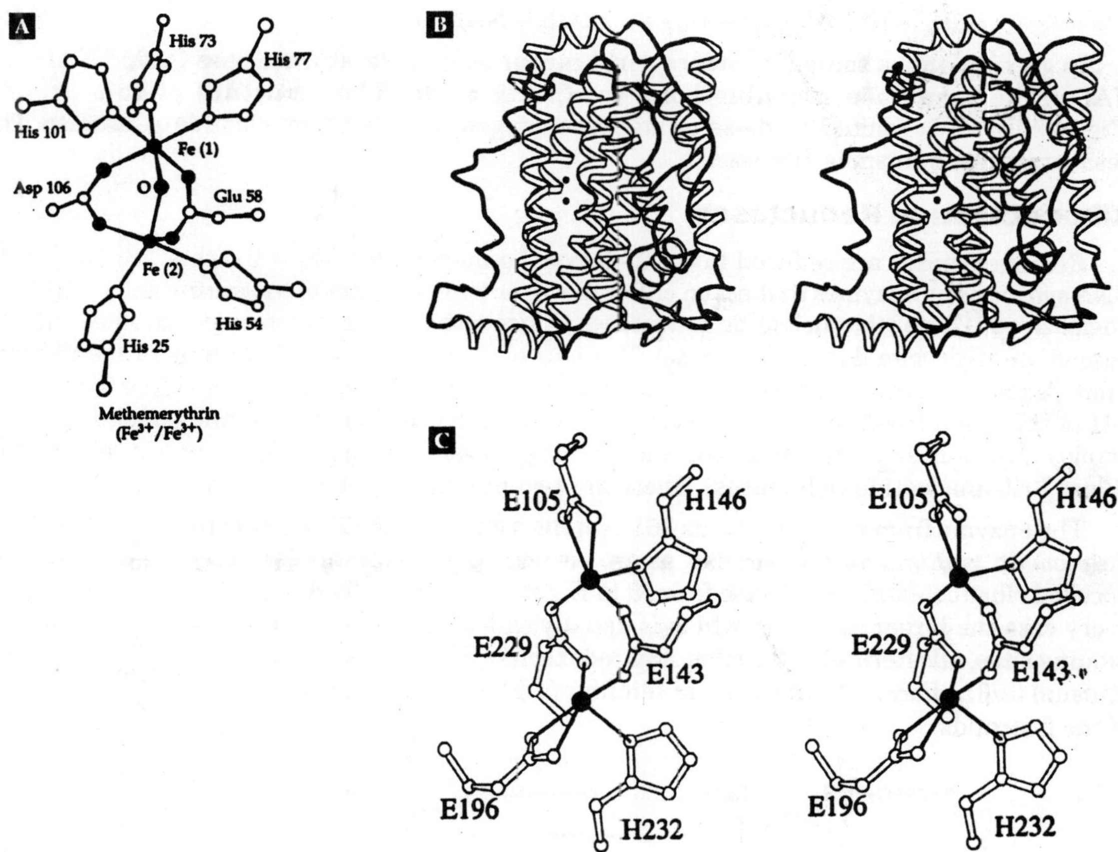

Fig. 4.20. (A) The active site of hemerythrin showing the two iron atoms (green) and their ligands which include the μ oxo bridge and two bridging carboxylate groups. From Lukat *et al.* The active site is between four parallel helices. (B) Stereoscopic view of the backbone structure of a Δ^9 stearoylacyl carrier protein desaturase which also contains a diiron center. Notice the nine-antiparallel-helix bundle. The diiron center is between four helices as in hemerythrin. (C) Stereoscopic view of the diiron center of the desaturase.

generates purple inactive forms which lack a detectable EPR signal. The Fe^{3+} of the active enzyme can be replaced with Ga^{3+} and the Fe^{2+} with Zn^{2+} with retention of activity; also, some plants contain phosphatases with Fe-Zn centers. The catalytic mechanism resembles those of other metallophosphatases the change of oxidation state of the Fe may play a regulatory role. On the other hand, a principal function of uteroferrin may be in transplacental transport of iron to the fetus.

Diiron Oxygenases ane Desaturases

In the (μ-oxo) diiron oxygenases O_2 is initially bound in a manner similar to that in hemerythrin but one atom of the bound O_2 is reduced to H_2O using electrons supplied by a cosubstrate such as NADPH. The other oxygen atom enters the substrate. This is illustrated methane monooxygenase. A toluene monooxygenase has similar properties.

$$CH_4 + NADH + H^+ + O_2 \rightarrow CH_3OH + NAD^+ + H_2O \qquad (4.21)$$

In green plants a soluble Δ^9 stearoylacyl carrier protein desaturase uses O_2 and NADH or NADPH to introduce a double bond into fatty acids. The structure of this protein (Fig. 4.20B, C) is related to those of methane oxygenase and ribonucleotide reductase. The desaturase mechanism is discussed.

Ribonucleotide Reductases

Ribonucleotides are reduced to the 2′-deoxyribo-nucleotides (Eq. 4.21) that are needed for DNA synthesis by enzymes that act on either the di- or triphosphates of the purine and pyrimidine nucleosides. These ribonucleotide reductases utilize either thioredoxin or glutaredoxin as the immediate hydrogen donors (Eq. 4.22). The pair of closely spaced $^-$SH groups in the reduced thioredoxin or glutaredoxin are converted into a disulfide bridge at the same time that the 2′-OH of the ribonucleotide di- (or tri-) phosphate is converted to H_2O. While some organisms employ a vitamin B_{12}- dependent enzyme for this purpose, most utilize iron-tyrosinate enzymes (Class I ribonucleotide reductases). These are two-protein complexes of composition $\alpha_2\beta_2$.

The enzyme from *E. coli* contains 761-residue a chains and 375-residue p chains. That from *Salmonella typhimurium* is similar, as are corresponding mammalian enzymes and a virus-encoded ribonucleotide reductase formed in *E. coli* following infection by T4 bacteriophage. In every case the larger α_2 dimer, which is usually called the *R1 protein*, contains the substrate binding sites, allosteric effector sites, and redoxactive SH groups. Each α chain is folded into an unusual $(\alpha/\beta)_{10}$ barrel. As in the more familiar $(\alpha/\beta)_8$ barrels, the active site is at the N termini of the β strands.

P—P—O—CH2 O Base H OH OH → P—P—O—CH2 O Base H OH H

R(SH)2 Thioredoxin or glutaredoxin → H_2O, RS_2 ...(4.22)

Each polypeptide chain of the β_2 dimer or *R2 protein* contains a diiron center which serves as a free radical generator. A few bacteria utilize a dimanganese center. Oxygenation of this center is linked to the uptake of both a proton and an electron and to the removal of a hydrogen atom from the ring of tyrosine 166 to form H_2O and an organic radical (Eq. 4.23).

$$2Fe(II)^{2+} + O_2 + H^+ + e^- + Y122\text{-}OH \rightarrow Fe(III)^{3+} -O^{2-} Fe(III)^{3+} - Y122\text{-}O\bullet + H_2O \qquad ...(4.23)$$

The tyrosyl radical is used to initiate the ribonucleotide reduction at the active site in the R1 protein ~3.5 nm away. The tyrosyl radical is very stable and was discovered by a characteristic EPR spectrum of isolated enzyme. Alteration of this spectrum when bacteria were grown in deuterated tyrosine indicated that the radical is located on a tyrosyl side chain and that the spin density is delocalized over the tyrosyl ring. Using protein engineering techniques the ring was located as Tyr 122 of the *E. coli* enzyme. A few of the resonance structures that can be used to depict the radical are the following:

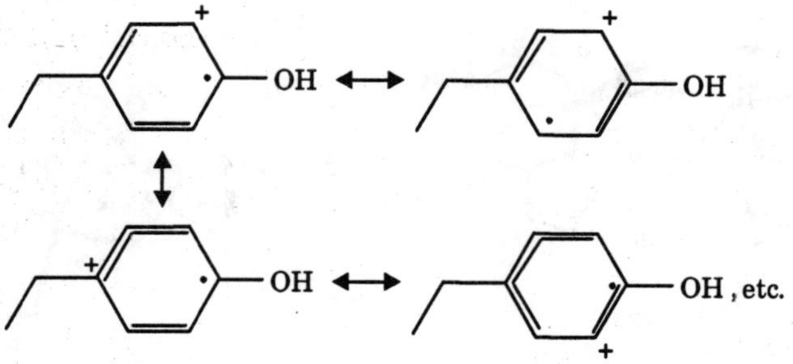

A chain of hydrogen-bonded side chains apparently provides a pathway for transfer of an unpaired electron from the active site to the Tyr 122 radical and from there to the radical generating center. The tyrosyl radical can be destroyed by removal of the iron by exposure to O_2 or by treatment of ribonucleotide reductases with hydroxyurea, which reduces the radical and also destroys catalytic activity:

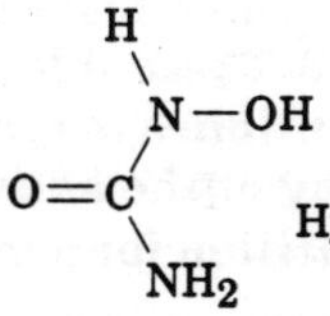

A second group of ribonucleotide reductases (Class II), found in many bacteria, depend upon the cobalt-containing *vitamin B_{12} coenzyme* which is discussed in Section B. These enzymes are monomeric or homodimeric proteins of about the size of the larger a subunits of the Class I enzymes. The radical generating center is the 5′-deoxyadenosyl coenzyme. Class III or *anaerobic nucleotide reductases* are used by various anaerobic bacteria including *E. coli* when grown anaerobically and also by some bacteriophages.

Like the Class I reductases, they have an $\alpha_2\beta_2$ structure but each β subunit contains an Fe_4S_4 cluster which serves as the free radical generator, that forms a stable glycyl radical at G580. In this respect the enzyme resembles pyruvate formatelyase. As with other enzymes using Fe_4S_4 clusters as radical generators, S-adenosylmethionine is also required. All ribonucleotide reductases may operate by similar radical mechanisms. When a 2′-Cl or -F analog of UDP was used in place of the substrate an irreversible side reaction occurred by which Cl^- or F^-, inorganic pyrophosphate, and uracil were released.

When one of these enzyme-activated inhibitors containing ^3H in the 3′ position was tested, the tritium was shifted to the 2′ position with loss of Cl^- and formation of a reactive 3′-carbonyl compound (Eq. 4.24) that can undergo β elimination at each end to give an unsarurated ketone which inactivates the enzyme. This suggested that the Fe-tyrosyl radical abstracts an electron (through a chain of intermediate groups) from an -SH group, now identified as C439 in the *E. coli* enzyme (Fig. 4.25). The resulting thiyl radical is thought to abstract a hydrogen atom from C′-3 of a true substrate to form a substrate radical (Eq. 4.25), with the help of the C462 -SH group, facilitates the loss of ^-OH from C-2 in step *b*.

The resulting C2 radical would be reduced by the nearby redox-active thiol pair C462 and C225 (Fig. 4.21) and Eq. 4.25 step *c*). In Eq. 4.25 the reaction is shown as a hydride transfer with

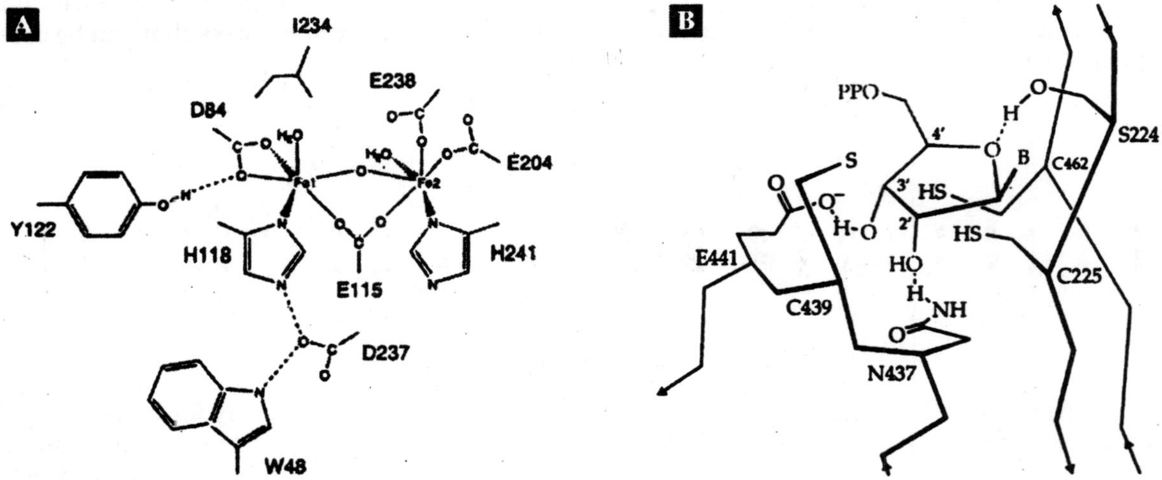

Fig. 4.21. (A) Scheme showing the diiron center of the R2 subunit of *E. coli* ribonucleotide Included are the side chains of tyrosine 122, which loses an electron to form a radical, and of histidine 118, aspartate 237, and tryptophan 48. These side chains provide a pathway for radical transfer to the Rl subunit where the chain continues to tyrosines 738 and 737 and cysteine 429. From Andersson *et al.* (B) Schematic drawing of the active site region of the *E. coli* class III ribonucleotide reductase with a plausible position for a model-built substrate molecule.

an associated one-electron shift but the mechanism is uncertain. In step *d* of Eq. 4.25 the thiyl radical is regenerated and continues to function in subsequent rounds

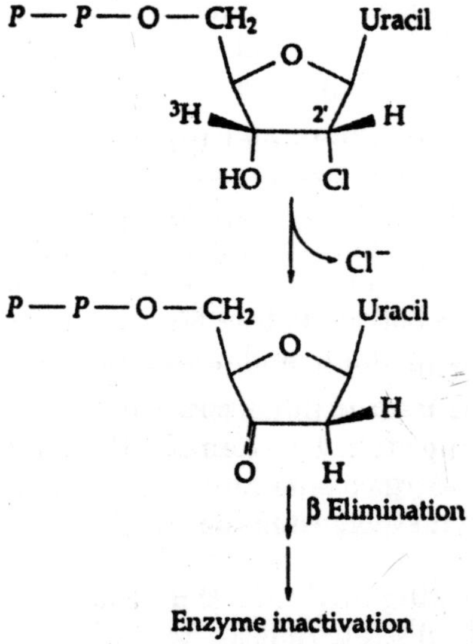

P—P—O—CH₂ … Uracil

β Elimination

Enzyme inactivation

...(4.24)

of catalysis. In the final step (step *e)* the redox active pair is reduced by reduced thioredoxin or glutaredoxin. The active site must open to release the product and to permit this reduction, which may involve participation of still other –SH groups in the protein.

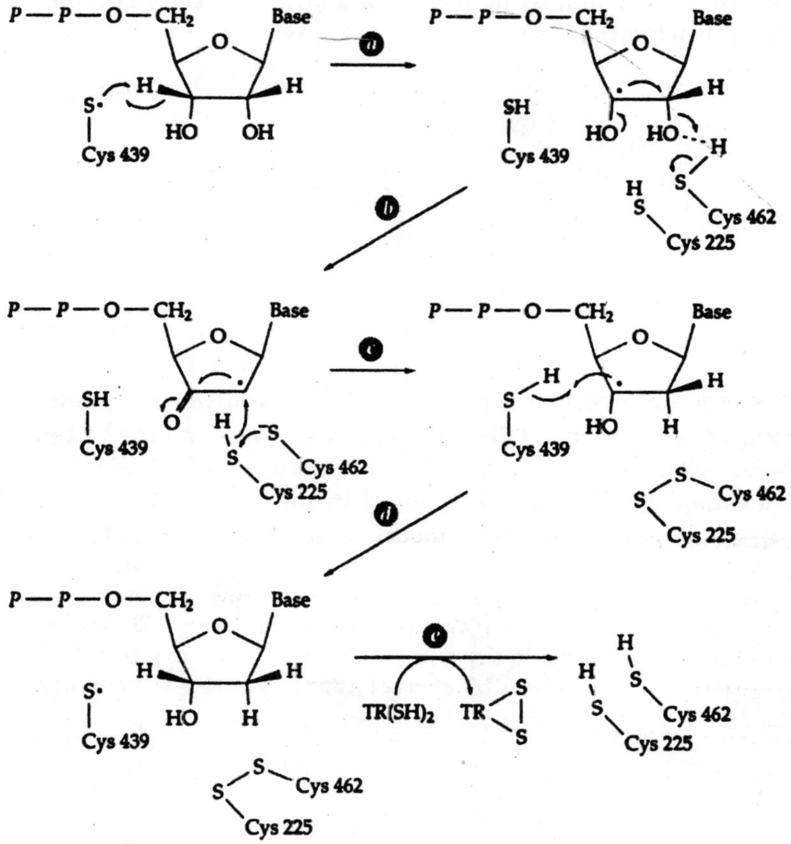

...(4.25)

Superoxide Dismutases

Metalloenzymes of at least three different types catalyze the destruction of superoxide radicals that arise from reactions of oxygen with heme proteins, reduced flavoproteins, and other metalloenzymes. These superoxide dismutases (SODs) convert superoxide anion radicals $^{\bullet}O_2^-$ into H_2O_2 and O_2 Eq. 4.26. The H_2O_2 can then be destroyed by catalase (Eq. 4.8).

$$2O_2^- + 2\,H \rightarrow H_2O_2 + O_2 \qquad ...(4.26)$$

The much studied Cu/Zn superoxide dismutase of eukaryotic cytoplasm is described in Section D. However, éukaryotic mitochondria contain manganese SOD and some eukaryotes also synthesize an iron-containing SOD. For example, the protozoan *Leishmania tropica,* which takes up residence in the phagolysosomes of a victim's macrophages, synthesizes an iron-containing SOD to protect itself against superoxide generated by the macrophages.

Mycobacterium tuber-culosis secretes an iron SOD which assists its survival in living tissues and is also a target for the immune response of human hosts. Iron and manganese SODs have ~ 20-kDa subunits in each of which a single ion of Fe or Mn is bound by three imidazole groups and a carboxylate group. The metal ion undergoes a cyclic change in oxidation state as illustrated by Eq. 4.27. Notice that two protons must be taken up for formation of H_2O_2. In Cu/Zn SOD the copper cycles between Cu^{2+} and Cu^{+}. The structure of the active site of an Fe SOD (is shown in Fig. 4.22). That of Mn SOD is almost identical. In addition to the histidine and carboxylate ligands, the metal binds a hydroxyl ion

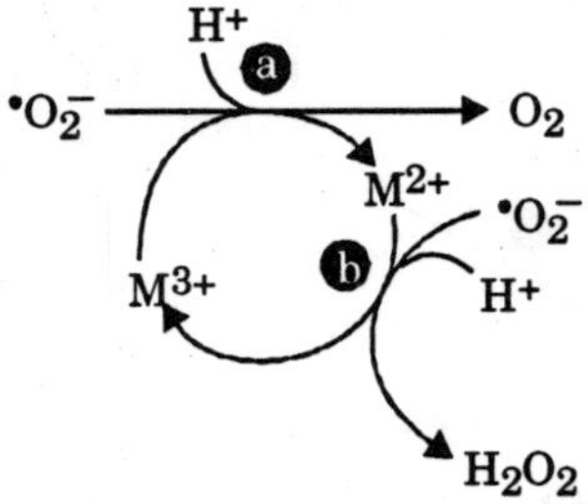

...(4.27)

~OH or H_2O and has a site open for binding of $^{\bullet}O_2^-$. As indicated, uptake of one proton is associated with each reaction step. As illustrated, the first proton may be taken up to convert the bound ~OH to H_2O. The enzyme in the Fe^{2+} form has a pK_a of 8.5 that has been associated with tyrosine 34. Perhaps this residue is involved in the proton uptake process. A similarly located Tyr 41 from *Sulfolobus* is covalently modified, perhaps by phosphorylation.

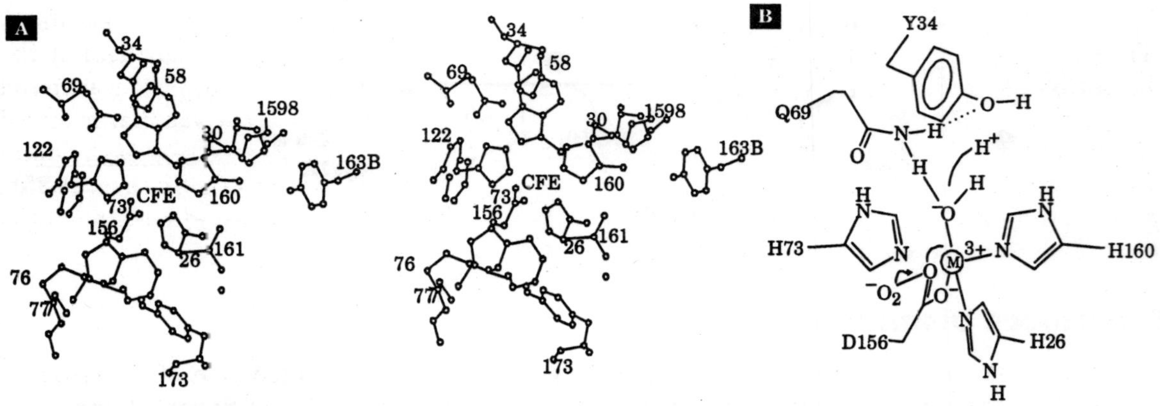

Fig. 4.22. (A) Structure of the active site of iron superoxide dismutase from *E. coli*. (B) Interpretive drawing illustrating the single-electron transfer from a superoxide molecule to the Fe^{3+} of superoxide dismutase and associated proton uptake.

B. COBALT AND VITAMIN B_{12}

The human body contains only about 1.5 mg of cobalt, almost all of it is in the form of *cobalamin*, vitamin B_{12}. Ruminant animals, such as cattle and sheep, have a relatively high nutritional need for cobalt and in regions with a low soil cobalt content, such as Australia, cobalt deficiency in these animals is a serious problem. This need for cobalt largely reflects the high requirement of the microorganisms of the rumen (paunch) for vitamin B_{12}. All bacteria require vitamin B_{12} but not all are able to synthesize it. For example, *E. coli* lacks one enzyme in the biosynthetic pathway and must depend upon other bacteria to complete the synthesis.

Coenzyme Forms

For several years after the discovery of cobalamin its biochemical function remained a mystery, a major reason being the extreme sensitivity of the coenzymes to decomposition by light. Progress came after Barker and associates discovered that the initial step in the anaerobic fermentation of glutamate by *Clostridium tetanomorphum* is rearrangement to β-methylaspartate (Eq. 4.28).

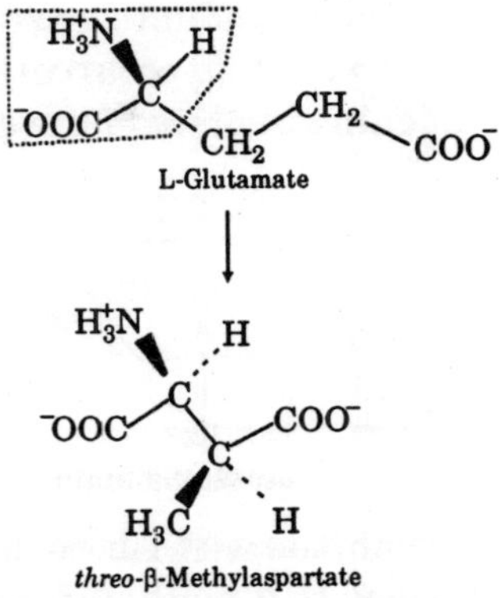

L-Glutamate

threo-β-Methylaspartate

...(4.28)

The latter compound can be catabolized by reactions that cannot be used on glutamate itself. Thus, the initial rearrangement is an indispensable step in the energy metabolism of the bacterium. A new coenzyme required for this reaction was isolated in 1958 after it was found that protection from light during the preparation was necessary. The coenzyme was characterized in 1961 by X-ray diffraction as *5′-deoxyadenosylcobalamin*. It is related to cyanocobalamin replacement of the CN group by a 5′-deoxyadenosyl group as indicated in the following abbreviated formulas. Here the planes represent the corrin ring system and Bz the dimethyl-benzimidazole that is coordinated with the cobalt from below the ring.

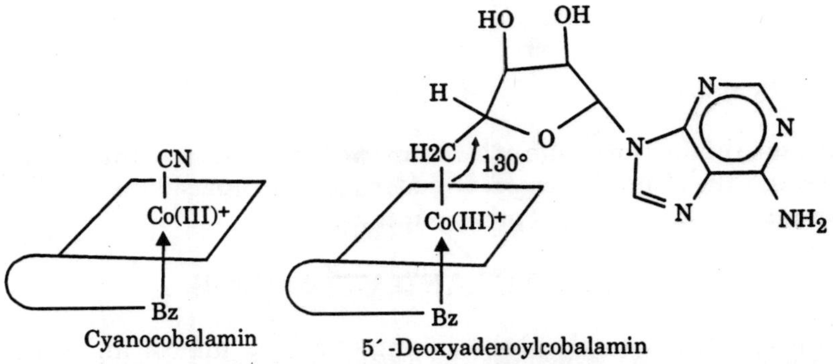

Cyanocobalamin

5′-Deoxyadenoylcobalamin

The most surprising structural feature is the Co-C single bond of length 0.205 nm. Thus, the coenzyme is an alkyl cobalt, the first such compound found in nature. In fact, alkyl cobalts were previously thought to be unstable. Vitamin B_{12} contains Co(III), and cyano-cobalamin can be imagined as arising by replacement of the single hydrogen on the inside of the corrin ring by Co^{3+} plus CN^-. However, bear in mind that three other nitrogens of the corrin ring and a nitrogen of dimethylbenzimidazole also bind to the cobalt. Each nitrogen atom donates an electron pair to form coordinate covalent linkages. Because of resonance in the conjugated double-bond system of the corrin, all four of the Co-N bonds in the ring are nearly equivalent and the positive charge is distributed over the nitrogen atoms surrounding the cobalt. The strength of the axial Co-C bond is directly influenced by the strength of bonding of the dimethylimidazole whose conjugate base has a microscopic pK_a of 5.5.

Protonation of this base breaks its bond to cobalt and may thereby strengthen the Co-C bond. Steric factors are also important in determining the strength of this bond. NMR techniques are now playing an important role in investigation of these factors. In both bacteria and liver, the 5′-deoxyadenosyl coenzyme is the most abundant form of vitamin B_{12}, while lesser amounts of *methylcobalamin* are present. Other naturally occurring analogs of the coenzymes include *pseudo vitamin* B_{12} which contains adenine in place of the dimethylbenzimidazole.

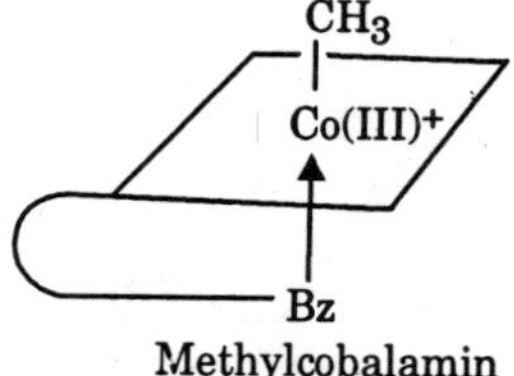

Methylcobalamin

Like dimethylbenzimidazole, it is combined with ribose in the unusual a linkage. A compound called factor A is the vitamin B_{12} analog with 2-methyladenine. Related compounds have been isolated from such sources as sewage sludge which abounds in anaerobic bacteria. It has been suggested that plants may contain vitamin B_{12}-like materials which do not support growth of bacteria. Thus, we may not have discovered all of the alkyl cobalt coenzymes.

Reduction of Cyanocobalamin and Synthesis of Alkyl Cobalamins

Cyanocobalamin can be reduced in two one-electron steps (Eq. 4.29). The cyanide ion is lost

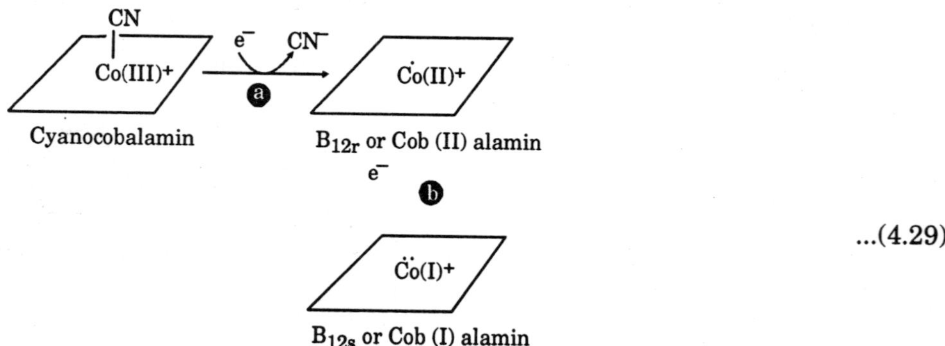

...(4.29)

in the first step (Eq. 4.29 step *a*), which may be accomplished with chromous acetate at pH 5 or by catalytic hydrogenation. The product is the brown paramag-netic compound B_{12r}, a tetragonal low-spin cobalt(II) complex. In the second step (Eq. 4.29 step *b*), an additional electron is added, *e.g.*, from sodium boro-hydride or from chromous acetate at pH 9.5, to give the gray-green exceedingly reactive B_{12s}. The latter is thought to be in equilibrium with cobalt(III) hydride as shown in Eq. 4.30 step *a*. The hydride is unstable and breaks down slowly to H_2 and B_{12r} (Eq. 4.30 step *b*).

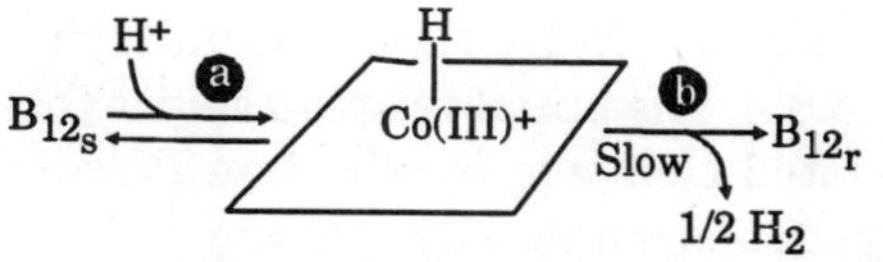

$$B_{12s} \xrightarrow[\;]{H^+ \;(a)} H\text{–}Co(III)^+ \xrightarrow[\text{Slow}]{(b)} B_{12r} + 1/2\, H_2 \quad ...(4.30)$$

Vitamin B_{12s} reacts rapidly with alkyl iodides (*e.g.*, methyl iodide or a 5′-chloro derivative of adenosine) via nucleophilic displacement to form the alkyl cobalt forms of vitamin B_{12} (Eq. 4.31). These reactions provide a convenient way of preparing isotopically labeled alkyl

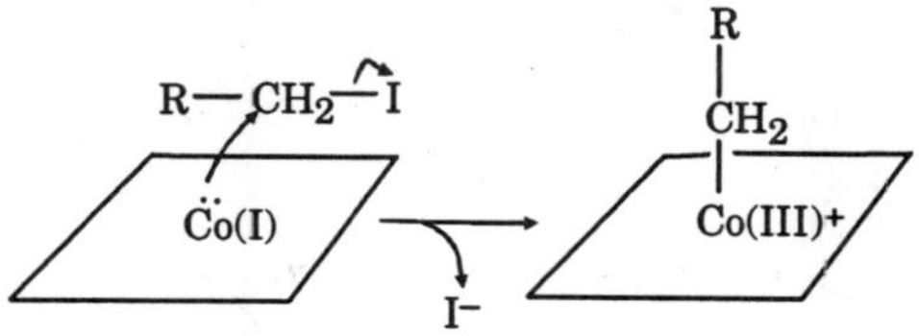

$$R\text{—}CH_2\text{—}I + \ddot{C}o(I) \xrightarrow{} R\text{—}CH_2\text{—}Co(III)^+ + I^- \quad ...(4.31)$$

cobalamins, including those selectively enriched in ^{13}C for use in NMR studies. The bio-synthesis of 5′-deoxyadenosylcobalamin utilizes the same type of reaction with ATP as a substrate. A B_{12s} *adenosyltransferase* catalyzes nucleophilic displacement on the 5′ carbon of ATP with formation of the coenzyme and displacement of inorganic tripolyphosphate PPP_i.

Three Nonenzymatic Cleavage Reactions of Vitamin B_{12} Coenzymes

The 5′-deoxyadenosyl coenzyme is easily decomposed by a variety of agents. Anaerobic irradiation with visible light yields principally vitamin B_{12r} and a cyclic 5′, 8-deoxyadenosine which is probably formed through an intermediate radical (Eq. 4.32). Irradiation in the presence of air gives a variety of products.

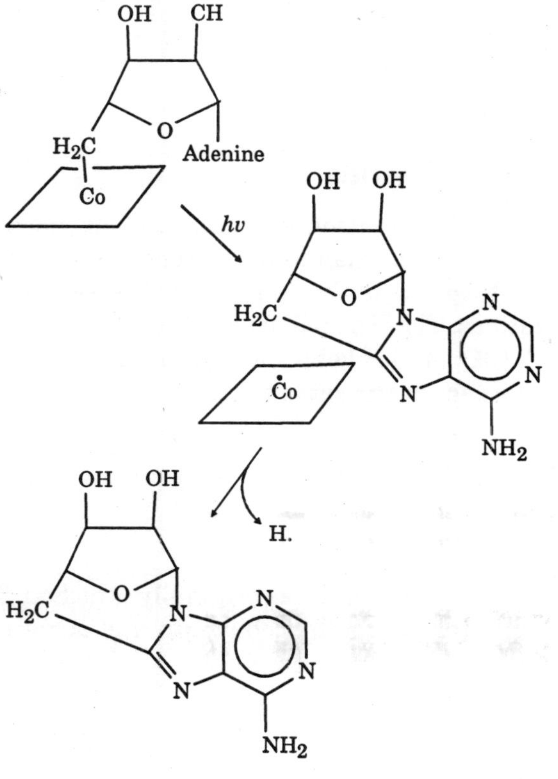

OH CH

H₂C Co Adenine

hv

OH OH

H₂C N N N N NH₂

Ċo

H.

OH OH

H₂C N N N N NH₂

...(4.32)

Hydrolysis of deoxyadenosylcobalamin by acid (1 M HC1,100°, 90 min) yields hydroxycobalamin, adenine, and an unsaturated sugar (Eq. 4.33). The initial reaction step is thought to be protonation of the oxygen of the ribose ring. A related cleavage by alkaline cyanide can be viewed as a nucleophilic displacement of the deoxyadenosyl anion by cyanide.

The end product is *dicyanocobalamin*, in which the loosely bound nucleotide containing dimethyl benzimidazole is replaced by a second cyanide ion. Methyl and other simple alkyl cobalamins are stable to alkaline cyanide. A number of other cleavage reactions of alkyl cobalamins are known.

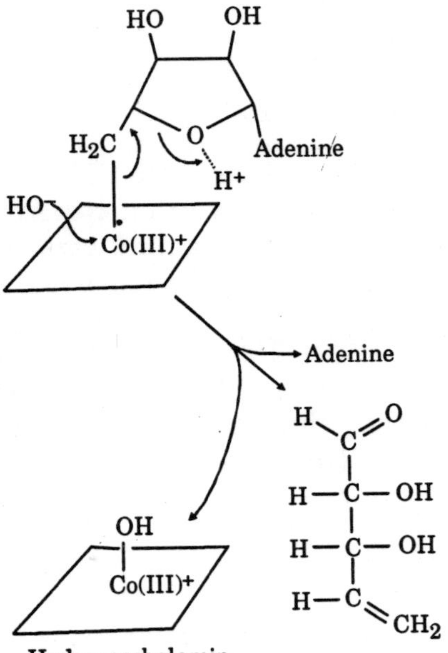

HO OH

H_2C O Adenine

H^+

HO Co(III)$^+$

Adenine

H C O

H—C—OH

OH H—C—OH

Co(III)$^+$ H—C CH_2

Hydroxocobalamin

...(4.33)

Enzymatic Functions of B_{12} Coenzymes

Three types of enzymatic reactions depend upon alkyl corrin coenzymes. The first is the reduction of ribonucleotide triphosphates by cobalamin-dependent ribonucleotide reductase, a process involving *intermolecular* hydrogen transfer (Eq. 4.21). The second type of reaction encompasses the series of isomerizations. These can all be depicted as in Eq. 4.34, Some group X, which may be attached by a C-C, C-O, or C-N bond shown in table 4.1, is transferred to an adjacent carbon atom bearing a hydrogen. At the same time,

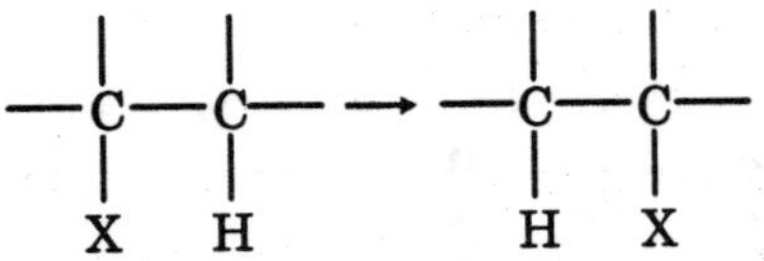

$$\begin{array}{cc} | & | \\ -C- & C- \\ | & | \\ X & H \end{array} \longrightarrow \begin{array}{cc} | & | \\ -C- & C- \\ | & | \\ H & X \end{array} \qquad ...(4.34)$$

the hydrogen is transferred to the carbon to which X was originally attached. The third type of reaction is the transfer of methyl groups via methylcobalamin and some related bacterial metabolic reactions.

Cobalamin-dependent Ribonucleotide Reductase

Lactic acid bacteria such as *Lactobacillus leichmanni* and many other bacteria utilize a 5′-deoxyadenosylcobalamin-containing enzyme to reduce nucleoside triphosphates according to Eq. 4.21. Thioredoxin or dihydrolipoic acid can serve as the hydrogen donor. Early experiments showed that protons from water are reversibly incorporated at C-2′ of the reduced nucleotide with retention of configuration.

A more important finding was a large kinetic isotope effect of 1.8 when 3′-^{3}H-containing UTP was reduced by the enzyme. Reaction of the reductase with dihydrolipoic acid in the presence of deoxy-GTP, which apparently serves as an allosteric activator, leads to formation, within a few milliseconds, of a radical with a characteristic EPR spectrum that can be studied when the reaction mixture is rapidly cooled to 130°K. When GTP (a true substrate) is used instead of dGTP, the radical signal reaches a maximum in about 20 ms and then decays. Of the various oxidation states of cobalt (3+, 2+, and 1+) only the 2+ state of vitamin B_{12r} paramagnetic and gives rise to an EPR signal.

The electronic absorption spectrum of the coenzyme of ribonucleotide reductase is also changed rapidly by substrate in a way that suggests formation of B_{12r}. Thus, it was proposed that a hemolytic cleavage occurs to form B_{12r} and a stabilized 5′-deoxyadenosyl radical (Eq. 4.35). However, H3 is not transferred from the 3′ position of the substrate into the deoxyadenosyl part of the coenzyme. The enzyme has many properties in common with the previsouly discussed iron-tyrosinate ribonucleotide reductases including limited peptide sequence homology. Stubbe and coworkers suggested that the deoxyadenosyl radical is formed as a radical chain initiator and that the mechanism of ribonucleotide reduction is as shown in Eq. 4.25. Studies of enzyme-activated inhibitors support this mechanism.

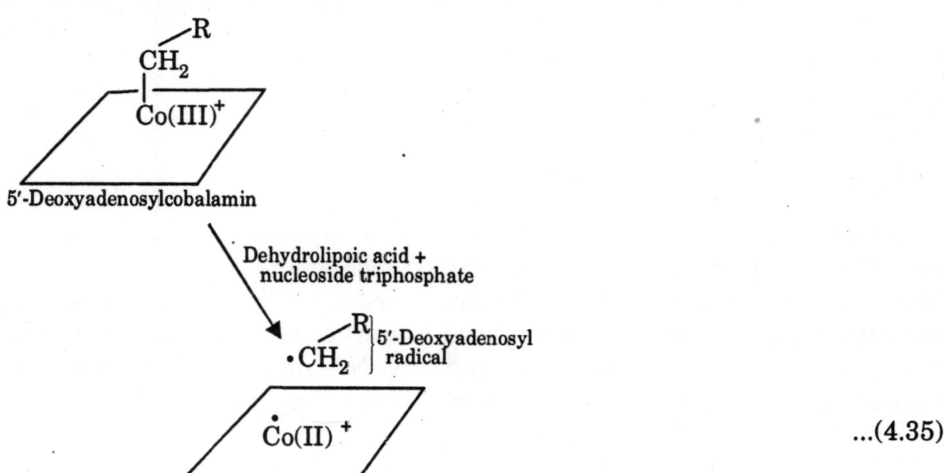

...(4.35)

The Isomerization Reactions

At least 10 reactions of the type described by Er. 4.34 are known (Table 4.1). They can be subdivided into three groups. First, X = OH or NH_2 in Eq. 4.34 isomerization gives a *geminal*-diol or aminoalcohol that can eliminate H_2O or NH_3 to give an aldehyde. All of these enzymes, which are called *hydro-lyases* or *ammonia-lyases*, specifically required K^+ as well as the vitamin B_{12} coenzyme. Second, X = NH_2 in Eq. 4.34. For this group of *aminomutases* PLP is required as a second coenzyme. Third, X is attached via a carbon atom; the enzymes are called *mutases*. *Methylmalonly-Co mutase* is required for catabolism of propionate in the human body, and is one of only two known vitamin B^{12-} dependent enzymes. The related isolbutyryl-CoA mutase participates in the microbial synthesis of such polyether antibiotics as monensin A. Other mutases are involved in anaerobic bacterial fermentations.

Table 4.1. Isomerization Reactions Involving Hydrogen Transfer and Dependent upon a Vitamin B_{12} Coenzyme.

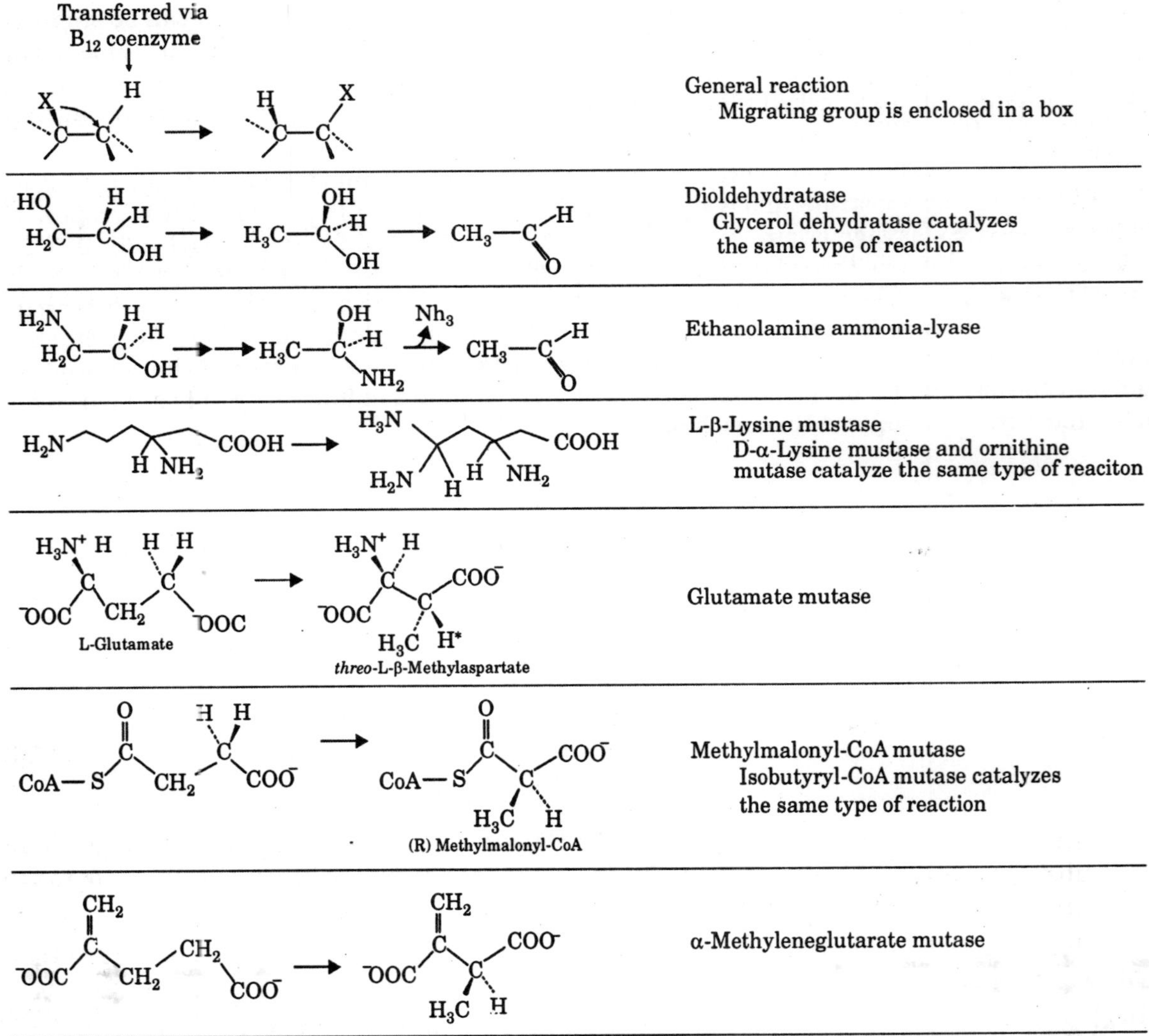

In these reactions the hydrogen is always transferred via the B_{12} coenzyme. No exchange with the medium takes place. Since X may be an electronegative group such as OH, the reactions could all the treated formally as hydride ion transfers but it is more likely that they occur via hemolytic cleavages. Such cleavage is indicated by the observation of EPR signals for several of the enzymes in the presence of their substrates. Abeles and associates showed that when *dioldehydratase* (Tabel 4.1) catalyzes the conversion of 1, 2-[1-^{3}H]propanediol to propionaldehyde, tritium appears in the coenzyme as well as in the final product. When ^{3}H-containing coenzyme is incubated with unlabeled propanediol, the product also contains ^{3}H, which was shown by chemical degradation to be exclusively on C-5′. Synthetic 5′-deoxyadenosyl coenzyme containing ^{3}H in the 5′ position transferred ^{3}H to product.

Most important, using a mixture of propanediol and ethylene glycol, a small amount of *intermolecular* transfer was demonstrated; that is, ^{3}H was transferred into acetaldehyde, the product of dehydration of ethylene glycol. Similar results were also obtained with ethanolamine ammonia-lyase.Another important experiment showed that ^{18}O from [2-^{18}O]propanediol was transferred into the 1 position without exchange with solvent. Furthermore, ^{18}O from (*S*)-[1-^{18}O]propanediol was retained in the product while that from the (*R*) isomer was not. Thus, it appears that the enzyme stereo specifically dehydrates the final intermediate.

From these and other experiments, it was concluded that initially a 5′-deoxyadenosyl radical is formed via Eq. 4.35. This radical then abstracts the hydrogen atom marked by a shaded box in Eq. 4.36 to form a substrate radical and 5′-deoxyadenosine. One proposal, illustrated in Eq. 4.36 is that the substrate radical immediately recombines with the Co(II) of the coenzyme to

(5′-deoxyadenosyl coenzyme)

HCH_2—R
(5′-deoxyaenosine)

CH_3—CH_2—CH

...(4.36)

from an organo-cobalt substrate compound. The 5′-deoxyadenosine now contains hydrogen from the substrate; because of rotation of the methyl group this hydrogen becomes equivalent to the two already present in the coenzyme (step *a* or Eq. 4.36).

The substrate-cobalamin compound formed in this step then undergoes isomerization, which, in the case of dioldehydratase, leads to intramolecular transfer of the OH group (step *b*). In step *c* the hydrogen atom is transferred back from the 5′-deoxyadenosine to its new location in the product and in step *d* the resulting *gem*-diol is dehydrated to form the aldehyde product. Carbocation, carboanion, and and free radical intermediated have all been proposed for the isomerizaton step in the reaction. A carbocation would presumably be cyclized to an epoxide which could react with the B_{12s} (Eq. 4.37, step *b*) to complete the isomerization.

$Co(III)^+$ —(a)— $Co(I)$ (B_{12s}) —(b)— $Co(III)^+$ ...(4.37)

At present most evidence favors, for the isomerization reactions, an enzyme-catalyzed rearrangement of the substrate radical produced initially during formation of the 5′-deoxyadenosine (Eq. 4.38).

...(4.38)

According to this mechanism the Co(II) of the B_{12r} formed in Eq. 4.35 has no active role in the isomerization and does not form an organocobalt intermediate. Its only role is to be available to recombine with the 5′-deoxyadenosyl radical at the end of the reaction sequence. Support for this interpretation has been obtained from study of model reactions and of organic radicals generated in other ways. Recently, EPR spectroscopy with ^2H- and 13C-labeled glutamates as substrates for glutamate mutase permitted identification of a 4-glutamyl radical as a probable intermediate for that enzyme.

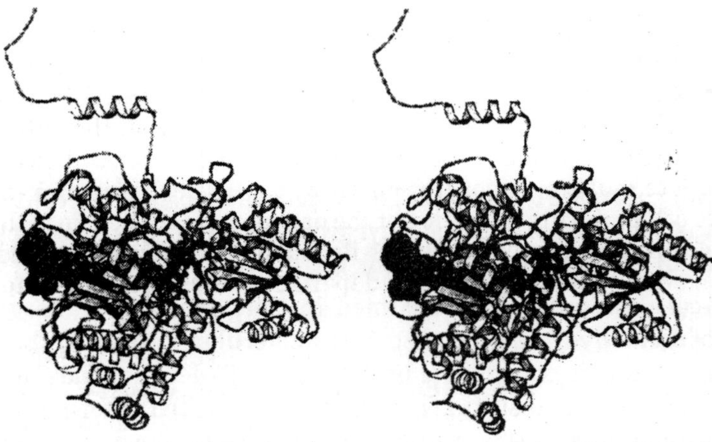

Fig. 4.23. Three-dimensional structure of methylmalonyl-CoA mutase from *Propionobacterium shermanii*. The B_{12} coenzyme is deeply buried, as is the active site. A molecule of bound desulfocoenzyme A, a substrate analog, blocks the active site entrance on the left side.

On the other hand, for methylmalonyl-CoA mutase Ingraham suggested cleavage of the Co-C bond of an organocobalt intermediate to form a carbanion, the substrate-cobalamin compound serving as a sort of "biological Grignard reagent." The carbanion would be stabilized by the carbonyl group of the thioester forming a "homoenolate anion". The latter coul break up in either of two ways reforming a C-Co bond and causing the isomerization. Some experimental results also favor an ionic or organocobalt pathway.

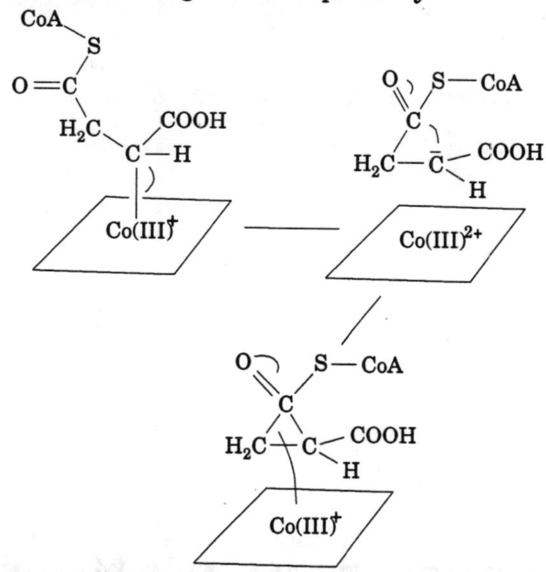

...(4.39)

The three-dimensional structure of methylmalonyl-CoA mutase from *Propionibacterium shermanii* shows that the vitamin B_{12} is bound in a base-off conformation with the dimethlbenzimidazole group bound to the protein far from the corrin ring (Fig. 4.23). A histidine of the protein coordinates the cobalt, as also in methionine synthase. The entrance to the deeply buried active site is blocked by the coenzyme part of the substrate. The buried active site may be favorable for free radical rearrangement reaction.

The structure of substrate complexes shows that the coenzyme is in the cob(II)alamin (B_{12r}) form with the 5′-deoxyadenosyl group detached from the cobalt, rotated, shifted, and weakly bound to the protein. Side chains from neighboring Y89, R207, and H244 all hydrogen bond to the substrate. The R207 guanidinium group makes an ion pair with the substrate carboxylate, and the phenoic and imidazole groups may have catalytic functions. Studies are also in progress on crystalline glutamate mutase. According to all of these mechanism, 5′-deoxyadenosine is freed from its bond to cobalt during the action of the enzymes. Why then does the deoxyadenosine not escape from the coenzyme entirely, leading to its inactivation? Substrate-induced inactivation is not ordinarily observed with coenzyme B_{12}– dependent reactions, but some quasi-substrates do inactivate their enzymes.

Thus, glycolaldehyde converts the coenzyme of dioldehydratase to 5′-deoxyadenosine and ethylene glycol does the same with ethanolamine deaminase. When 5′-deoxyinosine replaces the 5′-deoxyadenosine of the normal coenzyme in dioldehydratase, 5′-deoxyinosine is released quantitatively by the substrate. This suggests that the dehydratase may normally hold the adenine group of 5′-through hydrogen bonding to the amino group. Because the OH group of inosine tautomerizes to C = O, inosine may not be held as tightly. The deeply buried active site (Fig. 4.23) may also prevent escape of the deoxyadenosine.

Despite the evidence in its favor, there was initially some reluctance to accept 5′-deoxyadenosine as an intermediate in vitamin B_{12}-dependent isomerization reactions. It was hard to believe that a methyl group could exchange hydrogen atoms so rapidly.

It was suggested that protonation of the oxygen of the ribose ring as in Fig. 4.33 might facilitate release of a hydrogen atom. However, substitution of the ring oxygen by CH_2 in a synthetic analog did not destroy the coenzymatic activity. Another possibility is that the methyl group has an unusual reactivity if the cobalt is reduced to Co(II).

Stereochemistry of the Isomerization Reactions

Dioldehydratase acts on either the (*R*) or (*S*) isomers of 1, 2-propanediol (Eqs. 4.40 and 4.41; asterisks and daggers mark positions of labeled atoms in specific experiments).

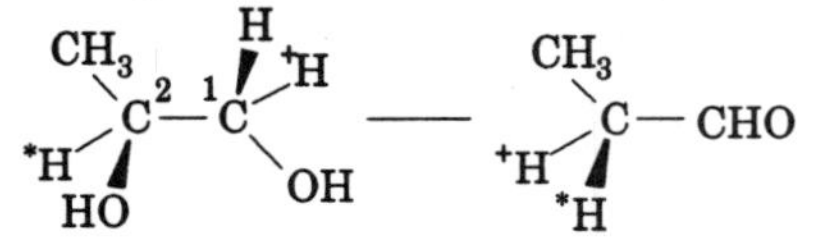

$$CH_3\text{-}\overset{2}{C}(^{*}H)(HO)\text{-}\overset{1}{C}(H)(^{+}H)(OH) \longrightarrow CH_3\text{-}C(^{+}H)(^{*}H)\text{-}CHO \qquad ...(4.40)$$

(S)-1, 2-Propanediol

$$CH_3\text{-}C(HO)(^{*}H)\text{-}C(^{+}H)(H)(OH) \longrightarrow H_3C\text{-}C(^{*}H)(H)\text{-}CHO \qquad ...(4.41)$$

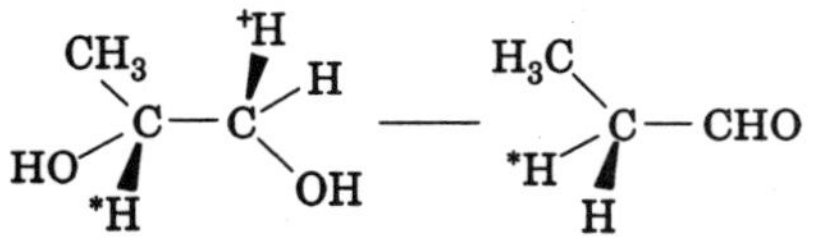

(R)-1, 2 Propanediol

In both cases the reaction proceeds with retention of configuration at C-2 and with stereochemical specificity for one of the two hydrogens at C-1. The reaction catalyzed by methylmalonyl-CoA mutase likewise proceeds with retention of configuration at C-2 (Table 4.1) but the glutamate mutase reaction is accompanied by inversion (Eq. 4.28).

Aminomutases

The enzymes *L-β-lysine mutase* (which is also *D-α-lysine mutase*) and *D-ornitine mutase* catalyze the transfer of an ω-amino group to an adjacent carbon atom (Table 4.1). Two proteins are needed for the reaction; pyridoxal phosphate is required and is apparently directly involved in the amino group migration. In the β-lysine mutase the 6-amino group of L-β-lysine replaces the pro-*S* hydrogen at C-5 but with inversion at C-5 to yield (3*S*, 5*S*)-3, 5-diaminohexanoic acid. A bacterial D-lysine 5, 6-aminomutase interconverts D-lyisne with 2,5-diaminohexanoic acid.

Another related enzyme is *L-leucine* 2, 3-*aminomutase*, which catalyzes the reversible interconversion of L-leucine and β-leucine. It was reported to be present in plants and also in the human body, but the latter could not be confirmed. The interconversion of L-α-lysine and L-β-lysine is catalyzed by a lysine 2, 2-aminomutase found in certain clostridia. This enzyme also required pyridoxal phosphate and catalyzes a reaction with the same stereochemistry as that of β-lysine mutase. However, it does not contain vitamin B_{12} but depends upon *S*-adenosylmethionine (AdoMet) and an iron-sulfur cluster. The adenosyl group of AdoMet may function in the same manner as does the deoxyadenosyl group of the adenosylcobalamin coenzyme. In these mutases a 5′-deoxyadenosyl radical may abstract a hydrogen atom from the β position of a Schiff base of PLP with the amino acid substrate.

The radical isomerizes (Eq. 4.42) then accepts a hydrogen atom back from the 5′-deoxyadenosine to complete the reaction. Its properties suggest that lysine 2, 3-aminomutase

R—ĊH—CH—COO⁻ ... (reaction scheme with pyridoxal phosphate intermediates, steps a and b)

...(4.42)

is related to the pyruvate formate-lyase of *E.coli* class I ribonucleotide reductases, and other enzymes that act by hemolytic mechanism.

Transfer Reactions of Methyl Groups

The generation and utilization of methyl groups is a quantitatively important aspect of the metabolism of all cells. As we have seen, methyl groups can be created by the reduction of one-carbon compounds attached to tetrahydrofolic acid. Methyl groups of methyltetrahydrofolic acid (N_5-CH_3THF) can then be transferred to the sulfur atom of homocysteine to form methionine (Eq. 4.43). The latter is converted to *S*-adenosylmethionine, the nearly universal methyl group donor for transmethylation reactions. In some bacteria, fungi, and higher plants, the methyl-THF-homocysteine transmethylase does not depend upon vitamin B_{12} but is a metalloenzyme with zinc at the active center.

However, human beings share with certain strains of *E. coli* and other bacteria the need for methylcobalamin. The structure of the *E. coli* enzyme (Fig. 4.24) shows methylcobalamin bound in a base-off conformation, with histidine 759 of the protein replacing dimethylbenzimidazole in the distal coordination position on the cobalt. This histidine is part of a sequence Asp-X-His-X-X-Gly that is found not only in methionine symthase but also in methylamalonyl-CoA mutase, glutamate mutase, and 2-methyleneglutarate mutase.

However, diol dehydratase lacks this sequence and binds adenosylcobalamin with the dimethylbenzimidazole-cobalt bond intact.The coenzyme evidently functions in a cyclic process. The cobalt alternates between the +1 and +3 oxidation states as shown in Eq. 4.43. The first indication of such a cyclic process was the report by Weissabch that ^{14}C-labeled methylcobalamin could be isolated following treatment of the enzyme with such methyl donors as AdoMet and

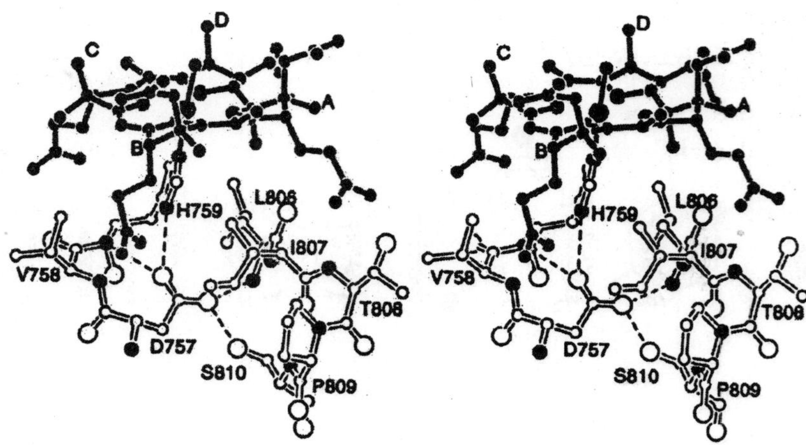

Fig. 4.24. Stereoscopic views of the active site of methionine synthase from *E. coli*. Methylcobalamin (black) is in the active site with His 759 of the protein in the distal position of the coenzyme in a base-off conformation. The dimethylbenzimidole nucleotide has been omitted for clarity. Notice the hydrogen-bonded His 759-D757-S810 triad.

$H_3\overset{+}{N}$ H
HS — COO^-
Homocysteine
H_3C—S — COO^-
a
H^+
CH_3
$Co(III)^+$
Co(I)
b
THF $N^5 - CH_3 - THF$
d
$Co(II)^+$
c
e^-
Ado-Hcy e^- $AdoMet^+$

...(4.43)

methyl iodide after reduction (*e.g.*, with reduced riboflavin phosphate). The sequence parallels for the laboratory synthesis of methylcobalamin. Nevertheless, the transmethylase demonstrates some complexities. Initially, it must be "activated" by AdoMet or methyl iodide, after which it cycles according, steps *a* and *b*, but is gradually inactivated. This apparently happens by oxidation to a Co(II) form of the enzyme that must be reductively methylated with AdoMet and reduced flavodoxin regenerate the active form. It is also possible that the methyl groups is not transferred as a formal CH_3^+ as pictured as a $\bullet CH_3$ radical generated by hemolytic cleavage of methylcobalamin to cob(II)alamin. Whatever the mechanism, a chiral methyl group is transferred from 5-methyl-THF to homocysteine with overall retention f configuration. Another important group of methyl transfer reactions are those from methyl corrinoids to mercury, tin, arsenic, selenium, and tellurium.

For example, describes the methylation of Hg^{2+}. These reactions are of special interest because of the generation of toxic methyl and dimethyl mercury and dimethylarisne. Notice that whereas methyl group is transferred as CH_3^+ with no valence change occurring in the cobalt. However, it is also possible that transfer occurs as methyl radical. Methyl corrinoids are able to undergo this type of reaction nonenzymatically, and the ability to transfer a methyl anion is a property of methyl corrinoids not shared by other transmethylating agents such as AdoMet. At the conclusion of the reaction cobalt is in the +3 state. To be remethylated, it must presumably be reduced to Co(II). A second methyl group can be transferred by the same type of reaction to form $(CH_3)_2Hg$.

Hg^{2+} ··· CH_3–Co(III)$^+$ + H_2O ⟶ HO–Co(III)$^+$ + CH_3Hg^+ + H^+ ...(4.44)

Methylation of arsenic is an important pollution problem because of the widespread use of arsenic compounds in insecticides and because of the presence of arsenate in insecticides and because of the presence of arsenate in the phosphate used in household detergents. After reduction to arsenite, methylation occurs. Additional reduction steps result in the formation of *dimethylarsine*, one of the principal products of action of methanogenic bacteria on arsenate.

The methyl transfer is shown as occurring through CH_3^+, with an accompanying loss of a proton from the substrate. However, a CH_3 radical may be transferred with formation of a cobalt(II) corrinoid.

Arsenate $\xrightarrow{2e^-}$ $(HO)_2As–O^-$ $\underset{H_2O}{\rightleftharpoons}$ O=As–O$^-$ (Arsenite)

$\xrightarrow{^+CH_3,\ -H^+}$ Methylarsonate

$\xrightarrow{^+CH_3,\ H^+,\ 2e^-,\ -H_2O}$ Dimethylarsinate

$\xrightarrow{4e^-,\ 5H^+,\ -2H_2O}$ $(H_3C)_2As–H$ Dimethylarsine ...(4.45)

Corrinoid-dependent Synthesis of Acetyl-AoA

The anaerobic bacterium *Clostridium thermoaceticum* obtains its energy for growth by reduction of CO_2 with hydrogen. One of the CO_2 molecules is reduced to formate which is converted via 5-methyl-THF to the methyl corrinoid 5-methoxybenzimidazolyl-cobamide. The methyl group of the latter

$$4H_2 + 2CO_2 \longrightarrow CH_3COO^- + H^+ + 2H_2O \quad \Delta G' \text{ (pH 7)} = -94.9 \text{ kJ mol}^{-1} \qquad ...(4.46)$$

combines with another CO_2 to form acetyl-CoA. The process, which requires the nickel-containing carbon monoxide dehydrogenase, is discussed in Section C.

C. NICKEL

It was not until the 1970s that nickel was first recognized as a dietary essential for animals. Nickel-deficient chicks grew poorly, had thickened legs, and developed dermatitis. Tissues of deficient animals contained swollen mitochondria and swollen perinuclear space suggesting a function for Ni in membranes. However, doubts have been raised about the conclusions based on these experiments. Within tissues the nickel content ranges from 1 to 5 μg/1. Some of the metal in serum is present as complexes of low molecular mass and some is bound to serum albumin and to a specific nickel-containing protein of the macroglobulin class, known as *nickeloplasmin*.

Nickel is also present in plants; in some, *e.g., Allysum,* it accumulates to high concentrations. It is essential for legumes and possibly for all plants. Nickel uptake proteins have been identified in bacteria and fungi. Because of its ubiquitous occurrence it is difficult to prepare a totally Ni-free diet. The acute toxicity of orally ingested Ni(II) is low, and homeostatic mechanisms exist in the animal body for regulating its concentration. However, the volatile nickel carbonyl $Ni(CO)_4$ is very toxic and Ni from jewelry is a common cause of dermatitis. In its compounds nickel usually has the +2 oxidation state but the +3 and +4 states occur rarely in complexes.

The Ni^{2+} ion contains eight 3d electrons, a configuration that favours square-planar coordination of four ligands. However, the ion is also able to form a complex with six ligands and an octahedral geometry. It has been suggested that this "ambivalence" may be of biochemical significance. Nickel is found in at least four enzymes: *urease*, certain *hydrogenases*, *methyl-*CoM *reductase* (in its *cofactor* F_{430}) of methanogenic bacteria, and *carbon monoxide dehydrogenase* of acetogenic and methanogenic bacteria.

Urease

Urease, which was first isolated from the jack bean has a special place in biochemical history as the first enzyme to be crystallized. This was accomplished by J. B. Sumner in 1926, and although Sumner eventually obtained the Nobel Prize, his first reports were greeted with skepticism and outright disbelief. The presence of two atoms of nickel in each molecule of urease was not discovered until 1975. The metal ions had been overlooked previously, despite the fact that the absorption spectrum of the purified enzyme contains an absorption "tail" extending into the visible region with a shoulder at 425 nm and weak maxima at 725 and 1060 nm.

Urease catalyzes the hydrolytic cleavage of urea to two molecules of ammonia and one of bicarbonate and is useful in the analytical determination of urea. Jack bean urease is a trimer or hexamer of identical 91-kDa subunits while that of the bacterium *Klebsiella* has an $(\alpha\beta_2\gamma_2)_2$ stoichiometry. Nevertheless, the enzymes are homologous and both contain the same binickel catalytic center (Fig. 4.25). The three-dimensional structure of the *Klebsiella* enzyme revealed that the two nickel ions are bridged by a carbamyl group of a carbamylated lysine. Like ribulose bisphosphate carboxylase, urease also requires CO_2 for formation of the active enzyme.

Formation of the metallocenter also requires four additional proteins, including a chaperonin and a nickel-binding protein. The mechanism of urease action is probably related to those of metalloproteases such as carboxypeptidase A zinc-dependent carbonic anhydrase. In Fig. 4.25

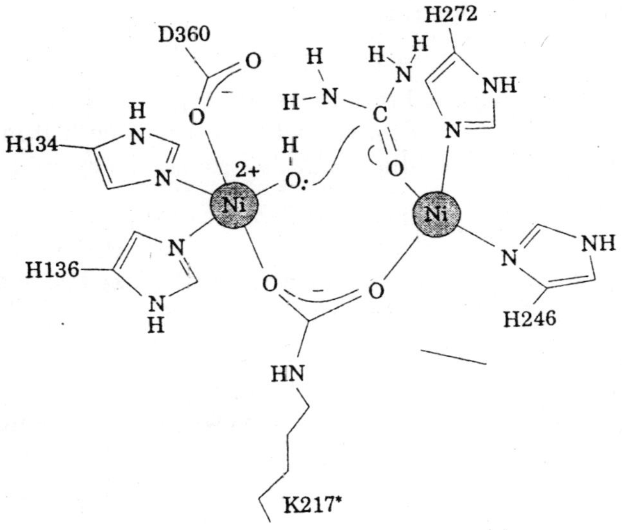

Fig. 4.25. The active site of urease showing the two Ni⁺ ions held by histidine side chains and bridged by a carbamylated lysine (K217*). A bound urea molecule. It has been placed in an open coordination position on one nickel and is shown being attacked for hydrolytic cleavage by a hydroxyl group bound to the other nickel.

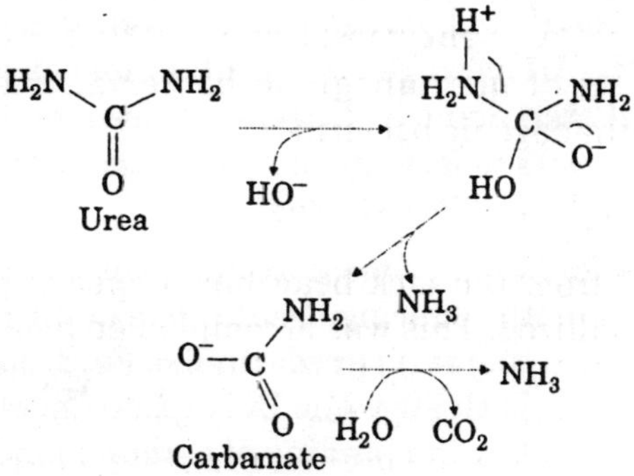

...(4.47)

one nickel ion is shown polarizing the carbonyl group, while the second provides a bound hydroxyl ion that serves as the attacking nucleophile. A probable intermediate product is a carbamate ion (Eq. 4.47).

Urease is an essential enzyme for bacteria and other organisms that use urea as a primary source of nitrogen. The peptic ulcer bacterium *Helicobacter pylori* uses urease to hydrolyze urea in order to defend against the high acidity of the stomach. The enzyme is also present in plant leaves and may play a necessary role in nitrogen metabolism. In nitrogen-fixing legumes urea derivatives, the *ureides*, have an important function but urease may not be involved in the catabolism of these compounds.

Hydrogenases

Many plants, animals, and microorganisms are able to evolve H_2 by reduction of hydrogen ions oxidize H_2 by the reverse of this reaction.

$$2H^+ + 2e^- \rightarrow H_2 \qquad ...(4.48)$$

Hydrogenases have been classified into two main types: *Fe-hydrogenases*, which contain iron as the only metal, and *Ni-hydrogenases*, which contain both iron and nickel. In a few Ni-hydrogenases a selenocysteine residue replaces a conserved cysteine side chain.

H₂C S Fe(II) Ni(III) S—CH₂—Cys 530 S—CH₂—Cys 65 (a) e⁻ → Fe(II) S Ni(II) S—CH₂—Cys 530 S—CH₂—Cys 65 (b) H₂ → H⁺ Fe(II) S H Ni(II) S—CH₂—Cys 530 H S⁺—CH₂—Cys 65 (c) e⁻ H⁺ Fe(II) S H Ni(II) H S⁺—CH₂—Cys 530 S—CH₂—Cys 65 (d) e⁻ 2H⁺

...(4.49)

Fe-hydrogenases are often extremely active and are utilized to rid organisms of an excess of electrons by evolution of H_2. Since they may also be used to acquire electrons by oxidation of H_2, they are often described as *bidirectional*. Ni-hydrogenases, as well as some Fe-hydrogenases, are involved primarily in *uptake* of H_2. All hydrogenases contain one or more Fe-S centers in addition to the H_2-forming catalytic center. Some hydrogenases are membrane bound and are often coupled through unidentified carriers to formate dehydrogenase. In the strict anaerobes such as clostridia, hydrogenases are linked to ferredoxins.

Hydrogenases are inactivated readily by O_2, which oxidizes the catalytic centers but can sometimes be reactivated by treatment with reducing agents. The 60-kDa all-iron monomeric hydrogenase I of *Clostridium pasteurianum* contains ferredoxin-like Fe_4S_4 clusters plus additional Fe and sulfur atoms organized as a special H cluster. The EPR spectrum of the catalytic center, recognized because the spectrum is altered by the binding of carbon monoxide, is unusual. Its *g* values of 2.00, 2.04, and 2.10 are similar to those of oxidized high-potential iron proteins.

The *C. pasteurianum* hydrogenase II is a 53-kDa monomer containing eight Fe and eight S^{2-} ions. These are organized into one ferredoxin-like Fe_4S_4 cluster plus a three-Fe cluster and one iron ion in a unique environment.

In contrast to these iron-only hydrogenases, the large periplasmic hydrogenase of *Desulfovibrio gigas* consists of one 28-kDa subunit and one 60-kDa subunit and contains two Fe_4S_4 clusters, one Fe_3S_4 cluster, and another dimetal center containing a single atom of Ni. There are seen clearly in the three-dimensional structure depicted in Fig. 4.26. The three Fe-S clusters, at 5- to 6-nm intervals, form a chain from the external surface to the deeply buried nickel-iron center. A plausible pathway for transport of protons to the active center can also be seen.

The nickel center also contains an atom of Fe. Four cysteine side chains participate in forming the Ni-Fe cluster, two of them provide sulfur atoms that bridge between the metals while two others are ligands to Ni (Fig. 42.6). A hydrogenase from *chromatium vinosum* has a similar structure. There are still uncertainties about other nonprotein ligands such as H_2O.

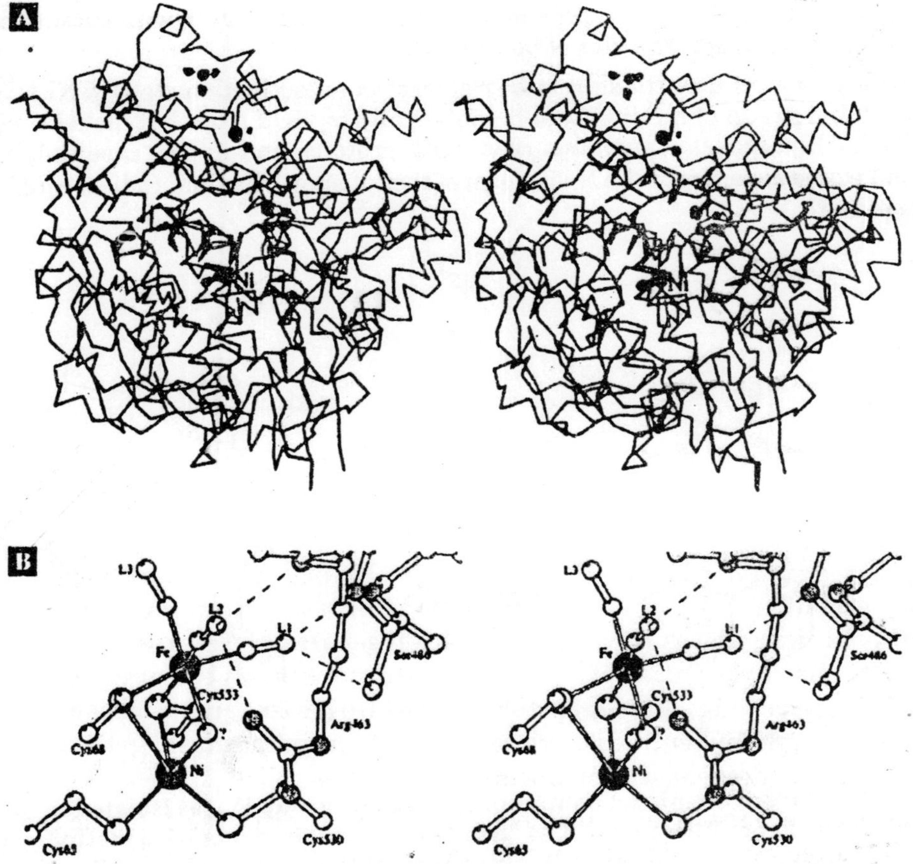

Fig. 4.26 (A) Stereo-scopic view of the structure of the *Desulfovibrio gigas* hydrogenase as an α-carbon plot. The electron density map at the high level of 8Σσ is superimposed and consists of dark spheres representing the Fe and Ni atoms. The iron atoms of the two Fe_4S_4 and one Fe_3S_4 clusters are seen clearly forming a chain from the surface of the protein to the Ni-Fe center. (B) The structure of the active site Ni-Fe pair. The two metals are bridged by two cysteine sulfur atoms and an uniden-tified atom, perhaps O, and the nickel is also coordinated by two additional cysteine sulfurs. Unidentified small molecules L1, L2, and L3 are also present.

All of the Ni-hydrogenases display an EPR signal that can be assigned to Ni(III). However, the active enzyme from D. *gigas* contains Ni(II). A proposed mechanism is indicated Eq. 4.49. Step *a* of this equation is a reductive activation. In step *b* a molecule of H_2 is bound as a hydride ion on Ni and a proton on a nearby sulfur. Protonation of a second sulfur ligand to Ni is needed to promote the cleavage of H_2 prior to the two-step oxidation of the bound H^-. One of two Ni-containing hydrogenases of *Methanobacteriuni thermoautotrophicum* contains FAD as well as Fe-S clusters. It specifically reduces the 5-deazaflavin cofactor F_{420}. A major function of this deazaflavin is reduction of the nickel-containing cofactor F_{430}.

Cofactor F_{430} and Methyl-Coenzyme M Reductase

Cofactor F_{430} is a nickel tetrapyrrole with a structure (Fig. 4.27) similar to those of vitamin B_{12} and of siroheme. The tetrapyrrole ring is the most highly reduced in cofactor F_{430}, which functions in reduction of methyl-CoM to methane in methanogens. The methyl CoM reductase of *Methanobacterinm thermoautotrophicum* is a large 300-kDa protein with subunit composition $\alpha_2\beta_2\gamma_2$ and containing two molecules of bound F_{430}.

The nickel in F_{430} is first thought to be reduced, in an activation step, to Ni(I), which may attack the methyl group of methyl-CoM homolytically to yield a methyl nickel complex and a sulfur radical. Alkyl nickel compounds react with protons, and in this case they would yield methane and would regenerate the Ni(II) form of the cofactor. The CoM radical could be reduced back to free CoM.

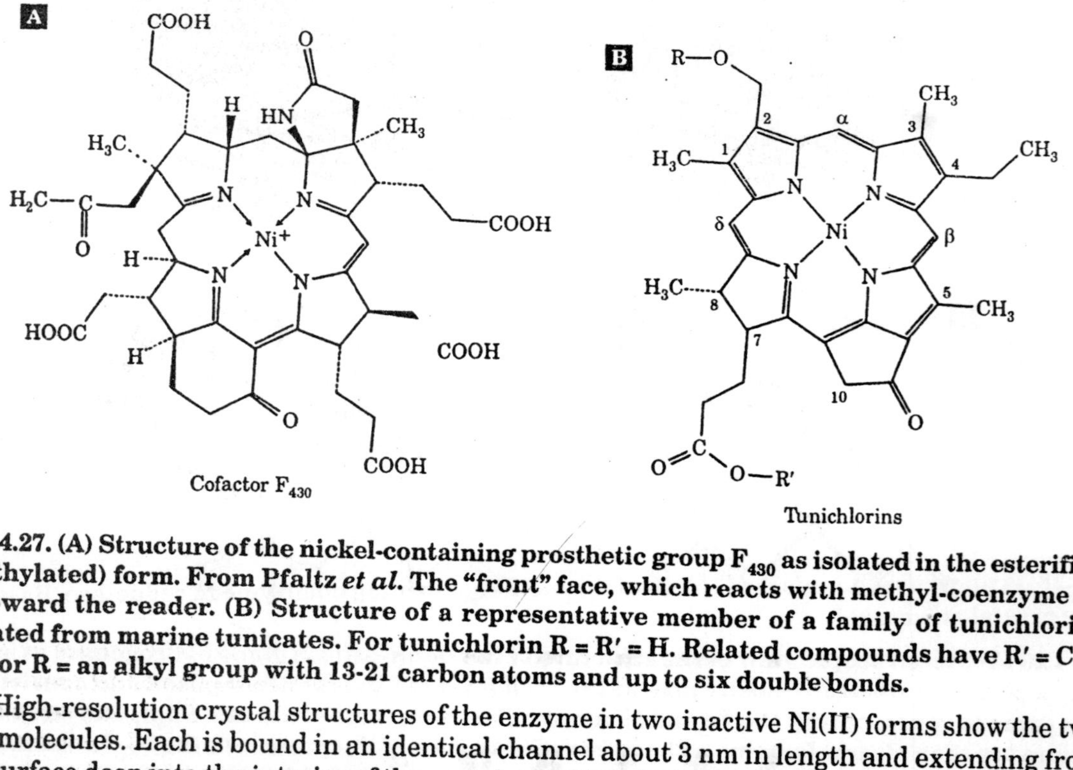

A Cofactor F_{430}

B Tunichlorins

Fig. 4.27. (A) Structure of the nickel-containing prosthetic group F_{430} as isolated in the esterified (methylated) form. From Pfaltz *et al.* The "front" face, which reacts with methyl-coenzyme M, is toward the reader. (B) Structure of a representative member of a family of tunichlorins isolated from marine tunicates. For tunichlorin R = R′ = H. Related compounds have R′ = CH_3 and/or R = an alkyl group with 13-21 carbon atoms and up to six double bonds.

High-resolution crystal structures of the enzyme in two inactive Ni(II) forms show the two F_{430} molecules. Each is bound in an identical channel about 3 nm in length and extending from the surface deep into the interior of the protein. The F_{430} lies at the bottom of this channel with its nickel atom coordinated with the oxygen atom of a glutamine side chain. In one form CoM lies directly above the nickel, with its thiolate sulfur providing the sixth ligand for the nickel. The long -SH-containing side chain of *heptanoyl threonine phosphate* lies within the channel with its amino acid head group blocking the entrance Fig. 4.28. In a second crystal form the mixed disulfide HTP-S-S-CoM, an expected product of the reaction, is present in the channel. Because of the distance from the -SH group of HTP and the nickel atom it is clear that there must be some motion of the methyl-CoM and that the methane formed may stay trapped in the active site until the HTP-S-S-CoA product leaves.

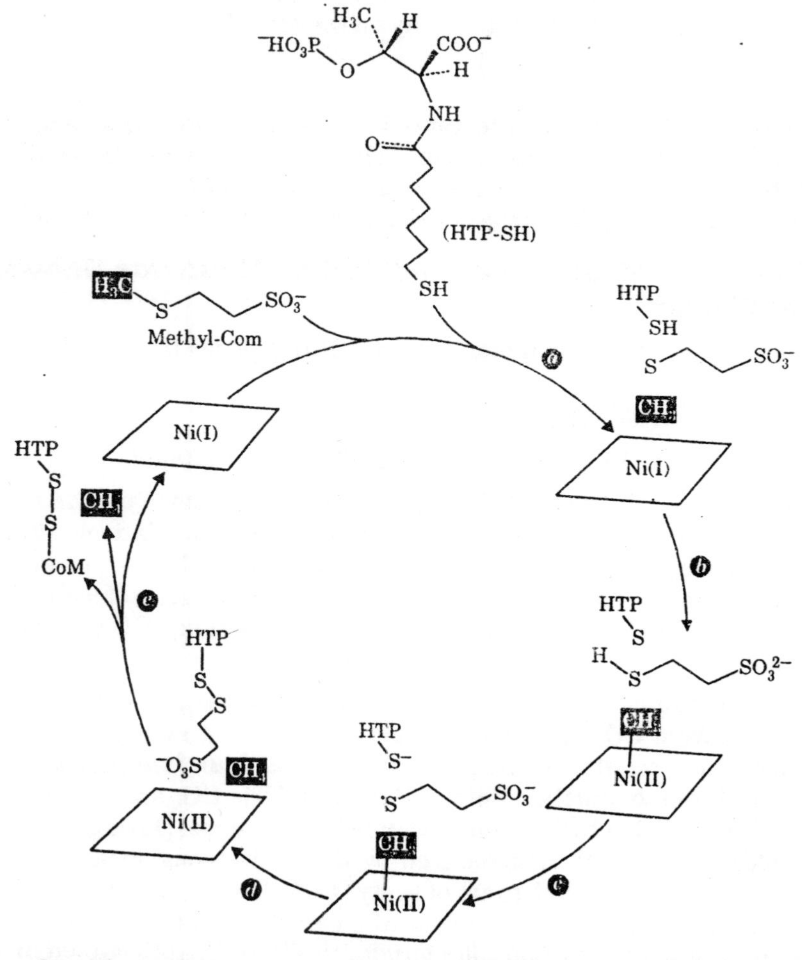

Fig. 4.28. Proposed mechanism of action of the methane-forming coenzyme M reductase. Based on the crystal structure.

A proposed mechanism for the catalytic cycle based on the X-ray results as well as previous chemical studies and EPR spectroscopy as shown in Fig. 4.28. The substrates enter. The position of the HTP, with its extended side chain, is probably the same as that seen in the X-ray structures of the Ni(II) complexes but the conformation of the methyl-CoM is different. The methyl transfer is reminiscent of that of methionine synthase (Eq. 4.43). Although the distance from the CoM sulfur and the HTP-SH is too great for direct proton transfer between the two, as indicated for there are two tyrosine hydroxyls that could provide a pathway for proton transfer. The region around the surface of the nickel coenzyme is largely hydrophobic and could facilitate formation of the thiyl radical in step *c*.

In the structure with the bound HTP-S-S-CoM heterodisulfide an oxygen atom of the sulfonate group of CoM is bonded tightly to the Ni(II). However, in the active Ni(I) form the nickel is nucleophilic and would probably repel the sulfonate, perhaps assisting the product release. The enzyme contains five posttranslationally modified amino acids near the active site: N-methyl-histidine, 5-methylarginine, 2-methylglutamine, 2-methyl-cysteine, and thioglycine in a thiopeptide bond. The latter may be the site of radical formation.

Tunichlorins

Nickel is found in various marine invertebrates. In the tunicates (sea squirts and their relatives) it occurs in a fixed ratio with cobalt, suggesting a metabolic role. A new class of nickel chelates called tunichlorins have been isolated. An example is shown in Fig. 4.27B. The function of tunichlorins is unknown but their existence suggests the possibility of unidentified biochemical roles for nickel.

Carbon Monoxide Dehydrogenases and Carbon Monoxide Dehydrogenase/ Acetyl-CoA Synthase

There are several bacterial carbon monoxide dehydrogenases that catalyze the reversible oxidation of CO to CO_2:

$$CO + H_2C \rightarrow CO_2 + H^+ + 2e^- \qquad ...(4.50a)$$

$$CO + H_2O \rightarrow CO_2 + H_2 \ \Delta G° = -38.1 \text{ kJ/mol} \qquad ...(4.50b)$$

Some bacteria use CO as both a source of energy and for synthesis of carbon compounds. The purple photosynthetic bacterium *Rhodospirillum rubrum* employs a relatively simple monomeric Ni-containing CO dehydrogenase containing one atom of Ni and seven or eight iron atoms, apparently arranged in Fe_4S_4 clusters. These bacteria can grow anaerobically with CO as the sole source of both energy and carbon. Some aerobic bacteria oxidize CO using a molybdenum-containing enzyme.

However, the most studied CO dehydrogenase is a complex enzyme that also synthe-sizes, reversibly, acetyl-CoA from CO and a methyl corrin. Employed by methanogens, acetogens, and sulfate-reducing bacteria, it is at the heart of the *Wood-Ljundahl pathway* of autotrophic metabolism, which is discussed further. Both oxidation of CO to CO_2 and reduction of CO_2 to CO are important activities of CO dehydrogenase/acetyl-CoA synthase. During growth on CO, some CO must be oxidized to CO_2 and then reduced by the pathways a methyl-tetrahydropterin which can be used to form the methyl group of acetyl-CoA. During growth on any other carbon compound CO_2 must be reduced to CO to form the carbonyl group of acetyl-CoA which can serve as a precursor to all other carbon compounds.

Native CO dehydrogenase/acetyl CoA synthase was isolated from cells of *Clostridium thermoaceticum* grown in the presence of radioactive ^{63}Ni. The protein is a 310-kDa $\alpha_2\beta_2$ oligomer. Each $\alpha\beta$ dimer contains 2 atoms of Ni, 1 of Zn, ~12 of Fe and –12 sulfide ions, which are organized into three metal clusters referred to as A, B, and C. Each cluster contains 4 Fe atoms and clusters A and C also contain 1 Ni each. Oxidation of CO occurs in the β subunits, each of which contains both cluster B, an Fe_4S_4 ferredoxin-type cluster, and cluster C, where the oxidation of CO is thought to occur. Cluster C contains 1 nickel ion as well as an Fe_4S_4 cluster that resembles that of aconitase. Cluster A, which is in the a subunit, also contains 1 atom of Ni and 4 Fe ions and is probably the site of synthesis of acetyl-CoA.

Oxidation of CO may require cooperation of the nickel ion and the Fe_4S_4 group within the C cluster.

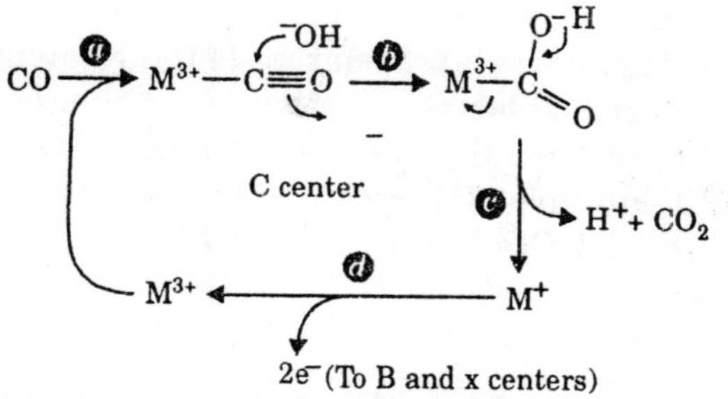

...(4.51)

CO probably binds to one of these metals and is attacked by a hydroxyl ion (Eq. 4.51 step *b*) which may be donated by the other metal of the pair. CO_2 and a proton are released rapidly (step *c*), after which the reduced metal center is reoxidized (step *d*). One electron is thought to be transferred directly to the Fe_4S_4 cluster B and the second by an alternative route. A multienzyme complex isolated from the methanogen *Methanosarcina* has an $(\alpha\beta\gamma\delta\varepsilon)_6$ structure, with the subunits having masses of 89, 60, 50, 48, and 20 kDa, respectively.

In this complex the CO dehydrogenase/acetyl-CoA synthase activity appears to reside in the $\alpha_2\varepsilon_2$ complex, the $\gamma\delta$ complex has a tetrahydropteridine: cob(I)amide-protein transferase, and the subunit has an acetyltransferase that binds acetyl-CoA and transfers the acetyl group to a group on the β subunit. Although the *Clostridium* and *Methanosarcina* systems are not identical, similar mechanisms are presumably involved. To generate acetyl-CoA a methyl group is first transferred from a tetrahydropterin such as tetrahydrofolate or, in methanogens, tetrahydromethanopterin or

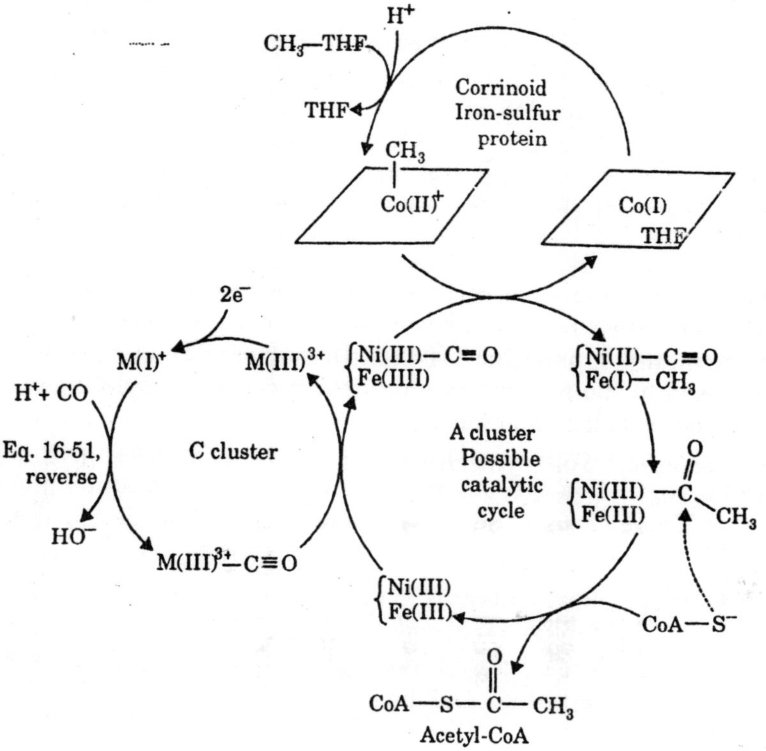

...(4.52)

tetrahydrosarcinapteri a methylcorrinoid. At the A center of the CO dehydrogenase a molecule of CO, which may be bound to the Ni, equilibrates with the methyl group, and with acetyl-CoA. As depicted in Eq. 4.52, acetyl-Ni may be an intermediate. Other details shown here are hypothetical. It is possible that the methyl group is transferred to the Ni atom in the M cluster before reaction with the CO, which might be bound to either Ni of cluster C or to Fe. This reaction of two transition-metal-bound ligands parallels a proposed industrial process for synthesis of acetic acid from methanol and CO and involving catalysis by rhodium metal and methyl iodide. It is thought that rhodium-bound CO is inserted into bound Rh-CH_3 to form an intermediate $Rh-\overset{\overset{\Large O}{\|}}{C}-CH_3$. The acetyl group is released as acetyl iodide which is hydrolyzed to acetic acid. An acetyl nickel intermediate may be involved in the corresponding biological reaction of Eq. 4.52. The stereochemistry of the sequence has been investigated using methyl-THF containing a chiral methyl group. Overall retention of the configuration of the methyl group in acetyl-CoA was observed.

D. COPPER

Copper was recognized as nutritionally essential by 1924 and has since been found to function in many cellular proteins. Copper is so broadly distributed in foods that a deficiency has only rarely been observed in humans. However, animals may sometimes receive inadequate amounts because absorption of Cu^{2+} is antagonized by Zn^{2+} and because copper may be tied up by molybdate as an inert complex. There are copper-deficient desert areas of Australia where neither plants nor animals survive. Copper-deficient animals have bone defects, hair color is lacking, and hemoglobin synthesis is impaired.

Cytochrome oxidase activity is low. The protein elastin of arterial walls is poorly crosslinked and the arteries are weak. Genetic defects in copper metabolism can have similar effects. An adult human ingests ~2-5 mg of copper per day, about 30% of which is absorbed. The total body content of copper is ~100 mg (~ 2×10^{-4} mol/kg), and both uptake and excretion (via the bile) are regulated. Since an excess of copper is toxic, regulation is important. Because Cu^{2+} is the most tightly bound metal ion in most chelating centers almost all of the copper present in living cells is complexed with proteins.

Copper is transported in the blood by a 132-kDa, 1046-residue sky-blue glycoprotein called *ceruloplasmin*. This one protein contains 3% of the total body copper. Regulation of copper uptake has been studied in most detail in the yeast *Saccharomyces cerevisiae*. Uptake of Cu^{2+} is similar to that of Fe^{3+}. The same plasma membrane reductase system, consisting of proteins Frelp and Fre2p (encoded by genes *FREI* and *FRE* 2), acts to reduce both Fe^{3+} and Cu^{2+}. These two genes are controlled in part by a transcriptional activator that responds to the internal copper concentration. Similar regulation is thought to occur in both plants and animals. The human hereditary disorders *Wilson's disease* and *Menkes' disease* have provided further insight into copper metabolism.

In Wilson's disease the ceruloplasmin content is low and copper gradually accumulates to high levels in the liver and brain. In Menkes syndrome, there is also a low ceruloplasmin level and an accumulation of copper in the form of copper metallothionein. Persons with this disease have abnormalities of hair, arteries, and bones and die in childhood of cerebral degeneration.

Similar symptoms are seen in some patients with Ehlers-Danlos syndrome. Genes for the proteins that are defective in both Wilson's and Menkes' diseases have been cloned and both proteins have been identified as P-type ATPase cation transporters. The two proteins must be similar in structure as indicated by a 55 per cent sequence identity.

Homologous genes involved in copper homeostasis have been located in both yeast and the cyanobacterium *Synechococcus.* The transporter encoded by this yeast gene, designated *Ccc2,* apparently functions to export copper from the cytosol into an extracytosolic compartment. In a similar way the Wilson and Menkes disease proteins, which reside in the *trans*-Golgi network, are thought to export copper or to provide copper for incorporation into essential proteins. The Wilson disease protein is also found in a shortened form in mitochondrial membranes. Other proteins associated with intracellular copper metabolism seem to be chaperones for Cu(I). The ability of copper ions to undergo reversible changes in oxidation state permits them to function in a variety of oxidation-reduction processes. Like iron, copper also provides sites for reaction with O_2, with superoxide radicals, and with nitrite ions.

Electron-Transferring Copper Proteins

A large group of small, intensely blue copper proteins function as single-electron carriers within bacteria and plants. Best known is *plastocyanin*, which is ubiquitous in green plants and functions in the electron transport chain between the light-absorbing photosynthetic centers I and II of chloroplasts. The bacterial *azurins* are thought to carry electrons between cytochrome c_{441} and cytochrome oxidase. *Amicyanin* accepts electrons from the coenzyme TTQ of methylamine dehydrogenase of methylotrophic bacteria and passes them to a cytochrome *c*. A basic blue copper protein *plastocyanin* of uncertain function occurs in cucumber seeds. The 10.5-kDa peptide chain of plastocyanin is folded into an eight-stranded β barrel, which contains a single copper atom.

In popular plastocyanin, the Cu is coordinated by the side chains of His 37, His 87, Met 92, and Cys 84 in a tetrahedral but distorted toward a trigonal bipyramidal geometry. Since copper-free apoplastocyanin has essentially the same structure, this geometry may be imposed by the protein onto the Cu^{2+}, which usually prefers square-planar or tetrahedral coordination. Calculations suggest that there is little or no strain and that the "Franck-Condon barrier" to electron transfer is low. The three-dimensional structure and copper environment of azurin are similar to those of plastocyanin. Messerschmidt *et al.*, suggested that the copper site in these proteins is perfectly adapted to its function because its geometry is a compromise between the optimal geometries of the Cu(I) and Cu(II) states between which it alternates. *Stellacyanin*, present in the Japanese lac tree and some other plants, is a mucoprotein; the 108-residue protein is over 40% carbohydrate. While its spectrum resembles that of plastocyanins and azurins, Stellacyanin contains no methionine and this amino acid cannot be a ligand to copper.

Rusticyanin functions in the periplasmic space of some chemolithotropic sulfur bacteria to transfer electrons from Fe^{2+} to cytochrome *c* as part of the energy-providing reaction for these organisms. *Halocyanin* functions in membranes of the archaeobacterium *Natronobacterium,* and *aurocyanin* functions in green photosynthetic bacteria. The blue colour of these "type 1" copper proteins is much more intense than are the well known colors of the hydrated ion $Cu(H_2O)_4^{2+}$ or of the more strongly absorbing $Cu(NH_3)_4^{2+}$. The blue colour of these simple complexes arises from a transition of an electron from one *d* orbital to another within the copper atom.

The absorption is somewhat more intense in copper peptide chelates of the type. However, the ~600 nm absorption bands of the blue proteins are an order of magnitude more intense, as is illustrated by the absorption spectrum of azurin. The intense blue is thought to arise as a result of transfer of electronic charge from the cysteine thiolate to the Cu^{2+} ion. A third type of copper center, first recognized in cytochrome *c* oxidase called Cu_A or *purple CuA*. Each copper ion is bonded to an imidazole and two cysteines serve as bridging ligands. The two copper ions are about 0.24 nm apart, and the two Cu^{2+} ions together can accept a single electron from an external donor such as cytochrome *c* or azurin to give *a*. half-reduced form.

Copper, Zinc-Superoxide Dismutase

Although of similar topology to the blue electron-transferring proteins, Cu, Zn-superoxide dismutase, has a different function. This dimeric 153-residue protein has been demonstrated in the cytoplasm of virtually all eukaryotic cells and in the periplasmic space of some bacteria where it converts superoxide ions $^{\bullet}O_2^-$ to O_2 and H_2O_2. The enzyme, which has a major protective role against oxidative damage to cells, presumably functions in a manner similar to that indicated iron or manganese. However, copper cycles between Cu^{2+} and Cu^+, alternately accepting and donating electrons. The active site of cytosolic superoxide dismutase (SOD) contains both Cu^{2+} and Zn^{2+}.

The copper ion is of "type 2": nonblue and paramagnetic. It is surrounded by four imidazole groups with an irregular square planar geometry. One of these imidazole groups (that of His 61) is shared with the Zn^{2+}, which is also bonded to two additional imidazole groups and a side chain carboxylate. The metal ions have evidently replaced the hydrogen atom that would otherwise be present on the imidazole of His 61. It has been suggested, as is also indicated in the diagram, that when the bound superoxide donates an electron to the Cu(II) to become O_2, a proton becomes attached to the bridging imidazole with breakage of its linkage to the Cu(I). The structure of a new crystalline form of reduced yeast SOD shows that the Cu(I) has moved 0.1 nm away from the bridging imidazole in agreement with this possibility. In the second half reaction the imidazole proton, together with a second proton from the medium and an electron from the Cu(I), would react to convert the second O_2^- into H_2O_2. The role of the Zn^{2+} may be in part structural but it may also serve to ensure that His 61 is protonated on the correct nitrogen atom. Arg 141 may assist in binding the O_2^- as is shown in the diagram. However, the fact that a mutant containing leucine in place of the active site arginine has over 10% of the activity of the native enzyme shows that the arginine is not absolutely essential.

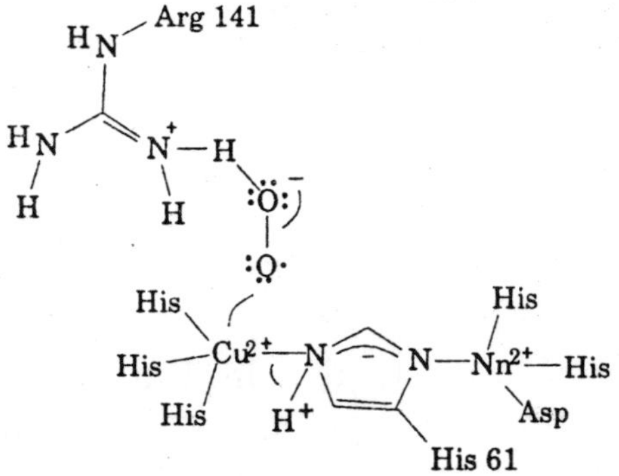

Additional nearby positively charged arginine and lysine side chains may provide "electrostatic guidance" that increases the velocity of reaction of superoxide ions. Cu, Zn-SOD is one of the fastest enzymes known. In addition to the cytosolic SOD there is a longer ~222-residue extracellular form that binds to the proteoglycans found on cell surfaces. Manganese SODs are found in mitochondria and in bacteria and iron SODs in plants and bacteria. They all appear to be important in protecting cells from superoxide radicals. This importance was dramatically emphasized when it was found that a defective SOD is present in persons (about 1 in 100,000) with a hereditary form of *amylotrophic lateral sclerosis* (ALS), which is also called Lou Gehrig's disease after the baseball hero who was stricken with this terrible disease of motor neurons in 1939 at the age of 36.

Nitrite and Nitrous Oxide Reductases

Copper enzymes participate in two important reactions catalyzed by denitrifying bacteria. Nitrite reductases from species of *Achromobacter* and *Alcaligenes* are trimeric proteins made up of 37-kDa subunits, each of which contains one type 1 (blue) copper and one type 2 (nonblue) copper. The first copper serves as an electron acceptor from a small blue *pseudoazurin*. The second copper, which is in the active site, is thought to bind to nitrite throught its nitrogen atom and to reduce it to NO.

$$NO_2^- + 2H^+ + e^- \rightarrow NO + H_2O \quad ...(4.53)$$

Crystallographic studies on the 343-residue *Alcaligenes* enzyme reveal two (β barrel domains with the type 1 copper embedded in one of them and the type 2 copper in an interface between the domains. Studies of EPR and ENDOR spectra and of various mutant forms have shown that, as for other copper enzymes, the type 1 copper is an electron-transferring center, accepting electrons from the pseudoazurin and passing them to the type 2 copper which binds and reduces nitrite.

The immediate product of nitrite reductase is NO, which is reduced in two one-electron steps to N_2O, and then to N_2. The second of these steps is catalyzed by another copper enzyme, nitrous oxide reductase.

$$N_2O + 2H^+ + 2e^- \rightarrow N_2 + H_2O \quad ...(4.54)$$

The biological significance of these reactions is considered further. The 132-kDa dimeric N_2O reductase from *Pseudomonas stutzeri* contains four copper atoms per subunit. One of its copper centers resembles the Cu_A centers of cytochrome *c* oxidase. A second copper center consists of four copper ions, held by seven histidine side chains in a roughly tetrahedral array around one sulfide (S^{2-}) ion. Rasmussen *et al.* speculate that this copper-sulfide cluster may be an acceptor of the oxygen atoms of N_2O in the formation of N_2. There is also a cytochrome cd_1 type of nitrite reductase.

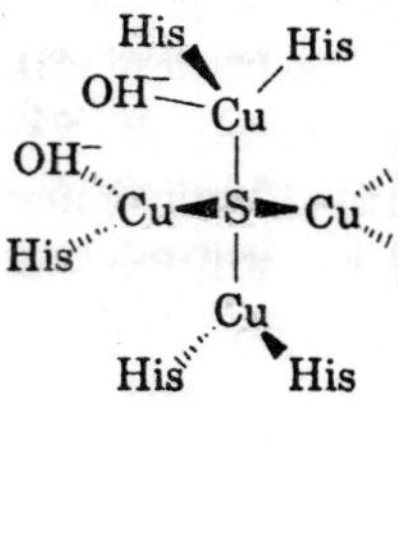

His His
OH⁻ Cu
OH⁻ His
Cu S Cu
His His
Cu
His His

Hemocyanins

While many copper proteins are catalysts for oxidative reactions of O_2, hemocyanin reacts with O_2 reversibly. This water-soluble O_2 carrier is found in the blue blood of many molluscs and arthropods, including snails, crabs, spiders, and scorpions. Hemocyanins are large oligomers ranging in molecular mass from 450 to 13,000 kDa. Molluscan hemocyanins are cylindrical oligomers which have a striking appearance under the electron microscope. Simpler hemocyanins, found in arthropods, are hexamers of 660-residue 75-kDa subunits. Each subunit of the hemocyanin from the spiny lobster is folded into three distinct domains, one of which contains a pair of Cu(I) atoms which bind the O_2. Each copper ion is held by three imidazole groups without any bridging groups between them, the Cu-Cu distance being 0.36-0.46 nm.

Octopus hemocyanin has a different fold and forms oligomers of ten subunits. However, the active sites are very similar. The O_2 is thought to bind between the two copper atoms. An allosteric mechanism may involve changes in the distance between the copper atoms. The oxygenated compound is distinctly blue with a molar extinction coefficient 5-10 times greater than that of cupric complexes. This fact suggests that the Cu(I) has been oxidized to Cu(II) and that the O_2 has been reduced to the peroxide dianion O_2^{2-} in the complex. Further support for this idea comes from the observation that treatment of oxygenated hemocyanin with glacial acetic acid leads to the formation of equal amounts of Cu^{2+} and Cu^+ and protonated superoxide.

$$(CuO_2Cu)^{2+} + H^+ \rightarrow Cu^{2+} + Cu^+ + HO_2 \quad ...(4.55)$$

Copper Oxidases

A large group of copper-containing proteins activate oxygen toward chemical reactions of dehydrogenation, hydroxylation, or oxygenation. *Galactose oxidase* (Fig. 4.29), from the mushroom *Polyporus,* is a dehydrogenase which converts the 6-hydroxymethyl group of galactose to an aldehyde while O_2 is reduced to H_2O_2.

$$R{-}CH_2OH \xrightarrow[O_2 \quad H_2O_2]{} R{-}CHO \quad ...(4.56)$$

Galactose oxidase has been used frequently to label glycoproteins of external cell membrane surfaces. Exposed terminal galactosyl or N-acetylgalactosaminyl residues are oxidized to the corresponding C-6 aldehydes and the latter are reduced under mild conditions with tritiated sodium borohydride. The single 639-residue polypeptide chain contains one type 2 copper ion. Neither oxygen nor galactose affects the absorption spectrum of the light murky green enzyme, but the combination of the two does, suggesting that both substrates bind to the enzyme before a reaction takes place. A side reaction releases superoxide ion and leaves the enzyme in the inactive Cu(II) state. EPR spectroscopic observations on the enzyme were puzzling.

The active enzyme shows no EPR signal but a one-electron reduction gives an inactive form with an EPR signal that arises from Cu(II). Experimental studies eventually pointed to the presence of a second reducible center which contains an organic free radical. In the active form this radical is *antiferromagnetically coupled* (spin-coupled) giving an "EPR-silent" enzyme able to accept two electrons. Another surprise was the discovery, from the X-ray structure, that a tyrosine side chain at the active site has been modified by addition of a thiolate group from a cysteine residue. This structure, which is also the site of the organic free radical, is shown in Fig. 4.29.

A possible mechanism of action is portrayed in Fig. 4.30. The substrate binds and its –OH group is deprotonated, perhaps by transfer of H^+ to the phertolate oxygen of tyrosine 495 (Fig. 4.29) in step *b* of Fig. 4.30. In step *c* a free radical hydrogen transfer occurs to form a ketyl radical which is immediately oxidized by Cu(II) in step *d*. In steps *c* and *f* the oxidant O_2 is converted to H_2O and the aldehyde product is also released. A fungal glyoxal oxidase has similar characteristics. The galactose oxidase proenzyme is self-processing. The Tyr-Cys cofactor arises as a result of copper-catalyzed oxidarive modification via a tyrosine free radical.

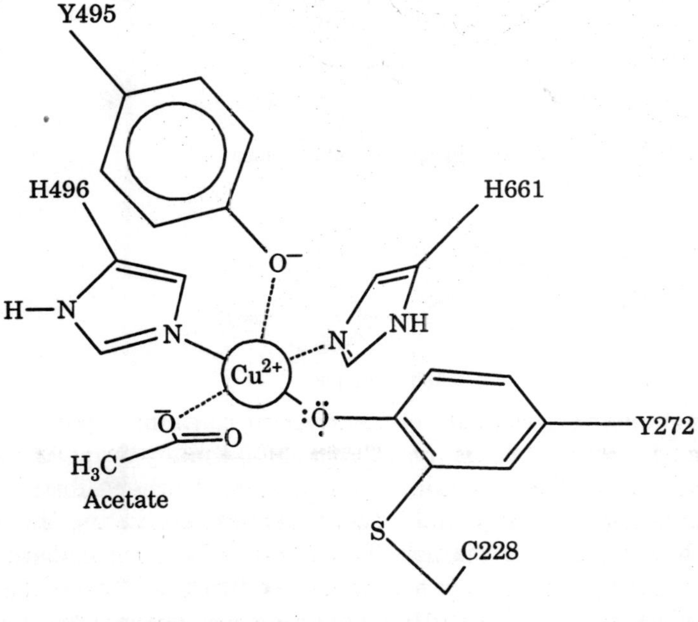

Y495

H496

H661

H—N

N

O^-

N

NH

Cu^{2+}

:Ö

O

O

Y272

H_3C

Acetate

S

C228

Fig. 4.29. Drawing of the active site of galactose oxidase showing both the Cu(II) atom and the neighbouring free radical on tyrosine 272, which has been modified by addition of the thiol of cysteine 228 and oxidation.

Several copper-containing amine oxidases convert amines to aldehydes and H_2O_2. They also contain one of the organic quinone cofactors. The dimeric *plasma amine oxidase* contains a molecule of the coenzyme TPQ and one Cu^{2+} per 90-kDa subunit. Whether the O_2 binds only to the single atom Cu(I) or also interacts directly with a cofactor radical in step *d* uncertain. *Lysyl oxidase*, which is responsible for conversion of ε-amino groups of side chains of lysine into aldehyde groups in collagen and elastin contains coenzyme LTQ as well as Cu.

The enzyme is specifically inhibited by β-aminopropionitrile, its activity is decreased in genetic diseases of copper metabolism. The *glycerol oxidase* of *Aspergillus* is a large 400 kDa protein containing one heme and two atoms of Cu. It converts glycerol + O_2 into glyceraldehyde and H_2O_2. Also containing copper is *urate oxidase*, whose action is indicated. *Tyrosinase* catalyzes hydroxylation followed by dehydrogenation. First identified in mushrooms, the enzyme has a widespread distribution in nature. It is present in large amounts in plant tissues and is responsible for the darkening of cut fruits.

In animals tyrosinase participates in the synthesis of *dihydroxyphenylalanine* (dopa) and in the formation of the black *melanin* pigment of skin and hair. Either a lack of or inhibition of this enzyme in the melanin-producing melanocytes causes *albinism*.

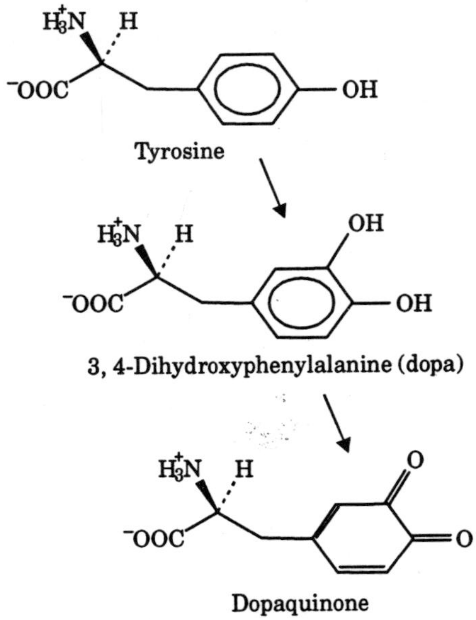

Tyrosine

3, 4-Dihydroxyphenylalanine (dopa)

Dopaquinone

...(4.57)

The 46-kDa monomeric tyrosinase of *Neurospora* contains a pair of spin-coupled Cu(II) ions. The structure of this copper pair (*type 3 copper*) has many properties in common with the copper pair in hemocyanin. For example, in the absence of other substrates, tyrosinase binds O_2 to form "oxytyrosinase", a compound with properties resembling those of oxyhemocyanin and containing a bound peroxide dianion. Tyrosinase is both an oxidase and a hydroxylase. Some other copper enzymes have only a hydroxylase function. One of the best understood of these is the *peptidylglycine α-hydroxylating monoxygenase*, which catalyzes the first step of the reaction. The enzyme is a colourless two-copper protein but the copper atoms are 1.1 run apart and do not form a binuclear center.

Ascorbate is an essential cosubstrate, with two molecules being oxidized to the semidehydroascorbate radical as both coppers are reduced to Cu(I). A ternary complex of reduced enzyme, peptide, and O_2 is formed and reacts to give the hydroxylated product. A related two-copper enzyme is *dopamine* (*β-monooxygenase*, which utilizes O_2 and ascorbate to hydroxylate dopamine to noradrenaline. These and other types of hydroxylases are compared. The *blue multicopper oxidases* couple the oxidation of substrates to the four-electron reduction of molecular oxygen to H_2O. In this respect they resemble cytochrome *c* oxidase, which also contains copper.

However, they do not contain iron. The best known member of this group is the plant enzyme *ascorbate oxidase*, which dehydrogenates ascorbic acid to dehydroascorbic acid. It is a dimeric blue copper with identical 70-kDa subunits. The three-dimensional structure revealed one type 1 copper ion held in a typical blue copper environment as in plastocyanin or azurin and also a *three-copper center*. In this center a pair of copper ions, each held by three imidazole groups, and bridged by a μ-oxo group as in hemerythrin, (Fig. 4.20) lie 0.51 nm apart and 0.41-0.44 nm away from the third copper, which is held by two other imidazoles. The type 1 copper shows typical intense 600-nm absorption and characteristic EPR signal, while the pair with the oxo bridge are antiferromagnetically coupled and EPR silent but with strong near ultraviolet light absorption. The additional metal ion in the trinuclear center is a type 2 copper which lacks characteristic spectroscopic features. Reduction of O_2 to 2 H_2O is thought to precede via superoxide radical intermediates. When substrate is added to the enzyme, the blue colour fades and it can be shown that the copper is reduced to the +1 state. The reduced enzyme then reacts with O_2, converting it into two molecules of H_2O. Similar to ascorbate oxidase in structure and properties are *laccase*, found in the latex of the Japanese lac tree and in the mushroom *Polyporus,* and the previously discussed *ceruloplasmin*.

Laccase is a catalyst for oxidation of phenolic compounds by a free radical mechanism involving the trinuclear copper center. Studied by Gabriel Bertrand in the 1890s, it was one of the first oxidative enzymes-investigated. In addition to its previously mentioned role in copper transport, ceruloplasmin is an amine oxidase, a superoxide dismutase, and a ferrooxidase able to catalyze the oxidation of Fe^{2+} to Fe^{3+}. Ceruloplasmin contains three consecutive homologous 350-residue sequences which may have originated from an ancestral copper oxidase gene. Like ascorbate oxidase, this blue protein contains copper of the three different types. Blood clotting factors V and VIII, and the iron uptake protein Fet3, are also closely related.

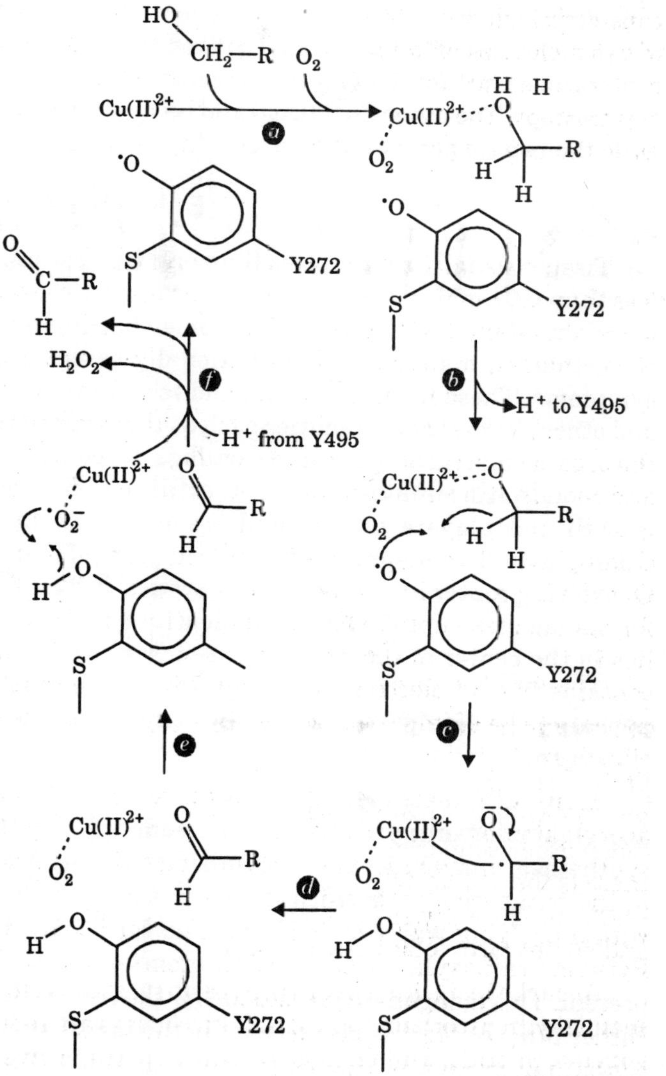

Fig. 4.30. Possible reaction cycle for catalysis of the 6-OH of D-galactose or other suitable alcohol substrate by galactose oxidase.

Cytochrome *c* Oxidase

The most studied of all copper-containing oxidases is cytochrome *c* oxidase of mitochondria. This multi-subunit membrane-embedded enzyme accepts four electrons from cytochrome *c* and uses them to reduce O_2 to 2 H_2O. It is also a proton pump. Its structure and functions are considered. However, it is appropriate to mention here that the essential catalytic centers consist of two molecules of heme *a* (*a* and a_3) and three Cu^+ ions. In the fully oxidized enzyme two metal centers, one Cu^{2+} (of the two-copper center Cu_A) and one Fe^{3+} (heme *a*), can be detected by EPR spectroscopy. The other Cu^{2+} (Cu_B) and heme a_3 exist as an EPR-silent exchange-coupled pair just as do the two copper ions of hemocyanin and of other type 3 binuclear copper centers.

E. MANGANESE

Tissues usually contain less than one part per million of manganese on a dry weight basis, less than 0.01 mM in fresh tissues. This compares with a total content in animal tissues of the more abundant Mg^{2+} of 10 mM. A somewhat higher Mn content (3.5 ppm) is found in bone. Nevertheless, manganese is nutritionally essential and its deficiency leads to well-defined symptoms. These include ovarian and testicular degeneration, shortening and bowing of legs, and other skeletal abnormalities such as the "slipped tendon disease" of chicks. In Mn deficiency the organic matrix of bones and cartilage develops poorly. The galactosamine, hexuronic acids, and chondroitin sulfates content of cartilage is decreased. Manganese is also essential for plant growth and plays a unique and essential role in the photosynthetic reaction centers of chloroplasts. Two magnetically coupled pairs of manganese ions bound in a protein act as the O_2 evolving center in photosynthetic system II. This function is considered. An ABC transporter for managanese uptake has been identified in the cyanobacterium *Synechocystis*. Manganese lies in the center of the first transition series of elements. The stable Mn^{2+} (manganous ion) contains five 3*d* electrons in a high-spin configuration. The less stable Mn^{3+} (manganic ion) appears to be of importance in some enzymes and is essential to the photosynthetic evolution of oxygen.

Many enzymes specifically require or prefer Mn^{2+}. These include galactosyl and N-acetylgalactosaminyltransferases needed for synthesis of mucopolysaccharides, lactose synthetase, and a muconate-lactonizing enzyme. *Arginase*, essential to the production of urea in the human body, specifically requires Mn^{2+} which exists as a spin-coupled dimetal center with a bridging water or $^-$OH ion. The Mn^{2+} may act much as does the Ni^{2+} ions of urease. Pyruvate carboxylase contains four atoms of tightly bound Mn^{2+}, one for each biotin molecule present. This manganese is essential for the transcarboxylation step in the action of this enzyme. Either Mn^{2+} or Mg^{2+} is also needed in the initial step of carboxylation of biotin. Another Mn^{2+}-containing protein is the lectin concanavalin A. The joining of *O*-linked oligosaccharides to secreted glycoproteins also seems to require manganese. Manganese is a component of a "pseudocatalase" of *Lactobacillus,* of lignin-degrading peroxidases, and of the wine-red superoxide dismutases found in bacteria and in the mitochondria of eukaryotes.

The dimeric dismutase from *E. coli* has a structure nearly identical to that of bacterial iron SOD. The manganese ions are presumed to alternate between the +3 and +2 states during catalysis. "Knockout" mice with inactivated Mn SOD genes live no more than three weeks, indicating that this enzyme is essential to life. However, mice lacking CuZn SOD appear normal in most circumstances. Some dioxygenases contain manganese. Many enzymes that require Mg^{2+} can utilize Mn^{2+} in its place, a fact that has been exploited in study of the active sites of enzymes. The highly paramagnetic Mn^{2+} is the most useful ion for EPR studies for investigations of paramagnetic relaxation of NMR signals. Manganese can also replace Zn^{2+} in some enzymes and may alter catalytic properties. Manganese may function in the regulation of some enzymes.

For example, glutamine synthetase one form requires Mg^{2+} for activity but upon adenylylation binds Mn^{2+} tightly.

Nucleases and DNA polymerases often show altered specificity when Mn^{2+} substitutes for Mg^{2+}. However, the significance of these differences *in vivo* is uncertain. Manganese is mutagenic in living organisms, apparently because it diminishes the fidelity of DNA replication. A striking accumulation of Mn^{2+} often occurs within bacterial spores, *Bacillus subtilus* absolutely requires Mn^{2+} for initiation of sporulation. During logarithmic growth the bacteria can concentrate Mn^{2+} from 1 μM in the external medium to 0.2 mM internally; during sporulation the concentrations become much higher.

F. CHROMIUM

Animals deficient in chromium grow poorly and have a reduced life span. They also have decreased "glucose tolerance," *i.e.,* glucose injected into the blood stream is removed only half as fast as it is normally. This is similar to the effect of a deficiency of insulin. Fractionation of yeast led to the isolation of a chromium-containing *glucose tolerance factor* which appeared to be a complex of Cr^{3+}, nicotinic acid, and amino acids. The chromium in this material is apparently well absorbed by the body but is probably not an essential cofactor. Nevertheless, dietary supplementation with chromium appears to improve glucose utilization, apparently by enhancing the action of insulin. Ingestion of glucose not only increases insulin levels in blood but also causes increased urinary loss of chromium, perhaps as a result of insulin-induced mobilization of stored chromium.

It has been suggested that a specific *chromium-binding oligopeptide* isolated from mammalian liver may be released in response to insulin and may activate a membrane phosphotyrosine phosphatase. Chromium concentrations in animal tissues are usually less than 2 μM but tend to be much higher in the caudate nucleus of the brain. High concentrations of Cr^{3+} have also been found in RNA-protein complexes. While several oxidation states, including +2, +3, and +6, are known for chromium, only Cr(III) is found to a significant extent in tissues. The Cr(VI) complex ions, chromate and dichromate, are toxic and chronic exposure to chromate-containing dust can lead to lung cancer.

Ascorbate is a principal biological reductant of chromate and can create mutagenic Cr(V) compounds that include a Cr(V)-ascorbateperoxo complex. However, Cr(III) compounds administered orally are not significantly toxic. Evidently, the Cr(VI) compounds can cross cell membranes and be reduced to Cr(III), which forms stable complexes with many constituents of cells including DNA. The use of such "exchange-inert" Cr(III) complexes of ATPin enzymology was considered. Most forms of Cr(III) are not absorbed and utilized by the body. For this reason, and because of the increased use of sucrose and other refined foods, a marginal human chromium deficiency may be wide-spread. This may result not only in poor utilization of glucose but also in other effects on lipid and protein metabolism. However, questions have been raised about the use of chromium picolinate as a dietary supplement. High concentrations have been reported to cause chromosome damage and there may be danger of excessive accumulation of chromium in the body.

G. VANADIUM

Vanadium is a dietary essential for goats and presumably also for human beings, who typically consume ~2 mg/day. However, because vanadium compounds have powerful pharmacological effects it has been difficult to establish the nutritional requirement for animals. The adult body contains only about 0.1 mg of vanadium.

Typical tissue concentrations are 0.1-0.7 μM and serum concentration may be 10 μM or less. Vanadium can assume oxidation states ranging from +2 to +5, the vanadate ion VO_4^{3-} being the predominant form of V(V) in basic solution and in dilute solutions at pH 7. However, at millimolar concentrations, $V_3O_9^{3-}$, $V_4O_{12}^{4}$, and other polynuclear forms predominate. In plasma most exists as metavanadate, VO_2–, but within cells it is reducted to the vanadyl cation VO^{2+} which is an especially stable double-bonded unit in compounds of V(IV). Only at very low pH is V(III) stable. The first suggestion of a possible biochemical function for vanadium came from the discovery that *vanadocytes*, the green blood cells of tunicates (sea squirts), contain – 1.0 M V(III) and 1.5-2 M H_2SO_4.

It was proposed that a V-containing protein is an oxygen carrier. However, the V^{3+} appears not to be associated with proteins and it does not carry O_2. It may be there to poison predators. The vanadium-accumulating species also synthesize several complex, yellow catechol-type chelating agents (somewhat similar to enterobactin (Fig. 4.1); which presumably complex V(V) and perhaps also reduce it to V(III). Vanadium is also accumulated by other marine organisms and by the mushroom *Amanita muscaria*. *Vanadoproteins* are found in most marine algae and seaweed and in some lichens. Among these are *haloperoxidases*, enzymes that are quite different from the corresponding heme peroxidases.

The vanadium is bound as hydrogen vanadate, HVO_4^{2-}, in trigonal bipyrimidal coordination with the three oxygens in equatorial positions and a histidine in one axial position. In the crystal structure an azide (N_3–) ion occupies the other axial position, but it is presumably the site of interaction with peroxide. The structure is similar to that of acid phosphatases inhibited by vanadate. Many nitrogen-fixing bacteria contain genes for a vanadium-dependent nitrogenase that is formed only if molyb-denum is not available. The nitrogenases are discussed. Much of current interest in vanadium stems from the discovery that vanadate (HVO_4^{2-} at pH 7) is a powerful inhibitor of ATPases such as the sodium pump protein (Na^+ + K^+)ATPase, of phosphatases, and of kinases. This can be readily understood from comparison of the structure of phosphate and vanadate ions.

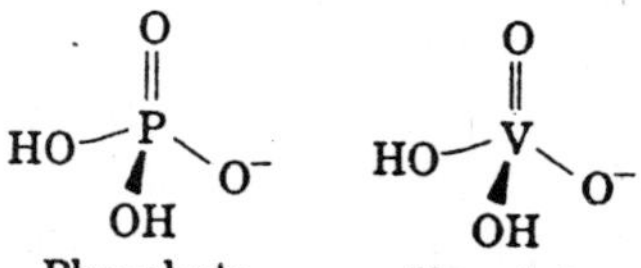

$HO-P(=O)(OH)-O^-$ Phosphate $HO-V(=O)(OH)-O^-$ Vanadate

Other enzymes such as the cyclic AMP-dependent protein kinase are *stimulated* by vanadium. Vanadate seems to inhibit most strongly those enzymes that form a phosphoenzyme intermediate. This inhibition may be diminished within cells because vanadate is readily reduced by glutathione and other intracellular reductants. The resulting vanadyl ion is a much weaker inhibitor and also stimulates several metabolic processes. Also of great interest is an insulin-like action of vanadium and evidence that vanadium may be essential to proper cardiac function. A role in lipid metabolism was suggested by the observation that in high doses vanadium inhibits cholesterol synthesis and lowers the phospholipid and cholesterol content of blood.

Vanadium is reported to inhibit development of caries by stimulating mineralization of teeth. Unlike tungsten, vanadium does not compete with molybdenum in the animal body. The sometimes dramatic effects of vanadate as an inhibitor, activator, and metabolic regulator are shared also by molybdate and tungstate. Even greater effects are observed with vanadate, molybdate, or tungstate plus H_2O_2. The resulting *pervanadate*, *permolybdate*, and *pertungstate* are often assumed to be monoperoxo compounds, *e.g.*, vanodyl hydroperoxide. However, there is some uncertainty.

H. MOLYBDENUM

Long recognized as an essential element for the growth of plants, molybdenum has never been directly demonstrated as a necessary animal nutrient. Nevertheless, it is found in several enzymes of the human body, as well as in 30 or more additional enzymes of bacteria and plants. *Aldehyde oxidases*, *xanthine oxidase* of liver and the related *xanthine* dehydrogenase, catalyze the reactions 4.58 and 4.59. Contain molybdenum that is essential for catalytic activity. Xanthine oxidase also contains two Fe_2S_2 clusters and bound FAD. The enzymes can also oxidize

$$R-CHO \xrightarrow[2e^-]{H_2O \quad 2H^+} R-COOH \qquad ...(4.58)$$

xanthine further (Eq. 4.59 step *b*) repetition of the same type of oxidation process at positions 8 and 9 to form *uric acid*. The much studied xanthine dehydrogenase has been isolated from milk, liver, fungi, and some bacteria. In the dehydrogenase NAD^+ is the electron acceptor that oxidizes the bound $FADH_2$ formed in Eq. 4.59. Xanthine dehydrogenase, in the absence of thiol compounds, is converted spontaneously into xanthine oxidase, probably as a result of a conformational change and formation of a disulfide bridge within the protein. Treatment with thiol compounds such as dithiothreitol reconverts the enzyme to the dehydrogenase. Evidently in the oxidase form the NAD^+ binding site has moved away from the FAD, permitting oxidation of $FADH_2$ by O_2 with formation of hydrogen peroxide.

Hypoxanthine —(a) H_2O, $2H^+$, $2e^-$ (to FAD)→ Xanthine —(b) H_2O, $2H^+ + 2e^-$ (to FAD)→ Uric acid ...(4.59)

A purine hydroxylase from fungi, bacterial quinoline and isoquinoline oxidoreductases, and a selenium-containing nicotinic acid hydroxylase from *Clostridium barberei* are members of the *xanthine oxidase family* (or molybdenum hydroxylase family). Also included in the family are aldehyde oxidoreductases from the sulfate-reducing *Desulfovibrio gigas* and from the tomato.

Two other families of molybdoenzymes are the *sulfite oxidase* family and the *dimethylsulfoxide reductase family*. *Nitrogenase* constitutes a fourth family. Sulfite oxidase essential human liver enzyme.

$$SO_3^{2-} + H_2O \rightarrow SO_4^{2-} + 2e^- + 2H^+ \qquad ...(4.60)$$

The *assimilatory nitrate reductase* (Eq. 4.61) of fungi and green plants also belongs to the sulfite oxidase family.

$$NO_3^{2} + 2e^- + 2H^+ \rightarrow NO_2- + H_2O \qquad ...(4.61)$$

DMSO reductase reduces dimethylsulfoxide to dimethylsulfide (Eq. 4.62) as part of the biological sulfur cycle.

$$(H_3C)_2S = O + 2H^+ + 2e^- \rightarrow (CH_3)_2S \qquad ...(4.62)$$

A number of other reductases and dehydrogenases, including *dissimilatory nitrate reductases* of *E. coli* and of denitrifying bacteria, belong to the DMSO reductase family. Other members are reductases for biotin *S*-oxide, trimethylamine *N*-oxide, and polysulfides as well as *formate dehydrogenases* (Eq. 4.63), formylmethanofuran dehydrogenase,

$$HCOO^- + 2e^- + 3H^+ \rightarrow CO_2 + 2H_2O \qquad ...(4.63)$$

and arsenite oxidase. Several other molyb-doenzymes, such as pyridoxal oxidase, had not been classified by 1996.

Molybdenum Ions and Coenzyme Forms

Molybdenum is a metal of the second transition series, one of the few heavy elements known to be essential to life. Its most stable oxidation state, Mo(VI), has 4*d* orbitals available for coordination with anionic ligands. Coordination numbers of 4 and 6 are preferred, but molybdenum can accommodate up to eight ligands. Most of the complexes are formed from the oxycation $Mo(VI)O_2^{2+}$. If two molecules of water are coordinated with this ion, the protons are so acidic that they dissociate completely to give $Mo(VI)O_4^{2-}$, the molybdate ion. Other oxidation states vary from Mo(III) to Mo(V). In these lower oxidation states, the tendency for protons to dissociate from coordinated ligands is less, *e.g.*, $Mo(III)(H_2O)_6^{3+}$ does not lose protons even in a very basic medium.

Molybdenum tends to form dimeric or polymeric oxygen-bridged ions. However, within the enzymes it exists as the unique *molybdenum coenzymes*. The Mocontaining enzymes usually also contain additional bound cofactors, including Fe-S clusters and flavin coenzymes or heme. The recognition that the Mo in the molybdoproteins exists in organic cofactor forms came from studies of mutants of *Aspergillus* and *Neurospora*. In 1964, Pateman and associates discovered mutants that lacked both nitrate reductase and xanthine dehydrogenase. Later, it was shown that acid-treated molybdoenzymes released a material that would restore activity to the inactived nitrate reductase from the mutant organisms. This new coenzyme, a phosphate ester of molybdopterin, was characterized by Rajagopalan and coworkers. A more complex form of the coenzyme, *molybdopterin cytosine dinucleotide* (Fig. 4.31) is found in the D. *gigas* aldehyde oxido-reductase.

Related coenzyme forms include nucleotides of adenine, guanine hypoxanthine. The structure of molybdopterin is related to that of *urothione*, a normal urinary constituent. The relationship to urothione was strengthened by the fact that several children with severe neurological and other symptoms were found to lack both sulfite oxidase and xanthine dehydrogenase as well as the molybdenum cofactor and urinary urothione.

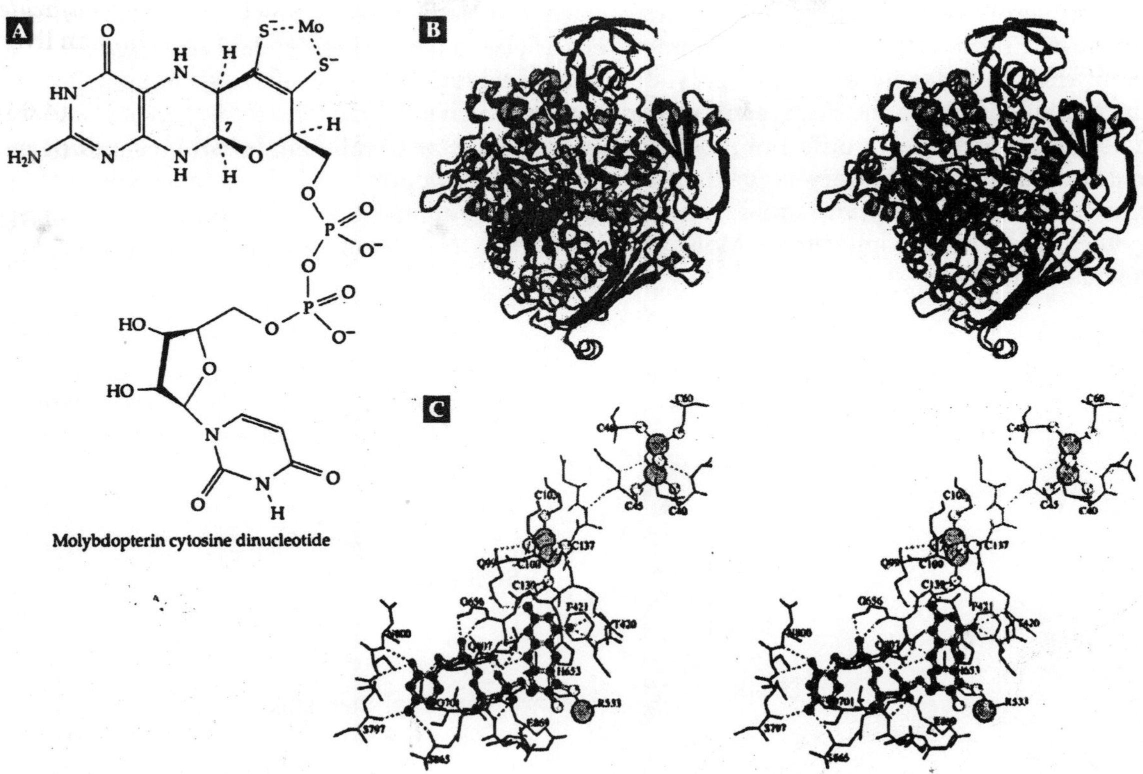

Fig. 4.31. (A) Structure of molybdopterin cytosine dinucleotide complexed with an atom of molybdenum. (B) Stereo-scopic ribbon drawing of the structure of one subunit of the xanthine oxidase-related aldehyde oxidoreductase from *Desulfovibrio gigas*. Each 907-residue subunit of the homodimeric protein contains two Fe_2S2 dusters visible at the top and the molybdenum-molybdopterin coenzyme buried in the center. (C) Alpha-carbon plot of portions of the protein surrounding the molybdenum-molybdopterin cytosine dinucleotide and (at the top) the two plant-ferredoxin-like Fe_2S_2 clusters. Each of these is held by a separate structural domain of the protein. Two additional domains bind the molybdopterin coenzyme and there is also an intermediate connecting domain. In xanthine oxidase the latter presumably has the FAD binding site which is lacking in the *D. gigas* enzyme.

Study by X-ray absorption spectroscopy of the extended *X-ray absorption fine structure* (EXAFS) has provided estimates of both the nature and the number of the nearest neighbouring atoms around the Mo. The EXAFS spectra of xanthine dehydrogenase and of nitrate reductase from *Chlorella* confirmed the presence of both the $Mo(VI)O_2$ unit with Mo-O distances of 0.17 run and two or three sulfur atoms at distances of 0.24 nm. The two sulfur atoms were presumed to come from the molybdopterin. A peculiarity of the xanthine oxidase family is the presence on the molybdenum of a "cyanolyzable" sulfur. This is a sulfide attached to the molybdenum, which is present as Mo(VI)OS rather than $Mo(VI)O_2$. Reaction with cyanide produces thiocyanate (Eq. 4.64).

$$\left[\text{O}{=}\text{Mo(VI)}{=}\text{S} \right]^{2+} \xrightarrow{CN^- \quad SCN^-} \left[\text{O}{=}\text{Mo(VI)}{=}\text{S} \right]^{2+} \qquad ...(4.64)$$

The active site structures of the three classes of molybdenum-containing enzymes are compared. In the DMSO reductase family there are two identical molybdopterin dinucleotide coenzymes complexed with one molybdenum. However, only one of these appears to be functionally linked to the Fe_2S_2 center. Nitrogenase, which catalyzes the reduction of N_2 to two molecules of NH_3, has a different *molybdenum-iron cofactor* (FeMo-co). It can be obtained by acid denaturation of the very oxygenlabile iron-molybdenum protein of nitrogenase followed by extraction with dimethylformamide. The coenzyme is a complex Fe-S-Mo cluster also containing *homocitrate* with a composition $MoFe_7S_9$-homocitrate. Nitrogenase and this coenzyme are considered further.

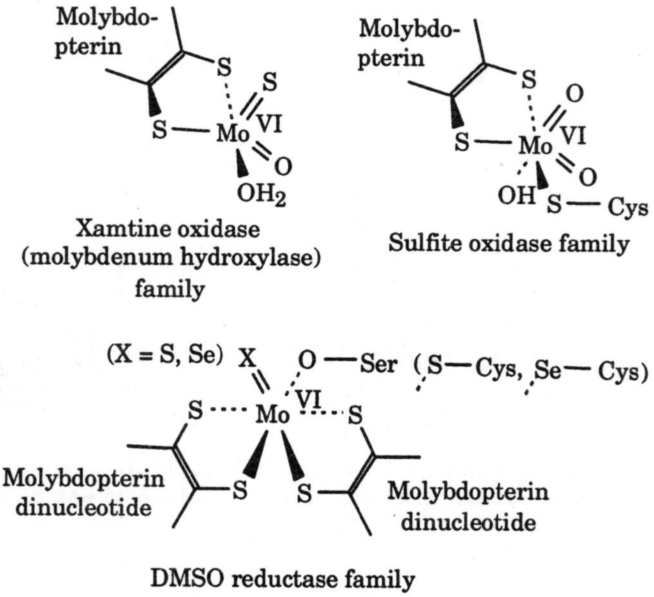

Xamtine oxidase (molybdenum hydroxylase) family

Sulfite oxidase family

DMSO reductase family

Fig. 4.32. Structures surrounding molybdenum in three families of molybdoenzymes..

Enzymatic Mechanisms

Although several of the reactions catalyzed by molybdoenzymes are classified as dehydrogenases, all of them except nitrogenase involve H_2O as either a reactant or a product. The EXAFS spectra suggest that the $Mo(VI)O_2$ unit is converted to Mo(IV)O during reaction with a substrate (Eq. 4.65 step *a*). Reaction of the Mo(IV)O with water completes the catalysis.

$$\overset{a}{[Mo(VI)O_2]^{2+} + Sub \rightarrow [Mo(IV)O]^{2+} + Sub\text{-}O}$$

$$\overset{b}{[Mo(IV)O]^{2+} + H_2O \rightarrow 2H^+ + 2e^- + [Mo(VI)]^{2+}O_2}$$

$$\text{SumSub} + H_2O \rightarrow Sub\text{-}O + 2H^+ + 2e^- \quad ...(4.65)$$

Step *a* of all of these reactions can be regarded as an *oxo-transfer*.

To complete the reaction, two electrons must be passed from Mo(IV) to a suitable acceptor, usually an Fe-S cluster or a bound heme group. FAD is also often present. Xanthine oxidase contains two Fe_2S_2 clusters and aiFAD for each of the two atoms of Mo in the dimer. Since this enzyme acts like a typical flavin oxidase that generates H_2O_2 from O_2, it may be that electrons pass from Mo to the Fe-S center and then to the flavin. Since the EPR signal of the paramagnetic Mo(V), with its characteristic six-line hyperfine structure, is seen during the action of xanthine oxidase and other molybdenum-containing enzymes, single-electron transfers are probably involved. In bacteria such as *E. coli* a dissimilatory nitrate reductase allows nitrate to serve as an oxidant in place of O_2.

An oxygen atom is removed from the nitrate to form nitrite as two electrons are accepted from a membrane-bound cytochrome *b*. The nitrate reductase consists of a 139-kDa Mo-containing catalytic subunit, a 58-kDa electron-transferring subunit that contains both Fe_3S_4 and Fe_4S_4 centers, and a 26-kDa heme-containing membrane anchor subunit. The assimilatory nitrate reductase of fungi, green algae, and higher plants contains both a *b*-type cytochrome and FAD and a molybdenum coenzyme in a large oligomeric complex. Formate dehydrogenases from many bacteria contain molybdopterin and also often selenium.

A membrane-bound Mocontaining formate dehydrogenase is produced by *E. coli* grown anaerobically in the presence of nitrate. Under these circumstances it is coupled to nitrate reductase via an electron-transport chain in the membranes which permits oxidation of formate by nitrate. This enzyme is also a multisubunit protein. Two other Mo- and Se- containing formate dehydrogenases are produced by *E. coli*. The three-dimensional structure is known for one of them, *formate dehydrogenase H*, a component of the anaerobic formate hydrogen lyase complex. The structure shows Mo held by the sulfur atoms of two molybdopterin molecules, as in DMSO reductase. The Se atom of SeCys 140 is also coordinated with the Mo atom, and the imidazole of His 141 is in close proximity. When ^{13}C-labeled formate was oxidized in ^{18}O-enriched water no ^{18}O was found in the released product, CO_2. This suggested that formate may be bound to Mo and dehydrogenated, with Mo(VI) being reduced to Mo(IV).

The formate hydrogen might be transferred as H^+ to the His 140 side chain. Mo(IV) could then be reoxidized by electron transfer in two one-electron steps. However, recent X-ray absorption spectra suggest the presence in the enzyme of a selenosulfide ligand to Mo. Mechanistic uncertainties remain! A flavin-dependent formate dehydrogenase system found in *Methanobacterium* passes electrons from dehydrogenation of formate to FAD and then to the deazaflavin coenzyme $F_{42}0$. In contrast to these Mocontaining enzymes, the formate dehydrogenase from *Pseudornonas oxalaticus,* which oxidizes formate with NAD^+ (Eq. 4.66), contains neither Mo or Se.

$$HCOO^- + NAD^+ \rightarrow CO_2 + NADH \qquad ...(4.66)$$

It is a large 315-kDa oligomer containing 2 FMN and ~20 Fe/S. Formate dehydrogenases of green plants and yeasts are smaller 70- to 80-kDa proteins lacking bound prosthetic groups. A key enzyme in the metabolism of carbon monoxide-oxidizing bacteria is CO oxidase, another membrane-bound molybdo-enzyme. It also contains selenium, which is attached to a cysteine side chain as *S-selanylcysteine*. A proposed reaction requence is shown in Eq. 4.67.

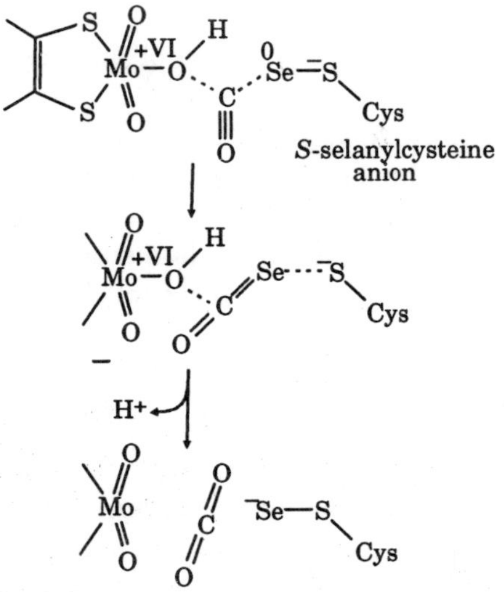

S-selanylcysteine anion

H^+

...(4.67)

Nutritional Need for Mo

The first hint of an essential role of molybdenum in metabolism came from the discovery that animals raised on a diet deficient in molybdenum had decreased liver xanthine oxidase activity. There is no evidence that xanthine oxidase is essential for all life, but a human genetic deficiency of sulfite oxidase or of its molybdopterin coenzyme can be lethal. The conversion of molybdate into the molybdopterin cofactor in *E. coli* depends upon at least five genes. In *Drosophila* the addition of the cyanolyzable sulfur (Eq. 4.64) is the final step in formation of xanthine dehydrogenase.

It is of interest that sulfur (S^0) can be transferred from rhodanese, or from a related mercaptopyruvate sulfurtransferase into the desulfo form of xanthine oxidase to generate an active enzyme. Uptake of molybdate by cells of *E. coli* is accom-plished by an ABC-type transport system. In some bacteria, *e.g.,* the nitrogen-fixing *Azotobacter,* molybdenum can be stored in protein-bound forms.

I. TUNGSTEN

For many years tungsten was considered only as a potential antagonist for molybdenum. However, in 1970 growth stimulation by tungsten compounds was observed for some acetogens, some methanogens, and a few hyperthermophilic bacteria. Since then over a dozen tungstoenzymes have been isolated. These can be classified into three categories: aldehyde oxidoreductases, formaldehyde oxidoreductases, and the single enzyme acetylene hydratase. In most cases the tungstoenzymes resemble the corresponding molybdoenzymes and in most instances organisms containing a tungsten-requiring enzyme also contain the corresponding molybdenum enzyme.

However, a few hyperthermophilic archaea appear to require W and are unable to use Mo. The aldehyde ferredoxin oxidoreductase from the hyperthermophile *Pyrococcus furiosus* was the first molybdopterin-dependent enzyme for which a three-dimensional structure became available. The tungstoenzyme resembles that of the related molybdoenzyme. A similar ferredoxin-dependent enzyme reduces glyceraldehyde-3-phosphate. Another member of the tungstoenzyme aldehyde oxidoreductase family is *carboxylic acid reductase*, an enzyme found in certain acetogenic clostridia. It is able to use reduced ferredoxin to convert unactivated carboxylic acids into aldehydes, even though $E^{\circ\prime}$ for the acetaldehyde/acetate couple is –0.58 V. Tungsten- and sometimes Se-containing formate dehydrogenases together with *N*-formylmethanofuran dehydrogenases a second family. Again, these appear to resemble the corresponding Mo-dependent enzymes.

The unique *acetylene hydratase* from the acetylene-utilizing *Pelobacter acetylenicus* catalyzes the hydration of acetylene to acetaldehyde.

$$H\text{–}C \equiv C\text{–}H + H_2O \rightarrow H_3C\text{–}CHO \qquad \text{...(4.68)}$$

In *Ttiermotoga maritima,* the most thermophilic organism known, tungsten promotes synthesis of an Fe-containing hydrogenase as well as some other enzymes but seems to have a regulatory rather than a structural role.

MAGNETIC IRON OXIDE IN ORGANISMS

An unusual form of stored iron is the magnetic iron oxide *magnetite* (Fe_3O_4). Honeybees, monarch butterflies, homing pigeons, migrating birds, and even magnetotactic bacteria' contain deposits of Fe_3O_4 that are suspected of being used in nagivation. Some bacteria have magnetic iron sulfide particles. Human beings have magnetic bones in their sinuses and in their brains' and may be able to sense direction magnetically. A set of possible magnetoreceptor cells, as well as associated nerve pathways, have been identified in trout. In the magnetotactic bacteria found in the Northern Hemisphere the magnetic domains are oriented parallel with the axis of motility of the bacteria which tend to swim toward the geomagnetic North and downward into sediments. Similar bacteria from the Southern Hemisphere prefer to swim south and downward.

The magnetic polarity of the bacterial magnetite crystals can be reversed by strong magnetic pulses, after which the bacteria swim in the direction opposite to their natural one. Magnetic ferritin can be produced artificially in the laboratory. The resulting particles may have practical uses, for example, in medical magnetic imaging. Magnetic materials in the human body are of interest not only in terms of a possible sensory function but also because of possible effects of electromagnetic fields on human health and behaviour.

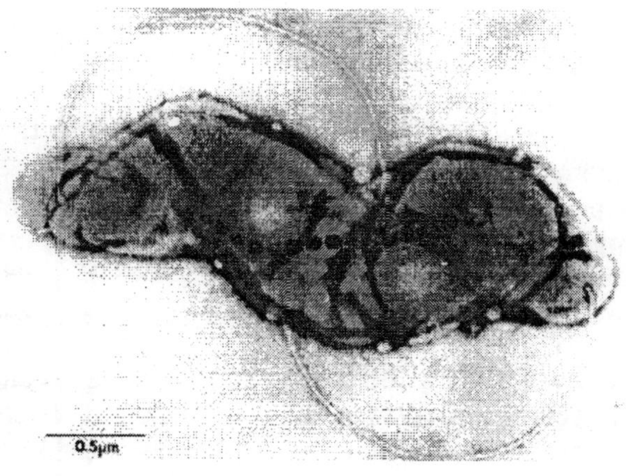

Magnetotactic soil bacterium containing 36 magnetite-containing magnetosomes.

COBALAMIN (VITAMIN B_{12})

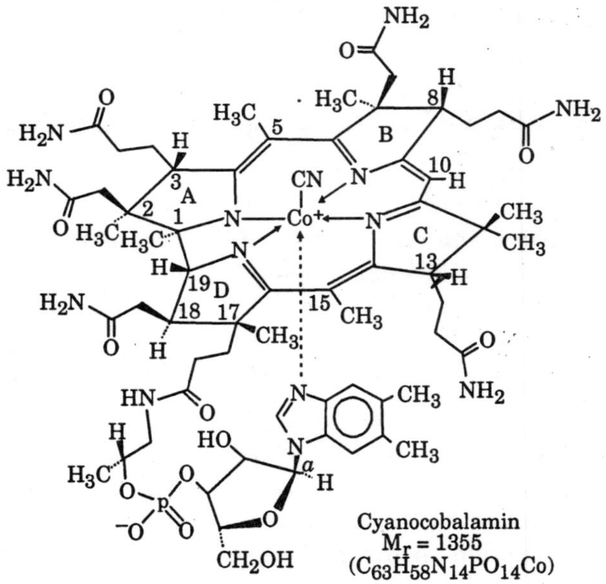

Cyanocobalamin
$M_r = 1355$
($C_{63}H_{58}N_{14}PO_{14}Co$)

The story of vitamin B_{12} began with pernicious anemia, a disease that usually affects only persons of age 60 or more but which occasionally strikes children. Before 1926 the disease was incurable and usually fatal. Abnormally large, immature, and fragile red blood cells are produced but the total number of erythrocytes is much reduced from $4 - 6 \times 10^6$ mm^{-3} to $1 - 3 \times 10^6$ mm^{-3}. Within the bone marrow mitosis appears to be blocked and DNA synthesis is suppressed. The disease also affects other rapidly growing tissues such as the gastric mucous membranes (which stop secreting HC1) and nervous tissues. Demyelination of the central nervous system with loss of muscular coordination (ataxia) and psychotic symptoms is often observed.

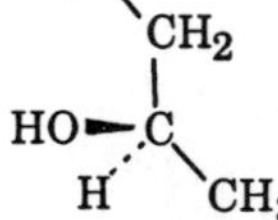

R-1-Amino-2-propanol   *α*-Ribofuranoside of dimethylimidazole

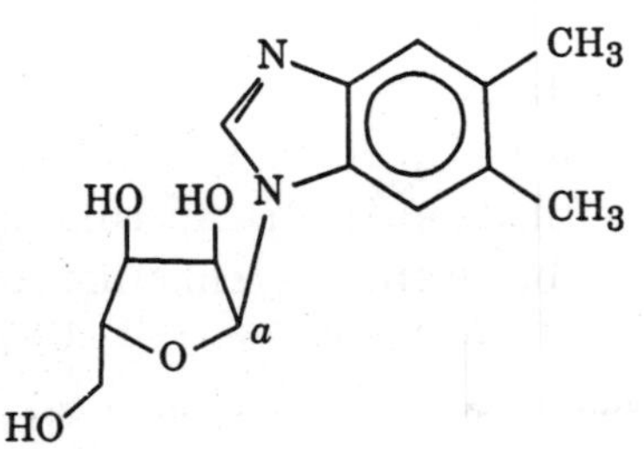

In 1926, Minot and Murphy discovered that pernicious anemia could be controlled by eating one-half pound of raw or lightly cooked liver per day, a treatment which not all patients accepted with enthusiasm. Twenty-two years later vitamin B_{12} was isolated (as the crystalline derivative cyanocobalamin) and was shown to be the curative agent. It is present in liver to the extent of

1 mg kg^{-1} or ~10^{-6} M. Although much effort was expended in preparation of concentrated liver extracts for the treatment of pernicious anemia, the lack of an assay other than treatment of human patients made progress slow. In the early 1940s nutritional studies of young animals raised on diets lacking animal proteins and maintained out of contact with their own excreta (which contained vitamin B_{12}) demonstrated the need for "animal protein factor" which was soon shown to be the same as vitamin B_{12}.

The animal feeding experiments also demonstrated that waste liquors from streptomyces fermentations used in production of antibiotics were extremely rich in vitamin B_{12}. Later this vitamin was recognized as a growth factor for a strain of *LactobaciHus lactis* which responded with half-maximum growth to as little as 0.013 μg/1(10^{-11}M). In 1948, red cobalt-containing crystals of vitamin B_{12} were obtained almost simultaneously by two pharmaceutical firms. Charcoal adsorption from liver extracts was followed by elution with alcohol and numerous other separation steps. Later fermentation broths provided a richer source.

Chemical studies revealed that the new vitamin had an enor-mous molecular weight, that it contained one atom of phosphorus which could be released as P_i, a molecule of aminopropanol, and a ribofuranoside of dimethyl benzimidazole with the unusual α configuration. Note the relationship of the dimethylbenzimidazole to the ring system of riboflavin. Several molecules of ammonia could be released from amide linkages by hydrolysis, but all attempts to remove the cobalt reversibly from the ring system were unsuccessful. The structure was determined in 1956 by Dorothy C. Hodgkin and coworkers using X-ray diffraction. At that time, it was the largest organic structure determined by X-ray diffraction.

The complete laboratory synthesis was accomplished in 1972. The ring system of vitamin B_{12}, like that of porphyrins, is made up of four pyrrole rings whose biosynthetic relationship to the corresponding rings in porphyrins is obvious from the structures. In addition, a number of "extra" methyl groups are present. A less extensive conjugated system of double bonds is present in the *corrin* ring of vitamin B_{12} than in porphyrins, and as a result, many chiral centers are found around the periphery of the somewhat nonplanar rings. Cyanocobalamin, the form of vitamin B_{12} isolated initially, contains cyanide attached to cobalt. It occurs only in minor amounts, if at all, in nature but is generated through the addition of cyanide during the isolation.

Hydroxocobalamin (vitamin B_{12a}) containing OH^- in place of CIST does occur in nature. However, the predominant forms of the vitamin are the coenzymes in which an alkyl group replaces the CN^- of cyanocobalamin. Intramuscular injection of as little as 3 – 6 μg of crystalline vitamin B_{12} is sufficient to bring about a remission of pernicious anemia and 1 μg daily provides a suitable maintenance dose (often administered as hydroxocobalamin injected once every 2 weeks). For a normal person a dietary intake of 2-5 μg/day is adequate.

There is rarely any difficulty in meeting this requirement from ordinary diets. Vitamin B_{12} has the distinction of being synthesized only by bacteria, and plants apparently contain none. Conseqxiently, strict vegetarians sometimes have symptoms of vitamin B_{12} deficiency. Pernicious anemia is usually caused by poor absorption of the vitamin. Absorption depends upon the *intrinsic factor*, a mucoprotein (or mucoproteins) synthesized by the stomach lining. Pernicious anemia patients often have a genetic predilection toward decreased synthesis of the intrinsic factor.

Gastrectomy, which decreases synthesis of the intrinsic factor, or infection with fish tapeworms, which compete for available vitamin B_{12} and interfere with absorption, can also induce the disease. Also essential are a plasma membrane receptor and two blood transport proteins *transcobalamin* and *cobalophilin*. The latter is a glycoprotein found in virtually every human biological fluid and which may protect the vitamin from photodegradation by light that penetrates tissues.

A variety of genetic defects involving uptake, transport, and conversion to vitamin B_{12} coenzyme forms are known. Normal blood levels of vitamin B_{12} are $\sim 2 \times 10^{-10}$ M or a little more, but in vegetarians the level may drop to less than one-half this value. A deficiency of folic acid can also cause megaloblastic anemia, and a large excess of folic acid can, to some extent, reverse the anemia of pernicious anemia and mask the disease.

5 Phosphorus Removal

Phosphorus is an essential macronutrient that spurs the growth of photosynthetic algae and cyanobacteria, leading to accelerated eutrophication of lakes. Wastewater discharges that reach lakes sensitive to eutrophication often require phosphorus removal over and above that normally taking place in primary and secondary treatment. For example, a typical municipal wastewater in the United States has a BOD_5 of 250 mg/1, a BOD_L of 370 mg/1, a COD of 500 mg/1, a TKN of 60 mg/1, and total P of 12 mg/1. Conventional primary sedimentation and secondary activated sludge reduce the effluent total P to about 6 mg/1. However, a typical effluent standard for a protected watershed is 1 mg P/l. Therefore, additional phosphorus removal is necessary. Phosphorus can be removed from wastewater prior to biological treatment, as part of biological treatment, or following biological treatment. The first and third approaches almost always are carried out by chemical precipitation of the phosphate anion (PO_4^{3-}) with Ca^{2+}, $A1^{3+}$, or Fe^{3+} cations. The second approach is the subject of this chapter, which shows how we can extend the capabilities of biotechnology processes already discussed for BOD and N removal.

Three phenomena can be exploited to remove phosphorus as part of a microbiological treatment process:

- Normal phosphorous uptake into biomass
- Precipitation by metal-salts addition to a microbiological process
- Enhanced biological phosphorus uptake into biomass

Each phenomenon and its application are described below. The latter two phenomena are the bases for so-called "advanced treatment" for phosphorus.

NORMAL PHOSPHORUS UPTAKE INTO BIOMASS

The biomass that develops normally in an aerobic biological process, such as activated sludge, contains 2 to 3% *P* in its dry weight. The stoichiometric formula for biomass can be modified to include this amount of *P*. The formula $C_5H_7O_2NP_{0.1}$ has a formula weight of 116 g/mol, of which *P* is 2.67 percent. Sludge wasted from the process removes *P* in proportion to the mass rate of sludge VSS wasted ($Q^w X_v^w$). A steady-state mass balance on total *P* is

$$0 = QP^0 - QP - Q^w X_v^w \text{ (0.0267 g P/g VSS)} \qquad ...(5.1)$$

in which P^0 and P are the influent and effluent total P concentrations, and Q is the influent flow rate. Equation 5.1 can be solved for the effluent total P concentration,

$$P = \frac{QP^0 - Q^w X_v^w (0.0267)}{Q} = P^0 - \frac{Q^w X_v^w (0.0267)}{Q} \quad ...(5.2)$$

The rate of sludge wasting is proportional to the BOD removal (ΔBOD_L) and the observed yield (Y_n):

$$Q^w X_v^w = Y_n Q (\Delta BOD_L) \quad ...(5.3)$$

The net yield depends on the SRT (θ_x), the true yield (Y), the endogenous decay rate (b), and the biodegradable fraction of the new biomass (f_d).

$$Y_n = Y \frac{1 + (1 - f_d) b\theta_x}{1 + b\theta_x} \quad ...(5.4)$$

Combining Equations 5.2 to 5.4 gives us the relationship describing how the effluent P concentration depends on the influent P concentration, the BOD_L, removal, and the SRT.

$$P = P^0 - \frac{(0.0267) Y (1 + (1 - f_d) b\theta_x)(\Delta BOD_L)}{1 + b\theta_x} \quad ...(5.5)$$

Table 5.1 compiles effluent P concentrations for a range of SRTs and BOD_L removals when $Y = 0.46$ mg VSS_a/mg BOD_L, $b = 0.1/d$, $f_d = 0.8$, and $P^0 = 10$ mg P/1. Only one scenario yields an effluent P concentration less than 1 mg/l: an SRT of 3 d and BOD_L, removal of 1,000 mg/1. Increasing SRT and decreasing BOD removal result is less P removal and failure to meet the effluent standard of 1 mg/1.

Table 5.1. Effects of SRT and BOD_L removal on the effluent P concentration when the influent P concentration is 10 mg P/1

BOD_L removal, mg BOD_L/l	*Effluent PO_4-P Concentration, mg/l*			
	3 d	*6 d*	*15 d*	*30 d*
100	9.0	9.1	9.4	9.5
300	7.0	7.4	8.1	8.5
500	5.0	5.7	6.8	7.5
1,000	0	1.4	3.6	5.1

Note: Y = 0.46 mg VSS_a/mg BOD_L, $f_d = 0.8$, $b = 0.1/d$, and biomass P content = 2.67 percent.

For domestic sewage, which is best represented by the BOD_L removal of 300 mg/1, the effluent P concentration is well above 1 mg P/l, and the need for advanced treatment is obvious. In many cases, N removal must accompany P removal. Thus, the large SRTs needed for nitrification reduce sludge wasting and normal P removal. For example, a conservative SRT of 30 d leaves 8.5 mg P/1 in the effluent, whereas a conventional SRT of 6 d leaves 7.4 mg P/1. These simple calculations for the normal removal of phosphorus are important to perform when phosphorus is regulated.

In most cases, they define the amount of additional P removal required by one of the advanced techniques. On the other hand, a wastewater having a low P:BOD_L ratio may not require advanced treatment if the SRT is carefully controlled to keep a low, but positive P concentration.

PRECIPITATION BY METAL-SALTS

Aluminum and ferric cations precipitate with the orthophosphate anion at pH values that are compatible with microbiological treatment. Therefore, salts of Al^{3+} or Fe^{3+} can be added directly to the wastewater as it enters or leaves the bioreactor. Precipitates form, are incorporated into sludge, and are removed from the system by sludge wasting.

The key precipitates are $AlPO_{4(S)}$ and $FePO_{4(S)}$. Their standard dissolution reactions and solubility products (pK_{so}) are

$$AlPO_{4(S)} = Al^{3+} + PO_4^{3-} \qquad pK_{so} = 21$$

$$FePO_{4(S)} = Fe^{3+} + PO_4^{3-} \qquad pK_{so} = 21.9 \text{ to } 23$$

Although the solubility products are very small, the conditional solubility of phosphates is not necessary miniscule, because Al^{3+}, Fe^{3+}, and PO_4^{3-} undergo competing acid/base and complexation reactions. For phosphate, the key competing reactions are acid/base ones that result in the formation of protonated species. The acid/base reactions and pK_a values are

$$HPO_4^{2-} = H^+ + PO_4^{3-} \qquad pK_{a,3} = 12.3$$

$$H_2PO_4^- = H^+ + HPO_4^{2-} \qquad pK_{a,2} = 7.2$$

$$H_3PO_4 = H^+ + H_2PO_4^- \qquad pK_a,1 = 2.1$$

At near neutral pH, HPO_4^{2-} and $H_2PO_4^-$, not PO_4^{3-}, are the dominant species. For example, at pH = 7.0, PO_4^{3-} comprises about 0.00025 percent of the total dissolved orthophosphates. Higher pH gives a greater fraction as PO_4^{3-}.

The aluminum and ferric cations form numerous complexes, and the hydroxy complexes always are very important. Key complex-formation reactions and stability constants ($pK_{1\text{-}4}$) include:

$$Fe^{3+} + OH^- = FeOH^{2+} \qquad pK_1 = 11.8$$

$$FeOH^{2+} + OH^- = Fe(OH)_2^+ \qquad pK_2 = 10.5$$

$$Fe(OH)_2^+ + OH^- = Fe(OH)_3^0 \qquad pK_3 = 7.7$$

$$Fe(OH)_3 + OH^- = Fe(OH)_4^- \qquad pK_4 = 4.4$$

$$Al^{3+} + OH^- = AlOH^{2+} \qquad pK_1 = 9.0$$

$$AlOH^{2+} + OH^- = Al(OH)_2^+ \qquad pK_2 = 9.7$$

$$Al(OH)_2^+ + OH^- = Al(OH)_3^0 \qquad pK_3 = 8.3$$

$$Al(OH)_3 + OH^- = Al(OH)_4^- \qquad pK_4 = 6.0$$

The acid/base nature of the hydroxy complexes means that the iron or aluminum speciation is pH dependent. At neutral pH, the dominant species are $Fe(OH)_2^+$, $Fe(OH)_3^0$, and $Al(OH)_3^0$.

Al^{3+} and Fe^{3+} are tiny fractions of the total aluminum and iron. Lower pH is required to give greater fractions of Al^{3+} or Fe^{3+}. Due to the opposing pH trends for the precipitating anion (PO_4^{3-}) versus the precipitating cations (Al^{3+} or Fe^{3+}), an optimal pH exists.

Based on the reactions shown above; the pHs of minimum solubility are near 6 for $AlPO_{4(S)}$ and near 5 for $FePO_{4(S)}$. In principle, stoichiometric additions of the cations can drive the total phosphate concentration to well below 1 mg P/l near these optimal pH values. In reality, the theoretical predictions only define a rough location for a "window" in which precipitation through metal-salts addition can work within a biological-treatment process. Empirical testing is used to define the best dosage in practice, and the metal-salts dosage always exceeds the stoichiometric amount defined by the solubility products for $A1PO_{4(S)}$ and $FePO_{4(S)}$. Several factors act to complicate the chemistry and increase the metal-salts addition.

1. Phosphate forms competing complexes, such as $CaHPO_4$, $MgHPO_4$, and $FeHPO_4^+$. Thus, the fraction of total phosphate that is present as PO_4^{3-} is less than predicted by acid-base chemistry alone.
2. Aluminum and iron form other complexes, particularly with organic ligands, or precipitate as $Al(OH)_{3(S)}$. These reactions reduce the available Al^{3+} and Fe^{3+}.
3. Some of the total phosphorus is not orthophosphate, but is tied up in organic compounds.
4. The optimal pH for precipitation may not be compatible with the optimal microbiological activity. The pH cannot be changed so much that metabolic activity is significantly inhibited.
5. The precipitation reaction may be kinetically controlled and not reach its maximum extent, which occurs at equilibrium.

Typically, the metal-salts dosage is 1.5 to 2.5 times the stoichiometric amount. Since the stoichiometric mole ratio is 1 mol metal to 1 mol P, the practical ratios, determined empirically, are 1.5 to 2.5 mol metal:mol *P*. This translates to 1.3 to 2.2 g Al/g *P* and 2.7 to 4.5 g Fe/g *P*. For example, a practical mole ratio of 2:1 requires an Fe dose of 36 mg Fe/1 to treat a wastewater with 10 mg *P*/1 and provide an effluent with l mg *P*/1.

An important consideration for metal-salts addition is that the added metals behave as acids and consume alkalinity as they precipitate with PO_4^{3-} or complex with hydroxide. For example, when Fe^{3+} is added (such as in FeCla) and reacts to form FePO4(s) or Fe(OH)3, it consumes three base equivalents per mole. The same occurs for $A1^{3+}$, such as from alum $(Al_2(SO_4)3 \cdot 14H_20)$. A dosage at a 2:1 mol ratio consumes 9.7 mg as CaCC3 of alkalinity per mg P. For low alkalinity waters, pH depression is a major risk. In such cases, a base, such as lime (CaO), can be added to supplement the alkalinity. Another,alternative is to dose sodium aluminate (NaAlO2), which dissolves in water to give

$$AlNaO_2 + 2H_2O = Na^+ + Al(OH)_3 + OH^-$$

This dissolution reaction is a net generator of alkalinity: one base equivalent/mol Al. Another alternative is to dose ferrous iron (Fe^{2+}), which is oxidized to Fe^{3+} in a reaction that produces 1 base equivalent per mole of Fe^{2+} oxidized:

$$Fe^{2+} + 0.25\ O_2 + 0.5\ H_2O = Fe^{3+} + OH^-$$

Precipitation of $AlPO_{4(S)}$ or $FePO_{4(S)}$ creates significant quantities of inorganic sludge that is retained within the biological process and ultimately wasted. The stoichiometric production ratios are 3.9 g $AlPO_{4(S)}$/g *P* and 4.9 g $FePO_{4(S)}$/g *P*. For example, removal of 8 mg P/2 produces 39 mg $FePO_{4(S)}$/1 of plant flow. If an activated sludge process has an SRT (θ_x) of 6 *d* and a hydraulic reaction time of 0.25 *d* (θ), the concentration factor (θ_x/θ) is 24, which means the concentration of $FePO_{4(S)}$ built up in the mixed liquor is 935 mg/1.

Clearly, phosphate precipitation increases the sludge to be wasted and significantly enriches it in inorganic solids. Most experience with metal-salts addition is with activated sludge treatment, in which the metal salt can be added with the influent wastewater, directly to the aeration basin, or to the mixed liquor before it goes to the settler. Phosphorus removal by metal-salt addition also is possible with biofilm processes that have excellent capability for capturing the precipitated solids and wasting them regularly. Large-granule, fixed-bed filters seem most amenable and have been used successfully.

Enhanced Biological Phosphorus Removal

Certain heterotrophic bacteria are capable of sequestering high levels of phosphorus as *intracellular polyphosphate* (poly *P*), which is an energy storage material. If such microorganisms are selected, induced to store poly *P*, and wasted when rich in poly *P*, the net removal of *P* through biomass uptake can be increased significantly. The goal of *enhanced biological phosphorus removal* is to create a process environment that attains each of those three goals. When enhanced biological phosphorus removal is successful, the biomass contains 2 to 5 times the *P* content of normal biomass. Equation 5.5 can be revised to account for the enrichment in poly *P*. For example, a three-fold enrichment gives

$$P = P^0 - \frac{(0.0801)Y(1+(1-f_d)b\theta_x)(\Delta \mathrm{BOD}_L)}{1+b\theta_x} \quad ...(5.6)$$

The effluent *P* concentration for a BOD_L removal of 300 mg/l and an influent *P* concentration of 10 mg/1 then becomes 4.2 mg/l for θ_x = 15 d and 2.2 mg/l for $\theta_X = 6\ d$. These values approach a typical *P* standard of 1 mg/1. Increased poly-*P* enrichment or a higher $BOD_L : P$ ratio in the influent make the effluent goal achievable. Fig. 5.1. Sketches the key components of an activated sludge system active in enhanced biological phosphorus removal.

Many technological innovations are used to optimize the performance, and they are summarized later. Fig. 5.1 presents the essential items and facilitates an explanation of how enhanced biological phosphorous removal works.

The four essential components for enhanced biological phosphorus removal are:

1. The influent wastewater and recycle sludge must mix and first enter an *anaerobic bioreactor.* Electron acceptors—particularly O_2 and NO_3^-—must be excluded to the maximum degree possible so that BOD oxidation is insignificant in this reactor. Due to the high mixed-liquor biomass level, hydrolysis and fermentation steps occur, but the electron equivalents in the influent BOD are not transferred to a terminal electron acceptor.

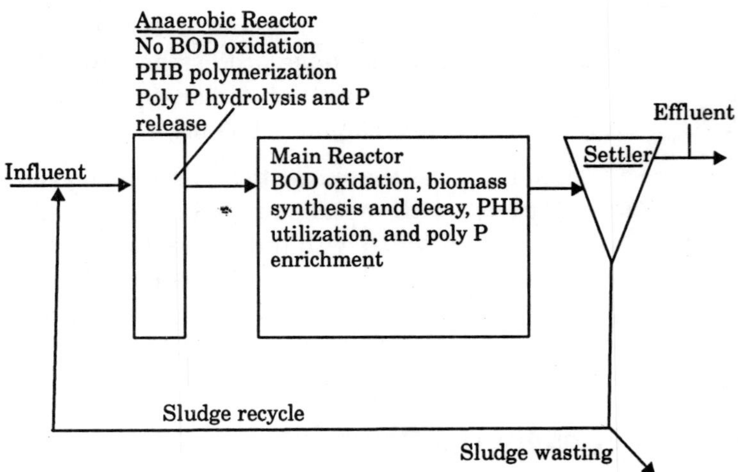

Fig. 5.1. Schematic of the required components of an activated sludge process active for enhanced biological phosphorus removal.

2. The mixed liquor flows from the anaerobic tank to the *main activated sludge bioreactor or bioreactors.* Depending on the system's SRT, nitrification and denitrification may or may not occur. Ample electron acceptors are available through aeration, which directly supplies O_2 and allows generation of NO_3^- if nitrification occurs. These electron acceptors make it possible for the heterotrophic bacteria to oxidize electron donors, gain energy, and grow.
3. The mixed liquor exiting the main bioreactor is settled, and most of it is *recycled* back to the head of the process, where it mixes with the influent and enters the anaerobic tank. This recycling ensures that all of the biomass *experiences alternating anaerobic and respiring conditions.*
4. The sludge that was most recently in the main bioreactor is *wasted* to control the SRT and to remove the biomass when it is enriched in poly P.

Each of the four essential components must be present in order that the biochemical and ecological mechanisms that drive enhanced biological phosphorus removal act. Figure 5.2 Summarizes the biochemical mechanisms acting in the anaerobic and main bioreactors. In the anaerobic phase (Figure 5.2*a*) certain heterotrophic bacteria are able to take up simple organic molecules produced by hydrolysis and fermentation. Because no electron acceptors are available, they sequester the electrons and carbon in insoluble intracellular solids, such *a spolyhydroxybutyrate (PHB).* To do the polymerization, the cells require an activated chemical form, acetyl coenzyme A (HSCoA).

Formation of HSCoA is an energy-consuming step, and the energy (ultimately transported as ATP) comes from the hydrolysis of poly *P*, which these microorganisms also contain and use as an energy-storage material. The hydrolysis of poly *P* releases phosphate (P_i) to the cellular P_i pool, and much of this P_i is released to the environment during the anaerobic phase.

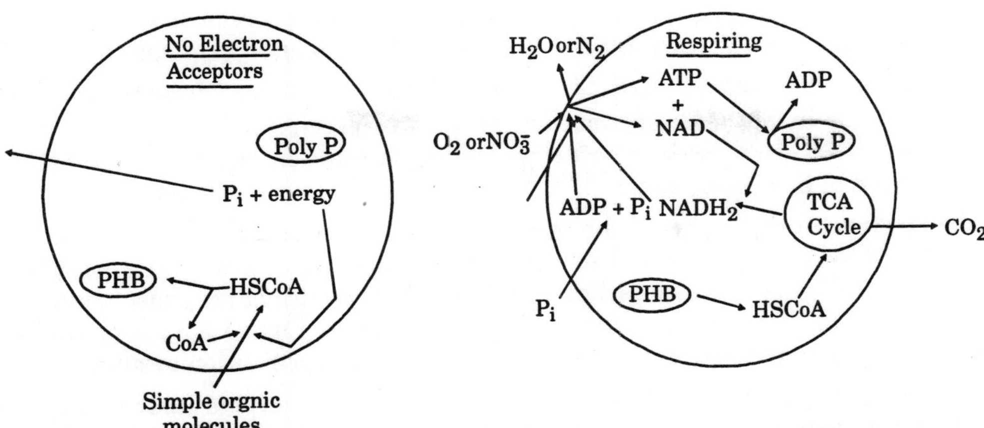

Fig. 5.2. Biochemical mechanisms operating in the anaerobic a. and main *b.* bioreactors of enhanced biological phosphorus removal. HSCoA = Acetyl coenzyme A, PHB = polyhydroxybutyrate.

When these heterotrophic bacteria move to the main reactor, they have an ample supply of electron acceptors, and the biochemical machinery essentially works in the opposite direction. The electron storage material (PHB) is hydrolyzed to HSCoA, which is then oxidized in the TCA cycle. The released electrons, carried on $NADH_2$, are used for ATP synthesis through respiration with O_2 or NO_3^- as the electron acceptor. Some of the ATP generated is "invested" in the synthesis of poly *P*, the energy-storage material. Inorganic phosphate, P_i, must be imported for poly *P* synthesis. Thus, these bacteria are significant sinks for environmental P_i when they are synthesizing poly *P*. They must be harvested and wasted when they are enriched in poly *P* and before they cycle to the anaerobic phase. Not all heterotrophic bacteria are capable of synthesizing poly *P* and PHB. In order to select for these bacteria, the so-called Bio-P *bacteria,* the process-must exert a strong ecological pressure to favor them.

The cycling, between the anaerobic and, respiratory phases creates that ecological pressure. During the anaerobic phase, the Bio-*P* bacteria take-up and sequester the available electron donors and carbon sources into PHB. They require poly *P* to fuel this "investment." During the respiratory phase, the Bio-*P* bacteria have rich reserves of electrons and carbon in PHB. Hydrolysis and oxidation of PHB gives them a major energy source for growth, even though the aqueous environment in the main bioreactor is oligotrophic.

Generating energy from storage reserves allows the Bio-*P* bacteria to outgrow other heterotrophic bacteria in the main bioreactor and gradually establish themselves as a major fraction of the biomass. In the past, Bio-*P* bacteria were identified as belonging to the genus *Acinetobacter.* Further research has shown that *Acinetobactor* is only one genus of Bio-*P* bacteria and often is not the dominant genus. Bio-*P* bacteria also are found among the *Pseudomonas, Arthrobacter, Nocardia, Beyerinkia, Ozotobacter, Aeromonas, Microlunatus,* and others. Although the biochemical and ecological foundation for enhanced biological phosphorus removal by Bio-*P* bacteria now seems to be firmly established, some evidence supports the concept that chemical precipitation plays a role, at least in some instances. The main candidate for the solid phase is $Ca_5(OH)(PO_4)_{3(S)}$, or hydroxyapatite. Its solubility product is pK_{so} = 55.9, which suggests that moderately hard waters could drive the orthophosphate below 1 mg *P*/1 for pH values greater than about 7.5. Struvite ($MgNH_4PO_4$) is another possibility when Mg^{2+} and NH_4^+ concentrations are elevated. Strong denitrification produces base that can increase the pH, particularly inside a floe or biofilm, and foster precipitation of hydroxyapatite or struvite. Very high aeration rates, which strip CO_2 from solution, also raise the pH and can foster precipitation of hydroxyapatite.

Chemical precipitation as part of enhanced biological phosphorous removal is accentuated by low alkalinity, which minimizes the buffering against these pH increases. Figure 5.3 summarizes schematically two approaches that link enhanced biological phosphorous removal to well-known predenitrification processes. The approach in Figure 5.3*a* sometimes called *Phoredox,* adds the initial anaerobic tank to a classical predenitrification system.

The process in figure 5.3*b* called *Bardenpho,* adds the anaerobic tank to the Barnard process, which has anoxic cell decay and aerobic polishing steps to effect added N removal. Burdick, Refling, and Stensel (1982) reported 93 percent removal of total N and 65 percent removal of total *P* with the Bardenpho process shown in Fig. 5.3. The sludge contained 4 to 4.5% *P*, which is about twice that of normal biomass. The effluent total P was 1.9 to 2.3 mg/l. Meganck and Faup (1988) summarized results from several other Bardenpho processes. In general, the effluent total ranged from about 0.5 to 6.3 mg/1, with the majority being close to 1 mg/1.

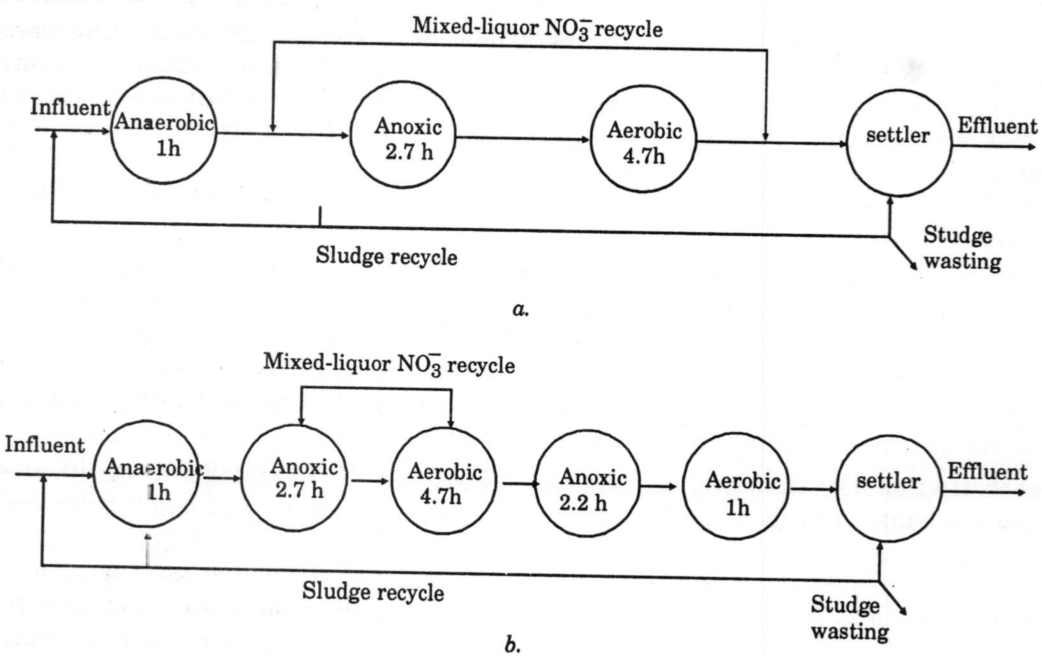

Fig. 5.3. Two approaches to incorporate enhanced biological phosphorus removal into classical predenitrification. The top a. links an anaerobic tank to classical predenitrification, while the bottom b. links to the Barnard process. Hydraulic detention times are typical and for general guidance only.

One possible drawback of the modification of predenitrification is that the sludge recycle returns NO_3^-; thus, some electron acceptor is applied to the anaerobic tank. Workers at the University of Cape Town (South Africa) developed the configurations shown in figure 5.4 to eliminate NO_3^- from the return sludge. The sludge recycle goes directly to the anoxic predenitrification tank. Then, biomass is recycled to the anaerobic tank from the anoxic tank, where active denitrification should keep the NO_3^- concentration very low. The modified UCT process further reduces NO_3^- transfer to the anaerobic tank by partitioning the anoxic tanking into two zones. The second zone receives the main mixed-liquor NO_3^- recycle, while the first zone receives only the small NO_3^- input from sludge recycle and is the source of mixed liquor sent to the anaerobic tank.

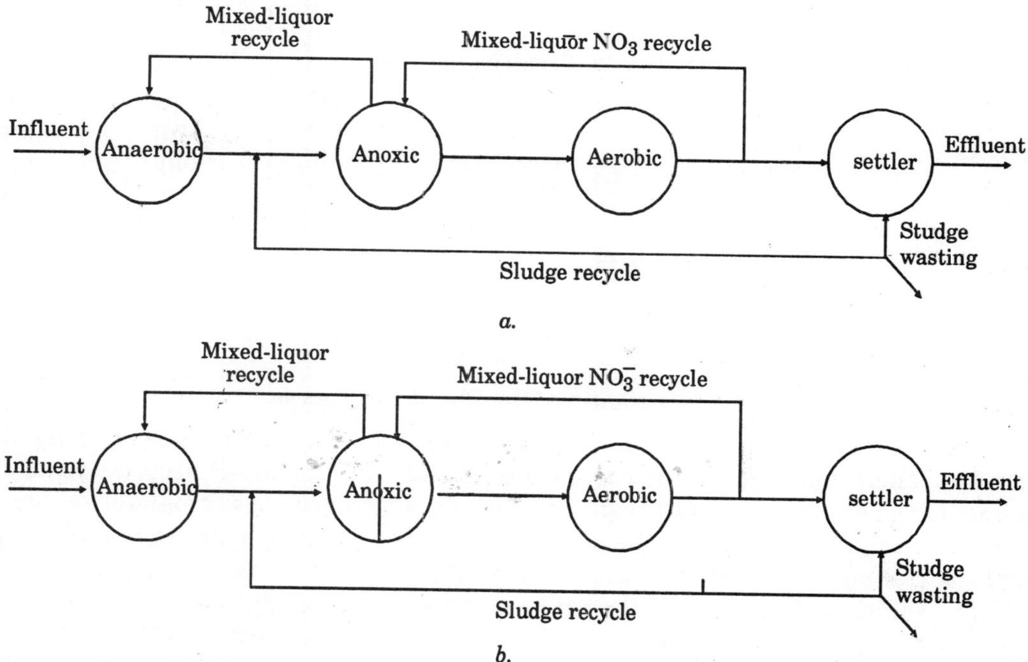

Fig. 5.4. Two University of Cape Town (UCT) processes minimize the recycling of NO_3^- to the anaerobic tank. In the modified UCT process *b.*, the anoxic tank is divided into two compartments. The mixed-liquor NO_3 recycle enters in the downstream part of the anoxic tank, while the mixed-liquor recycle to the anaerobic tank leaves from the upstream part of the anoxic tank. Mixing between the two parts is restricted by a physical barrier.

Sequencing batch reactors (SBRs) also can be used for enhanced biological phosphorous removal, as well as nitrogen removal. Carryover of settled sludge from the previous cycle always contains some NO_3^-. Once the NO_3^- is denitrified during an unaerated fill period, anaerobic conditions can be established and lead to poly *P* and PHB cycling.

Biofilm processes can be used for enhanced biological phosphorous removal if the biofilm is exposed to alternating anaerobic and respiratory periods. Characteristic responses of phosphate release during the anaerobic phase and phosphate uptake during the respirator phase occur. Oxygen or NO_3^- could be used as the terminal electron acceptor. Cycle times of around 6 *h* are workable, but further research is needed to optimize these processes for *N* and *P* removal.

Quite a unique process for enhanced biological phosphorus removal is *PhoStrip,* which combines Bio-*P* bacteria with chemical precipitation. Fig. 5.5 sketches the key features of PhoStrip. The aerobic bioreactor and settler form a typical activated sludge system, which

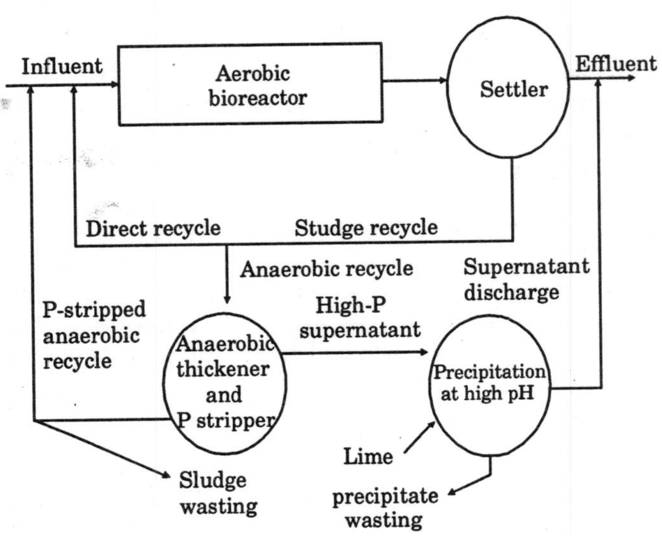

Fig. 5.5. Schematic of the PhoStrip process.

usually is operated with a conventional SRT, does not nitrify, and has a "plug flow" configuration. The ecological and biochemical pressures for the Bio-*P* bacteria come from routing part of the sludge recycle through a thickener, which becomes completely anaerobic. The thickened sludge is returned to the aeration basin after having hydrolyzed and released its poly *P* to the supernatant, which contains 30-70 mg *P*/1 and is sent to high-lime precipitation. Addition of lime raises the pH to around 11 and precipitates $Ca_3(PO4)_{2(S)}$. The $Ca_3(PO_4)_2$ sludge is wasted, while the clarified supernatant can be discharged to the effluent.

Field results show that PhoStrip can attain an effluent with less than 1 mg *P*/1 in most cases. PhoStrip is unique in that it does not rely on direct wasting of biomass enriched in poly *P*. Instead, it strips the poly *P* to soluble phosphates, which are then precipitated separately from the biomass. Despite this important difference from the other processes used for enhanced biological phosphorus removal, PhoStrip still relies on the same biochemical and ecological forces to select for and activate Bio-*P* bacteria.

6 Inorganic Pollutants

Inorganic air pollutants consist of many kinds of substances. The first to be addressed here are particulate pollutants. Many solid and liquid substances may become particulate air contaminants. Another important class of inorganic air pollutants consists of oxides of carbon, sulfur, and nitrogen. Carbon monoxide is a directly toxic material that is fatal at relatively small doses. Carbon dioxide is a natural and essential constituent of the atmosphere, and it is required for plants to use during photosynthesis. However, CO_2 may turn out to be the most deadly air pollutant of all because of its potential as a greenhouse gas that might cause devastating global warming.

Oxides of sulfur and nitrogen are acid-forming gases that can cause acid precipitation. Several other inorganic air pollutants, such as ammonia, hydrogen chloride, and hydrogen sulfide are also discussed in this chapter. A number of gaseous inorganic pollutants enter the atmosphere as the result of human activities. Those added in the greatest quantities are CO, SO_2, NO, and NO_2. (These quantities are relatively small compared to the amount of CO_2 in the atmosphere. The possible environmental effects of increased atmospheric CO_2 levels are discussed later in this chapter.) Other inorganic pollutant gases include NH_3, N_2O, N_2O_5, H_2S, Cl_2, HCl, and HF. Substantial quantities of some of these gases are added to the atmosphere each year by human activities.

Globally, atmospheric emissions of carbon monoxide, sulfur oxides, and nitrogen are of the order of one to several hundred million tons per year. As with most aspects of chemistry in the real world, it is somewhat artificial and arbitrary to divide air pollutants $\hat{q}$ between the inorganic and organic realms. For example, inorganic NO_2 undergoes photodissociation to start the processes that convert organic vapors to aldehydes, oxidants including inorganic O_3, and other substances characteristic of photochemical smog. Oxidants generated in such smog convert inorganic SO_2 to much more acidic sulfuric acid, the major contributor to acid precipitation. Numerous other examples could be cited to illustrate the interrelationships among air pollutants of various kinds.

MINUTE POLLUTANTS

Particles in the atmosphere, which range in size from about one-half millimeter (the size of sand or drizzle) down to molecular dimensions, are made up of an amazing variety of materials and discrete objects that may consist of either solids or liquid droplets. *Particulates* is a term that has come to stand for particles in the atmosphere, although *particulate matter* or simply *particles,* is preferred usage. Particulate matter makes up the most visible and obvious form of air pollution. Atmospheric aerosols are solid or liquid particles smaller than 100 μm in diameter. Pollutant particles in the 0.001 to 10 μm range are commonly suspended in the air near sources of pollution, such as the urban atmosphere, industrial plants, highways, and power plants. Very small, solid particles include carbon black, silver iodide, combustion nuclei, and sea-salt nuclei formed by the loss of water from droplets of seawater. Larger particles include cement dust, wind-blown soil dust, foundry dust, and pulverized coal. Liquid particulate matter, *mist*, includes raindrops, fog, and sulfuric acid mist. Some particles are of biological origin, such as viruses, bacteria, bacterial spores, fungal spores, and pollen. Particulate matter may be organic or inorganic; both types are very important atmospheric contaminants.

Formation of Inorganic Particles

Metal oxides constituted a major class of inorganic particles in the atmosphere. These are formed whenever fuels containing metals are burned. For example, particulate iron oxide is formed during the combustion of pyrite-containing coal:

Table 6.1. Important Terms Describing Atmospheric Particles

Term	*Meaning*
Aerosol	Colloidal-sized atmospheric particle
Condensation aerosol	Formed by condensation of vapors or reactions of gases
Dispersion aerosol	Formed by grinding of solids, atomization of liquids, or dispersion of dusts
Fog	Term denoting high level of water droplets
Haze	Denotes decreased visibility due to the presence of particles
Mists	Liquid particles
Smoke	Particles formed by incomplete combustion of fuel

$$3FeS_2 + 8O_2 \rightarrow Fe_3O_4 + 6SO_2 \qquad ...(6.1)$$

Organic vanadium in residual fuel oil is converted to particulate vanadium oxide. Part of the calcium carbonate in the ash fraction of coal is converted to calcium oxide and is emitted to the atmosphere through the stack;

$$CaCO_3 + heat \rightarrow CaO + CO_2 \qquad ...(6.2)$$

A common process for the formation of aerosol mists involves the oxidation of atmospheric sulfur dioxide to sulfuric acid, a hygroscopic substance that accumulates atmospheric water to form small liquid droplets:

$$2SO_2 + O_2 + 2H_2O \rightarrow 2H_2SO_4 \qquad ...(6.3)$$

In the presence of basic air pollutants, such as ammonia or calcium oxide, the sulfuric acid reacts to form salts:

$$H_2SO_4(\text{droplet}) + 2NH_3(g) \rightarrow (NH_4)_2SO_4\,(\text{droplet}) \qquad ...(6.4)$$

$$H_2SO_4(\text{droplet}) + CaO(s) \rightarrow CaSO_4(\text{ droplet}) + H_2O \qquad ...(6.5)$$

Under low-humidity conditions water is lost from these droplets and a solid aerosol is formed.

The preceding examples show several ways in which solid or liquid inorganic aerosols are formed by chemical reactions. Such reactions constitute an important general process for the formation of aerosols, particularly the smaller particles.

COMPOSITION OF PARTICLES

The proportions of elements in atmospheric particulate matter reflect relative abundances of elements in the parent material. The source of particulate matter is reflected in its elemental composition, taking into consideration chemical reactions that may change the composition. For example, particulate matter largely from ocean spray origin in a coastal area receiving sulfur dioxide pollution may show anomalously high sulfate and corresponding low chloride content. The sulfate comes from atmospheric oxidation of sulfur dioxide to form nonvolatile ionic sulfate, whereas some chloride originally from the NaCl in the seawater may be lost from the solid aerosol as volatile HCl:

$$2SO_2 + O_2 + 2H_2O \rightarrow 2H_2SO_4 \quad ..(6.6)$$

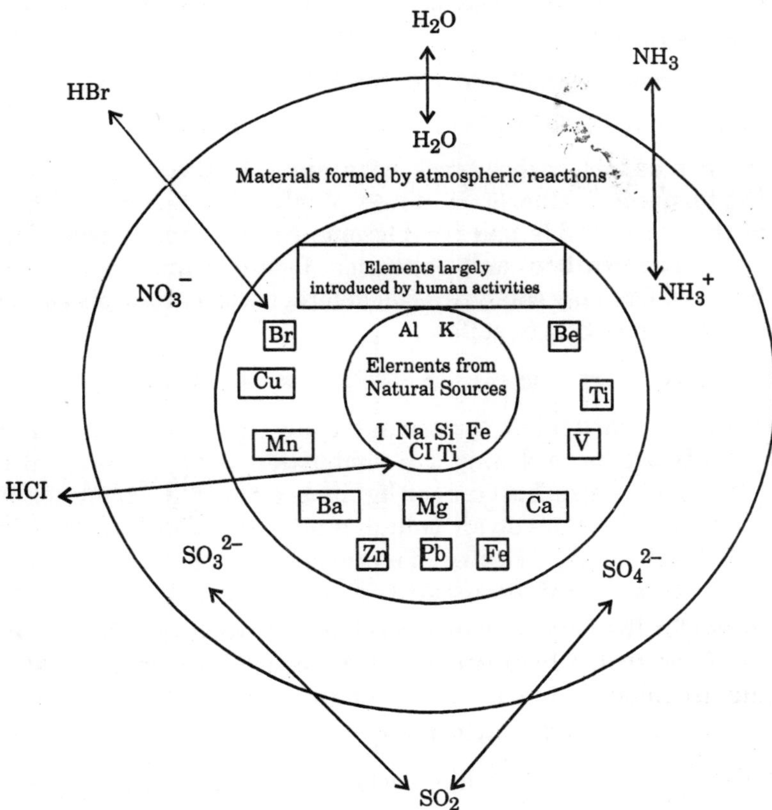

Fig. 6.1. Some of the components of inorganic particulate matter and their origins.

$$H_2SO_4 + 2NaCl(\text{particulate}) \rightarrow Na_2SO_4(\text{particulate}) + 2HCl \quad ...(6.7)$$

The chemical composition of atmospheric particulate matter is quite diverse. Among the constituents of inorganic particulate matter found in polluted atmospheres are salts, oxides, nitrogen compounds, sulfur compounds, various metals, and radionuclides. In coastal areas, sodium and chlorine get into atmospheric particles as sodium chloride from sea spray. The major trace elements that typically occur at levels above 1 $\mu g/m^3$ in particulate matter are aluminum, calcium, carbon, iron, potassium, sodium, and silicon; note that most of these tend to originate from terrestrial sources. Lesser quantities of copper, lead, titanium, and zinc and even lower levels of antimony, beryllium, bismuth, cadmium, cobalt, chromium, cesium, lithium, manganese, nickel, rubidium, selenium, strontium, and vanadium are commonly observed. The likely sources of some of these elements are given below:

(*i*) **Al, Fe, Ca, Si:** Soil erosion, rock dust, coal combustion

(*ii*) **C:** Incomplete combustion of carbonaceous fuels

(*iii*) **Na, Cl:** Marine aerosols, chloride from incineration of organohalide polymer wastes

(*iv*) **Sb, Se:** Very volatile elements, possibly from the combustion of oil, coal, or refuse

(*v*) **V:** Combustion of residual petroleum (present at very high levels in residues from Venezuelan crude oil)

(*vi*) **Zn:** Tends to occur in small particles, probably from combustion

(*vii*) **Pb:** Combustion of leaded fuels and wastes containing lead

Particulate carbon as soot, carbon black, coke, and graphite originates from auto and truck exhausts, heating furnaces, incinerators, power plants, and steel and foundry operations and composes one of the more visible and troublesome particulate air pollutants. Because of its good adsorbent properties, carbon can be a carrier of gaseous and other particulate pollutants. Particulate carbon surfaces may catalyze some heterogeneous atmospheric reactions, including the important conversion of SO_2 to sulfate.

Fly Ash

Much of the mineral particulate matter in a polluted atmosphere is in the form of oxides and other compounds produced during the combustion of high-ash fossil fuel. Much of the mineral matter in fossil fuels such as coal or lignite is converted during combustion to a fused, glassy bottom ash which presents no air pollution problems. Smaller particles of *fly ash* enter furnace flues and are efficiently collected in a properly equipped stack system. However, some fly ash escapes through the stack and enters the atmosphere. Unfortunately, the fly ash thus released tends to consist of smaller particles that do the most damage to human health, plants, and visibility. The composition of fly ash varies widely, depending upon the source of fuel.

The predominant constituents are oxides of aluminum, calcium, iron, and silicon. Other elements that occur in fly ash are magnesium, sulfur, titanium, phosphorus, potassium, and sodium. Elemental carbon (soot, carbon black) is a significant fly ash constituent.

Asbestos

Asbestos is the name given to a group of fibrous silicate minerals, typically those of the serpentine group, for which the approximate formula is $Mg_3P(Si_2O_5)(OH)_4$. The tensile strength, flexibility, and nonflammability of asbestos have led to many uses in the past including structural materials, brake linings, insulation, and pipe manufacture. Asbestos is of concern as an air pollutant because when inhaled it may cause asbestosis (a pneumonia condition), mesothelioma (tumor of the mesothelial tissue lining the chest cavity adjacent to the lungs), and bronchogenic carcinoma (cancer originating with the air passages in the lungs). Therefore, uses of asbestos have been severely curtailed and widespread programs have been undertaken to remove the material from buildings.

Toxic Metals

Some of the metals found predominantly as particulate matter in polluted atmospheres are known to be hazardous to human health. All of these except beryllium are so-called "heavy metals." Lead is the toxic metal of greatest concern in the urban atmosphere because it comes closest to being present at a toxic level; mercury ranks second. Others include beryllium, cadmium, chromium, vanadium, nickel, and arsenic (a metalloid). Atmospheric mercury is of concern because of its toxicity, volatility, and mobility. Some atmospheric mercury is associated with particulate matter. Much of the mercury entering the atmosphere does so as volatile elemental mercury from coal combustion and volcanoes. Volatile organomercury compounds such as dimethylmercury, $(CH_3)_2Hg$, and monomethylmercury salts, such as $(CH_3)HgBr$, are also encountered in the atmosphere.

With the reduction of leaded fuels, atmospheric leadis of less concern than it used to be. However, during the decades that leaded gasoline containing tetraethyllead was the predominant automotive fuel, particulate lead halides were emitted in large quantities. Lead halides are emitted from engines burning leaded gasoline through the action of dichloroethane and dibromoethane added as halogenated scavengers to form volatile lead chloride, lead bromide and lead chlorobromide, thereby preventing the accumulation of lead oxides inside engines.

Beryllium is used for the formulation of specialty alloys employed in electrical equipment, electronic instrumentation, space gear, and nuclear reactor components, so that distribution of beryllium is by no means comparable to that of other toxic metals such as lead or mercury. However, because of its "high tech" applications, consumption of beryllium may increase in the future. Because of its high toxicity beryllium has the lowest allowable limit in the atmosphere of all the elements. One of the main results of the recognition of beryllium toxicity hazards was the elimination of this element from phosphors (coatings which produce visible light from ultraviolet light) in fluorescent lamps.

Radioactive Particles

A significant natural source of radionuclides in the atmosphere is radon, a noble gas product of radium decay. Radon may enter the atmosphere as either of two isotopes, ^{222}Rn (half-life 3.8 days) and ^{220}Rn (half-life 54.5 seconds). The superscript numbers on the preceding symbols denote mass numbers, the sum of protons and neutrons in the isotope nucleus.) Both ^{220}Rn and ^{222}Rn are alpha emitters in decay chains that terminate with stable isotopes of lead. The initial decay products, ^{218}Po and ^{216}Po are nongaseous and adhere readily to atmospheric particulate matter. Therefore, some of the radioactivity detected in these particles is of natural origin. Furthermore, cosmic rays act on nuclei in the atmosphere to produce other radionuclides, including ^{10}Be, ^{14}Be, ^{14}C, ^{39}Cl, ^{3}H, ^{22}Na, ^{32}P, and ^{33}P.

The combustion of fossil fuels introduces radioactivity into the atmosphere in the form of radionuclides contained in fly ash. Large coal-fired power plants lacking ash-control equipment may introduce up to several hundred millicuries of radionuclides into the atmosphere each year, far more than either an equivalent nuclear or oil-fired power plant: The aboveground detonation of nuclear weapons can add large amounts of radioactive particulate matter to the atmosphere. Among tne radioisotopes that can be detected in rainfall falling after atmospheric nuclear weapon detonation are ^{91}Y, ^{144}Ce, ^{147}Ce, ^{147}Nd, ^{147}Pm, ^{149}Pm, ^{151}Sm, ^{155}Sm, ^{155}Eu, ^{156}Eu, ^{89}Sr, ^{90}Sr, ^{115m}Cd, ^{129m}Te, ^{131}I, ^{132}Te, and ^{140}Ba. (Note that "m" denotes a metastable state that decays by gamma-ray emission to an isotope of the same element.)

EFFECTS OF PARTICLES

Atmospheric particles have numerous effects. The most obvious of these is reduction and distortion of visibility. They provide active surfaces upon which heterogeneous atmospheric chemical reactions can occur and nucleation bodies for the condensation of atmospheric water vapor, thereby exerting a significant influence upon weather and air pollution phenomena. The most visible influence that aerosol particles have upon air quality results from their optical effects. Particles smaller than about 0.1 μm in diameter scatter light much like molecules; that is, Rayleigh scattering. Generally, such particles have an insignificant effect upon visibility in the atmosphere.

The light-scattering and intercepting properties of particles larger than 1 μm are approximately proportional to the particle's cross-sectional area. Particles of 0.1 μm-1 μm cause interference phenomena because they are about the same dimensions as the wavelengths of visible light, so their light-scattering properties are especially significant. Atmospheric particles inhaled through the respiratory tract may damage health. Relatively large particles are likely to be retained in the nasal cavity and in the pharynx, whereas very small particles are likely to reach the lungs and be retained by them. The respiratory system possesses mechanisms for the expulsion of inhaled particles.

In the ciliated region of the respiratory system, particles are carried as far as the entrance to the gastrointestinal tract by a flow of mucus. Macrophages in the nonciliated pulmonary regions carry particles to the ciliated region. The respiratory system may be damaged directly by particulate matter that enters the blood system or lymph system through the lungs. In addition, the soluble components of the particulate material may be transported to organs some distance from the lungs and have a detrimental effect on these organs.

Particles cleared from the respiratory tract are to a large extent swallowed into the gastrointestinal tract. A strong correlation has been found between increases in the daily mortality rate and acute episodes of air pollution. In such cases, high levels of particulate matter are accompanied by elevated concentrations of SO_2 and other pollutants, so that any conclusions must be drawn with caution.

EMISSIONS OF PARTICLES

The removal of particulate matter from gas streams is the most widely practiced means of air pollution control. A number of devices have been developed for this purpose, which differ widely in effectiveness, complexity, and cost. The selection of a particle removal system for a gaseous waste stream depends upon the particle loading, nature of particles (size distribution), and type of gas scrubbing system used.

Sedimentation Process

The simplest means of particulate matter removal is *sedimentation*, a phenomenon that occurs continuously in nature. Gravitational settling chambers may be employed for the removal of particles from gas streams by simply settling under the influence of gravity. These chambers take up large amounts of space and have low collection efficiencies, particularly for small particles. Gravitational settling of particles is enhanced by increased particle size which occurs spontaneously by coagulation. Thus, over time, the size of particles increases and the number of particles decreases in a mass of air that contains particles. Brownian motion of particles less than about 0. 1 μm in size is primarily responsible for their contact, enabling coagulation to occur. Particles greater than about 0.3 μm in radius do not diffuse appreciably and serve primarily as receptors of smaller particles.

Inertial Mechanisms

Inertial mechanisms are effective for particle removal. These depend upon the fact that the radius of the path of a particle in a rapidly moving, curving air stream is larger than the path of the stream as a whole. Therefore, when a gas stream is spun by vanes, a fan, or a tangential gas inlet, the paniculate matter may be collected on a separator wall because the particles are forced outward by centrifugal force. Devices utilizing this mode of operation are called *dry centrifugal collectors*.

Filtration Process

Fabric filters, as their name implies, consist of fabrics that allow the passage of gas but retain particulate matter. These are used to collect dust in bags contained in structures called *baghouses*. Periodically, the fabric composing the filter is shaken to remove the particles and to reduce back-pressure to acceptable levels. Typically, the bag is in a tubular configuration. Numerous other configurations are possible. Collected particulate matter is removed from bags by mechanical agitation, blowing air on the fabric, or rapid expansion and contraction of the bags.

Although simple, baghouses are generally effective in removing particles from exhaust gas. Particles as small as 0.01 μm in diameter are removed, and removal efficiency is relatively high for particles down to 0.5 μm in diameter.

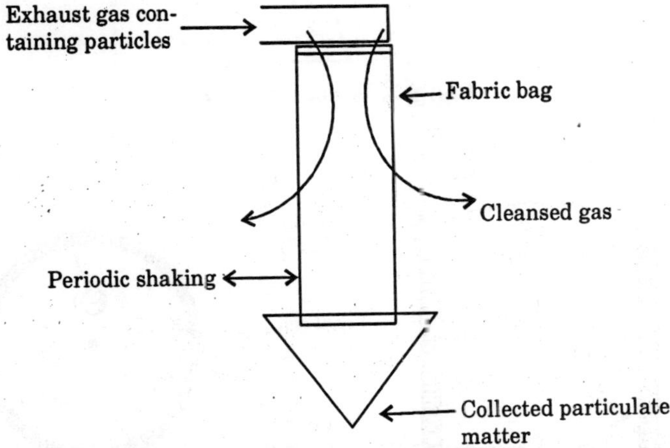

Fig. 6.2. Baghouse collection of particulate emissions.

Scrubbers

A venturi scrubber passes gas through a converging section, throat, and diverging section. Injection of the scrubbing liquid at right angles to incoming gas breaks the liquid into very small droplets, which are ideal for scavenging particles from the gas stream. In the reduced-pressure (expanding) region of the venturi, some condensation can occur, adding to the scrubbing efficiency. In addition to removing particles, Venturis may serve as quenchers to cool exhaust gas and as scrubbers for pollutant gases.

Electrostatic Removal

Aerosol particles may acquire electrical charges. In an electric field, such particles are subjected to a force, *F* (dynes) given by

$$F = Eq \quad ...(6.8)$$

where E is the voltage gradient (statvolt/cm) and q is the electrostatic charge charge on the particle (in esu). This phenomenon has been widely used in highly efficient *electrostatic precipitators*. The particles acquire a charge when the gas stream is passed through a high-voltage, direct-current corona. Because of the charge, the particles are attracted to a grounded surface from which they may be later removed. Ozone may be produced by the corona discharge. Similar devices used as household dust collectors may produce toxic ozone if not operated properly.

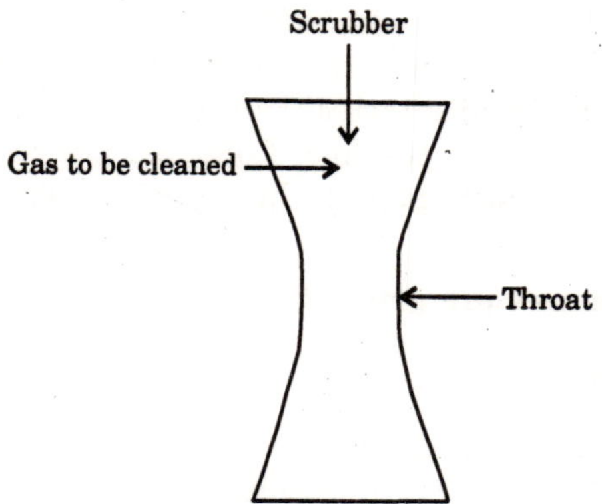

Fig. 6.3. Venturi scrubber.

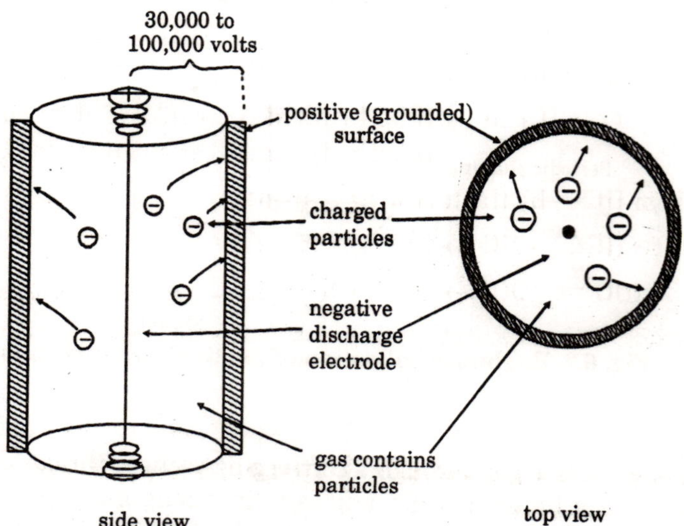

Fig. 6.4. Schematic diagram of an electrostatic precipitator

CARBON OXIDES

Carbon Monoxide

Carbon monoxide, CO, causes problems in cases of locally high concentrations. The overall atmospheric concentration of carbon monoxide is about 0.1 ppm. Much of this CO is present as an intermediate in the oxidation of methane by hydroxyl radical. Because of carbon monoxide emissions from internal combustion engines, highest levels of this toxic gas tend to occur in congested urban areas at times when the maximum number of people are exposed, such as during rush hours. At such times, carbon monoxide levels in the atmosphere may become as high as 50-100 ppm.

Regulation of Carbon Monoxide Emissions

Since the internal combustion engine is the primary source of localized pollutant carbon monoxide emissions, control measures have been concentrated on the automobile. Carbon monoxide emissions may be lowered by employing a leaner air-fuel mixture, that is, one in which the weight ratio of air to fuel is relatively high. At air-fuel (weight:weight) ratios exceeding approximately 16:1, an internal combustion engine emits virtually no carbon monoxide.

Modern automobiles use catalytic exhaust reactors to cut down on carbon monoxide emissions. Excess air is pumped into the exhaust gas, and the mixture is passed through a catalytic converter in the exhaust system, resulting in oxidation of CO to CO_2.

Fate of Atmospheric CO

The residence time of carbon monoxide in the atmosphere is of the order of 4 months. It is generally agreed that carbon monoxide is removed from the atmosphere by reaction with hydroxyl radical, HO.

$$CO + HO^{\bullet} \rightarrow CO_2 + H \qquad ...(6.9)$$

The reaction produces hydroperoxyl radical as a product:

$$O_2 + H + M \rightarrow HOO^{\bullet} + M \text{ (M is an energy-absorbing third body, usually a molecule of } O_2 \text{ or } N_2\text{)} \qquad ...(6.10)$$

$HO^{\bullet}$ is regenerated from HOO- by the following reactions:

$$HOO^{\bullet} + NO \rightarrow HO^{\bullet} + NO_2 \qquad ...(6.11)$$

$$HOO^{\bullet} + HOO^{\bullet} \rightarrow H_2O_2 + O_2 \qquad ...(6.12)$$

The latter reaction is followed by photochemical dissociation of H_2O_2 to regenerate $HO^{\bullet}$:

$$H_2O_2 + h\nu \rightarrow 2HO^{\bullet} \qquad ...(6.13)$$

Methane is also involved through the atmospheric CO-$HO^{\bullet}$-CH_4 cycle.

Soil microorganisms act to remove CO from the atmosphere. Therefore, soil is a sink for carbon monoxide.

Global Warming

Carbon dioxide and other infrared-absorbing trace gases in the atmosphere contribute to global warming—the "greenhouse effect"—by allowing incoming solar radiant energy to penetrate to the Earth's surface while reabsorbing infrared radiation emanating from it. Levels of these "greenhouse gases" have increased at a rapid rate during recent decades and are continuing to do so. Concern over this phenomenon has intensified since about 1980. This is because ever since accurate temperature records have been kept, the 1980s have been the warmest 10-year period recorded. On an annual basis, 1988 was the warmest year ever recorded, 1987 was second and 1981 third. To a degree, perhaps masked in part by the cooling effect of at least one major volcanic eruption, the global trend seemed to be continuing in the 1990s.

There are many uncertainties surrounding the issue of greenhouse warming. However, several things about the phenomenon are certain. It is known that CO_2 and other greenhouse gases, such as CH_4, absorb infrared radiation by which Earth loses heat. The levels of these gases have increased markedly since about 1850 as nations have become industrialized and as forest lands and grasslands have been converted to agriculture. Chlorofluorocarbons, which also are green-house gases, were not even introduced into the atmosphere until the 1930s. Although trends in levels of these gases are well known, their effects on global temperature and climate are much less certain. The phenomenon has been the subject of much computer modeling.

Most models predict global warming of 1.5 to 5°C, about as much again as has occurred since the last ice age. Such warming would have profound effects on rainfall, plant growth, and sea levels, which might rise as much as 0.5 to 1.5 meters. Carbon dioxide is the gas most commonly thought of as a greenhouse gas; it is responsible for about half of the atmospheric heat retained by trace gases. It is produced primarily by burning of fossil fuels and deforestation accompanied by burning and biodegradation of biomass. On a molecule-for-molecule basis, methane, CH_4, is 20-30 times more effective in trapping heat than is CO_2. Other trace gases that contribute are chlorofluorocarbons and N_2O.

Analyses of gases trapped in polar ice samples indicate that pre-industrial levels of CO_2 and CH_4 in the atmosphere were approximately 260 parts per million and 0.70 ppm, respectively. Over the last 300 years these levels have increased to current values of around 350 ppm, and 1.7 ppm, respectively; most of the increase by far has taken place at an accelerating pace over the last 100 years. (A note of interest is the observation based upon analyses of gases trapped in ice cores that the atmospheric level of CO_2 at the peak of the last ice age about 18,000 years past was 25 percent *below* preindustrial levels.)

About half of the increase in carbon dioxide in the last 300 years can be attributed to deforestation, which still accounts for approximately 20 percent of the annual increase in this gas. Carbon dioxide is increasing by about 1 ppm per year. Methane is going up at a rate of almost 0.02 ppm/year. The comparatively very rapid increase in methane levels is attributed to a number of factors resulting from human activities. Among these are direct leakage of natural gas, by-product emissions from coal mining and petroleum recovery, and release from the burning of savannas and tropical forests. Biogenic sources resulting from human activities produce large amounts of atmospheric methane. These include methane from bacteria degrading organic matter, such as municipal refuse in landfills; methane evolved from anaerobic biodegradation of organic matter in rice paddies; and methane emitted as the result of bacterial action in the digestive tracts of ruminant animals.

Both positive and negative feedback mechanisms may be involved in determining the rates at which carbon dioxide and methane build up in the atmosphere. Laboratory studies indicate that increased CO_2 levels in the atmosphere cause accelerated uptake of this gas by plants undergoing photosynthesis, which tends to slow buildup of atmospheric CO_2. Given adequate rainfall, plants living in a warmer climate that would result from the greenhouse effect would grow faster and take up more CO_2. This could be an especially significant effect of forests, which have a high CO_2-fixing ability. However, the projected rate of increase in carbon dioxide levels is so rapid that forests would lag behind in their ability to fix additional CO_2. Similarly, higher atmospheric CO_2 concentrations will result in accelerated sorption of the gas by oceans. The amount of dissolved CO_2 in the oceans is about 60 times the amount of CO_2 gas in the atmosphere. However, the times for transfer of carbon dioxide from the atmosphere to the ocean are of the order of years.

Because of low mixing rates, the times for transfer of oxygen from the upper approximately 100-meter layer of the oceans to ocean depths is much longer, of the order of decades. Therefore, like the uptake of CO_2 by forests, increased absorption by oceans will lag behind the emissions of CO_2. Severe drought conditions resulting from climatic warming could cut down substantially on CO_2 uptake by plants. Warmer conditions would accelerate release of both CO_2 and CH_4 by microbial degradation of organic matter. (It is important to realize that about twice as much carbon is held in soil in dead organic matter—necrocarbon—potentially degradable to CO_2 and CH_4 as is present in the atmosphere.) Global warming might speed up the rates at which biodegradation adds these gases to the atmosphere. It is certain that atmospheric CO_2 levels will continue to increase significantly.

The degree to which this occurs depends upon future levels of CO_2 production and the fraction of that production that remains in the atmosphere. Given plausible projections of CO_2 production and a reasonable estimate that half of that amount will remain in the atmosphere, projections can be made that indicate that sometime during the middle part of the next century the concentration of this gas will reach 600 ppm in the atmosphere. This is well over twice the levels estimated for pre-industrial times. Much less certain are the effects that this change will have on climate. It is virtually impossible for the elaborate computer models used to estimate these effects to accurately take account of all variables, such as the degree and nature of cloud cover. Clouds both reflect incoming light radiation and absorb outgoing infrared radiation, with the former effect tending to predominate.

The magnitude of these effects depends upon the degree of cloud cover, brightness, altitude, and thickness. In the case of clouds, too, feedback phenomena occur; for example, warming induces formation of more clouds, which reflect more incoming energy. Most computer models predict global warming of at least 3.0°C and as much as 5.5°C occurring over a period of just a few decades. These estimates are sobering because they correspond to the approximate temperature increase since the last ice age 18,000 years past, which took place at a much slower pace of only about 1 or 2°C per 1,000 years.

Drought is one of the most serious problems that could arise from major climatic change resulting from greenhouse warming. Typically, a 3-degree warming would be accompanied by a 10 percent decrease in precipitation. Water shortages would be aggravated, not just from decreased rainfall, but from increased evaporation, as well. Increased evaporation results in decreased run-off, thereby reducing water available for agricultural, municipal, and industrial use. Water shortages, in turn, lead to increased demand for irrigation and to the production of lower quality, higher salinity runoff water and wastewater.

In the U. S., such a problem would be especially intense in the Colorado River basin, which supplies much of the water used in the rapidly growing U. S. Southwest. A variety of other problems, some of them unforeseen as of now, could result from global warming. An example is the effect of warming on plant and animal pests—insects, weeds, diseases, and rodents. Many of these would certainly thrive much better under warmer conditions. Interestingly, another air pollutant, acid-rain-forming sulfur dioxide, may have a counteracting effect on greenhouse gases. This is because sulfur dioxide is oxidized in the atmosphere to sulfuric acid, forming a light-reflecting haze. Furthermore, the sulfuric acid and resulting sulfates act as condensation nuclei that increase the extent, density, and brightness of light-reflecting cloud cover.

CYCLE SULFUR

Sulfur cycle involves primarily H_2S, SO_2, SO_3, and sulfates. There are many uncertainties regarding the sources, reactions, and fates of these atmospheric sulfur species. On a global basis, sulfur compounds enter the atmosphere to a very large extent through human activities. Approximately 100 million metric tons of sulfur per year enter the global atmosphere through anthropogenic activities, primarily as SO_2 from the combustion of coal and residual fuel oil. The greatest uncertainties in the cycle have to do with nonanthropogenic sulfur, which enters the atmosphere largely as H_2S from volcanoes and from the biological decay of organic matter and reduction of sulfate.

The quantity added from biological processes may be as low as 1 million metric tons per year. Any H_2S that does get into the atmosphere is converted rapidly to SO_2 by processes that involve several intermediate steps, including reactions with hydroxyl radical.

Sulfur Dioxide in the Atmosphere

Many factors, including temperature, humidity, light intensity, atmospheric transport, and surface characteristics of paniculate matter, may influence the atmospheric chemical reactions of sulfur dioxide. Like many other gaseous pollutants, sulfur dioxide undergoes chemical reactions resulting in the formation of paniculate matter. Whatever the processes involved, much of the sulfur dioxide in the atmosphere ultimately is oxidized to sulfuric acid and sulfate salts, particularly ammonium sulfate and ammonium hydrogen sulfate.

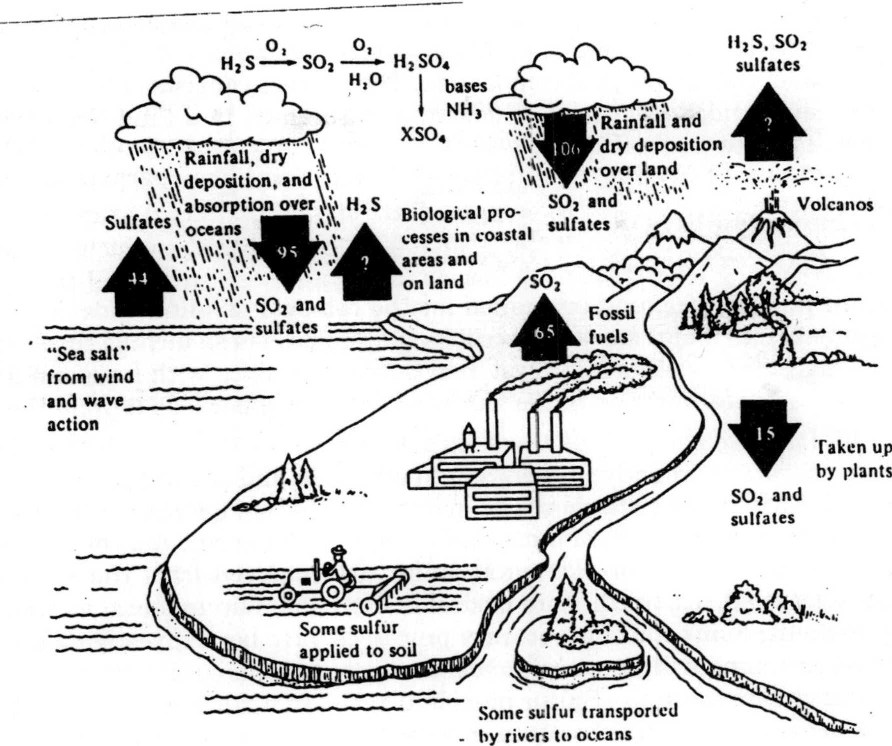

Fig. 6.5. The atmospheric sulfur cycle with quantities expressed in millions of metric tons per year.

Effects of Atmospheric Sulfur Dioxide

Though not terribly toxic to most people, low levels of sulfur dioxide in air do have some health effects. Sulfur dioxide's primary effect is upon the respiratory tract, producing irritation and increasing airway resistance, especially to people with respiratory weaknesses and sensitized asthmatics. Therefore, exposure to the gas may increase the effort required to breathe. Mucus secretion is also stimulated by exposure to air contaminated by sulfur dioxide. Atmospheric sulfur dioxide is harmful to plants. Acute exposure to high levels of the gas kills leaf tissue (leaf necrosis).

The edges of the leaves and the areas between the leaf veins are particularly damaged. Chronic exposure of plants to sulfur dioxide causes chlorosis, a bleaching or yellowing of the normally green portions of the leaf. Sulfur dioxide in the atmosphere is converted to sulfuric acid, so that in areas with high levels of sulfur dioxide pollution, plants may be damaged by sulfuric acid aerosols. Such damage appears as small spots where sulfuric acid droplets have impinged in leaves.

Removal of Sulfur Dioxide

A number of processes are being used to remove sulfur and sulfur oxides from fuel before combustion and from stack gas after combustion. Most of these efforts concentrate on coal, since it is the major source of sulfur oxides pollution. Physical separation techniques may be used to remove discrete particles of pyritic sulfur from coal. Chemical methods may also be employed for removal of sulfur from coal. Fluidized bed combustion of coal promises to eliminate SO_2 emissions at the point of combustion. The process consists of burning granular coal in a bed of finely divided limestone or dolomite maintained in a fluid-like condition by air injection. Heat calcines the limestone,

$$CaCO_3 \rightarrow CaO + CO_2 \quad ...(6.14)$$

and the lime produced absorbs SO_2:

$$CaO + SO_2 + 1/2O_2 \rightarrow CaSO_4 \quad ...(6.15)$$

Many processes have been proposed or studied for the removal of sulfur dioxide from stack gas. Table 6.2 summarizes major stack gas scrubbing systems. These include throwaway and recovery systems as well as wet and dry systems. Current practice with lime and limestone scrubber systems often uses injection of the slurry into the scrubber loop beyond the boilers.

Experience to date has shown that these scrubbers remove well over 90% of both SO_2 and fly ash when operating properly. In addition to corrosion and scaling problems, disposal of lime sludge poses formidable obstacles. The quantity of this sludge may be appreciated by considering that approximately 1 ton of limestone is required for each 5 tons of coal. Recovery systems in which sulfur dioxide or elemental sulfur are removed from the spent sorbing material, which is recycled, are much more desirable from an environmental viewpoint than are throwaway systems. Many kinds of recovery processes have been investigated, including those that involve scrubbing with magnesium oxide slurry, sodium sulfite solution, ammonia solution, or sodium citrate solution. Sulfur dioxide trapped in a stack-gas-scrubbing process can be converted to hydrogen sulfide by reaction with synthesis gas (H_2, CO, CH_4),

Table. 6.2. Major Stack Gas Scrubbing Systems[a]

Process	*Chemical Reactions*	*Major Advantages or Disadvantages*
Lime slurry scrubbing	$Ca(OH)_2 + SO_2 \rightarrow CaSO_3 + H_2O$	Up to 200 kg of lime are needed per metric ton of coal, producing huge quantities of waste product.
Limestone slurry scrubbing[b]	$CaCO_3 + SO_2 \rightarrow CaSO_3 + CO_2(g)$	Lower pH than lime slurry, and not so efficient.
Magnesium oxide scrubbing	$Mg(OH)_2(slurry) + SO_2 \rightarrow MgSO_3 + H_2O$	The sorbent can be regenerated, and this need not be done on site.
Sodium-base scrubbing	$Na2SO_3 + H_2O + SO_2 \rightarrow 2\ NaSHO_3$ $2\ NaHSO_3 + heat \rightarrow NaSO_3 + H_2O + SO_2$ (regeneration)	There are no major technological limitations. Annual costs are relatively high.
Double alkalib[b]	$2NaOH + SO_2 \rightarrow Na_2SO_3 + H_2P$ $Ca(OH)_2 + NaSO_3 \rightarrow$ $CaSO_3(s) + 2\ NaOH$ (regeneration of NaOH)	Allows for regeneration of expensive sodium alkali solution with inexpensive lime.

[b]These processes have also been adapted to produce a gypsum product by oxidation of $CaSO_3$ in the spent scrubber medium:

$$CaSO_3 + 1/2O_2 + 2\,H_2O \rightarrow CaSO_4 \cdot 2\,H_2(s)$$

Gypsum has some commercial value, such as in the manufacture of plasterboard, and makes a relatively settleable waste product.

$$SO_2 + (H_2, CO, CH_4) \rightleftarrows H_2S + CO_2 \quad ...(6.16)$$

The Claus reaction is then employed to produce elemental sulfur:

$$2H_2S + SO_2 \rightleftarrows 2H_2O + 3S \quad ...(6.17)$$

OXIDES OF NITROGEN

The three oxides of nitrogen normally encountered in the atmosphere are nitrous oxide (N_2O), nitric oxide (NO), and nitrogen dioxide (NO_2). Microbially generated nitrous oxide is relatively unreactive and probably does not significantly influence important chemical reactions in the lower atmosphere. Its concentration decreases rapidly with altitude in the stratosphere due to the photochemical reaction

$$N_2O + h\nu \rightarrow N_2 + O \quad ...(6.18)$$

and some reaction with singlet atomic oxygen:

$$N_2O + O \rightarrow N_2 + O_2 \quad ...(6.19)$$

$$N_2O + O \rightarrow NO + NO \quad ...(6.20)$$

These reactions are significant in terms of depletion of the ozone layer. Increased global fixation of nitrogen, accompanied by increased microbial production of N_2O, could contribute to ozone layer depletion. Colorless, odorless nitric oxide (NO) and pungent red-brown nitrogen dioxide (NO_2) are very important in polluted air. Collectively designated NO_X, these gases enter the atmosphere from natural sources, such as lightning and biological processes, and from pollutant sources.

The latter are much more significant because of regionally high NO_2 concentrations, which can cause severe air quality deterioration. Practically all anthropogenic NO_2 enters the atmosphere as a result of the combustion of fossil fuels in both stationary and mobile sources. The contribution of automobiles to nitric oxide production in the U.S. has become somewhat lower in the last decade as newer automobiles with nitrogen oxide pollution controls have become more common. Most NO_2 entering the atmosphere from pollution sources does so as NO generated from internal combustion engines. At the very high temperatures in an automobile combustion chamber, the following reaction occurs:

$$N_2 + O_2 \rightarrow 2NO \quad ...(6.21)$$

High temperatures favor both a high equilibrium concentration and a rapid rate of formation of NO. Rapid cooling of the exhaust gas from combustion "freezes" NO at a relatively high concentration because equilibrium is not maintained. Thus, by its very nature, the combustion process both in the internal combustion engine and in furnaces produces high levels of NO in the combustion products.

Reactions of No_x in Atmosphere

Atmospheric chemical reactions convert NO_X to nitric acid, inorganic nitrate salts, organic nitrates, and peroxyacetyl nitrate (see Chapter 18). The principal reactive nitrogen oxide species in the troposphere are NO, NO_2, and HNO_3. These species cycle among each other. Although NO is the primary form in which NO_X is released to the atmosphere, the conversion of NO to NO_2 is relatively rapid in the troposphere. Nitrogen dioxide is a very reactive and significant species in the atmosphere. It absorbs light throughout the ultraviolet and visible spectrum penetrating the troposphere. At wavelengths below 398 nm, photodissociation to oxygen atoms occurs,

$$NO_2 + h\nu \rightarrow NO + O \qquad ...(6.22)$$

giving rise to several significant inorganic reactions, in addition to a host of atmospheric reactions involving organic species. The reactivity of NO_2 to photo-dissociation is shown clearly by the fact that in direct sunlight the half-life of NO_2 is much shorter than that of any other atmospheric component (only 1 or 2 minutes).

Harmful Effects of Nitrogen Oxides

Nitric oxide, NO, is biochemically less active and less toxic than NO_2. Acute exposure to NO_2 can be quite harmful to human health. For exposures ranging from several minutes to one hour, a level of 50-100 ppm of NO_2 causes inflammation of lung tissue for a period of 6-8 weeks, after which time the subject normally recovers. Exposure of the subject to 150-200 ppm of NO_2 causes *bronchiolitis fibrosa obliterans,* a condition fatal within 3 – 5 weeks after exposure.

Death generally results within 2-10 days after exposure to 500 ppm or more of NO_2. "Silo-filler's disease," caused by NO, generated by the fermentation of ensilage (fermented corn plants used for cattle feed) containing nitrate, is a particularly striking example of nitrogen dioxide poisoning.

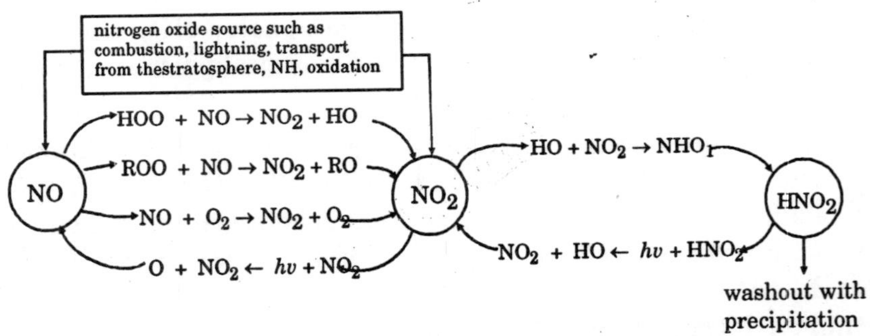

Fig. 6.6. Principal reactions among NO, NO_2, and HNO_3 in the atmosphere. ROO• represents an organic peroxyl radical, such as the methylperoxyl radical. CH_3OO.

Deaths have resulted from the inhalation of NO_2-containing gases from burning celluloid and nitrocellulose film and from leakage of NO_2 oxidant from missile rocket motors. Athough extensive damage to plants is observed in areas receiving heavy exposure to NO_2, most of this damage probably comes from secondary products of nitrogen oxides, such as PAN formed in smog. Exposure of plants to several parts per million of NO_2 in the laboratory causes leaf spotting and breakdown of plant tissue. Exposure to 10 ppm of NO causes a reversible decrease in the rate of photosynthesis.

Nitrogen oxides are known to cause fading of dyes and inks used in some textiles. This has been observed in gas clothes dryers and is due to NO_X formed in the dryer flame. Much of the damage to materials caused by NO_X, such as stress-corrosion cracking of electrical apparatus, comes from secondary nitrates and nitric acid.

Regulation of Nitrogen Oxides

The level of NO, emitted from stationary sources such as power plant furnaces generally falls within the range of 50-1000 ppm. NO production is favored both kinetically and thermodynamically by high temperatures and by high excess oxygen concentrations. These factors must be considered in reducing NO emissions from stationary sources. Reduction of flame temperature to prevent NO formation is accomplished by adding recirculated exhaust gas, cool air, or inert gases. Low-excess-air firing is effective in reducing NO_X emissions during the combustion of fossil fuels. As the term implies, low-excess-air firing uses the minimum amount of excess air required for oxidation of the fuel, so that less oxygen is available for the reaction

$$N2 + O_2 \rightarrow 2NO \quad ...(6.23)$$

in the high temperature region of the flame. Incomplete fuel burnout, with the emission of hydrocarbons, soot, and CO, is an obvious problem with low-excess-air firing. This may be overcome by a two-stage combustion process. In the first stage fuel is fired at a relatively high temperature with a substoichiometric amount of air, and NO formation is limited by the absence of excess oxygen. In the second stage fuel bürnout is completed at a relatively low temperature in excess air; the low temperature prevents formation of NO. Removal of NO_x trom stack gas presents some formidable problems. Possible approaches to NO_X removal are catalytic decomposition of nitrogen oxides, catalytic reduction of nitrogen oxides, and sorption of NO_X by liquids or solids.

Ammonia in the Atmosphere

Ammonia is present even in unpolluted air as a result of natural biochemical and chemical processes. Among the various sources of atmospheric ammonia are microorganisms, decay of animal wastes, sewage treatment, coke manufacture, ammonia manufacture, and leakage from ammonia-based refrigeration systems. High concentrations of ammonia gas in the atmosphere are generally indicative of accidental release of the gas. Ammonia is removed from the atmosphere by its affinity for water and by its action as a base. It is a key species in the formation and neutralization of nitrate and sulfate aerosols in polluted atmospheres. Ammonia reacts with these acidic aerosols to form ammonium salts:

$$NH_3 + HN0_3 \rightarrow NH_4NO_3 \quad ...(6.24)$$

$$NH_3 + H_2SO_4 \rightarrow NH_4HSO_4 \quad ...(6.25)$$

Ammonium salts are among the more corrosive salts in atmospheric aerosols.

ACID RAIN

As discussed in this chapter, much of the sulfur and nitrogen oxides entering the atmosphere are converted to sulfuric and nitric acids, respectively. When combined with hydrochloric acid 'arising from hydrogen chloride emissions, these acids cause acidic precipitation that is now a major pollution problem in some areas. Precipitation made acidic by the presence of acids stronger than $CO_2(aq)$ is commonly called *acid rain;* the term applies to all kinds of acidic aqueous precipitation, including fog, dew, snow, and sleet. In a more general sense, *acid deposition* refers to the deposition on the Earth's surface of aqueous acids, acid gases (such as SO_2), and acidic salts (such as NH_4HSO_4). According to this definition, deposition in solution form is *acid precipitation,* and deposition of dry gases and compounds is *dry deposition.* Sulfur dioxide, SO_2, contributes more to the acidity of precipitation than does CO_2 present at higher levels in the atmosphere because SO, is a stronger acid and it is more water soluble. Although acid rain can originate from the direct emission of strong acids, such as HCl gas or sulfuric acid mist, most of it is a secondary air pollutant produced by the atmospheric oxidation of acid-forming gases such as the following:

$$SO_2 + 1/2O_2 + H_2O \xrightarrow[\text{consisting of several steps}]{\text{Overall reaction}} \{2H^+ + SO_4^{2-}\}\,(aq) \qquad ...(6.26)$$

$$2NO_2 + 1/2O_2 + H_2O \xrightarrow[\text{consisting of several steps}]{\text{Overall reaction}} 2\{H^+ + N_3-\}\,(aq) \qquad ...(6.27)$$

Chemical reactions such as these play a dominant role in determining the nature, transport, and fate of acid precipitation. As the result of such reactions the chemical properties (acidity, ability to react with other substances) and physical properties (volatility, solubility) of acidic atmospheric pollutants are altered drastically. For example, even the small fraction of NO that does dissolve in water does not react significantly. However, its ultimate oxidation product, HNO_3, though volatile, is highly water-soluble, strongly acidic, and very reactive with other materials.

Therefore, it tends to be removed readily from the atmosphere and to do a great deal of harm to plants, corrodable materials, and other things that it contacts. Although emissions from industrial operations and fossil fuel combustion are the major sources of acid-forming gases, acid rain has also been encountered in areas far from such sources. This is due in part to the fact that acid-forming gases are oxidized to acidic constituents and deposited over several days, during which time the air mass containing the gas may have moved as much as several thousand km.

It is likely that the burning of biomass, such as is employed in "slash-and-burn" agriculture, evolves the gases that lead to acid formation in more remote areas. In arid regions, dry acid gases or acids sorbed to particles may be deposited, with effects similar to those of acid rain deposition.

Acid rain spreads out over areas of several hundred to several thousand kilometers, so it is classified as a *regional* air pollution problem compared to a *local* air pollution problem for smog and a *global* one for ozone-destroying chlorofluorocarbons and greenhouse gases. Analyses of the movements of air masses have shown a correlation between acid precipitation and prior movement of an air mass over major sources of anthropogenic sulfur and nitrogen oxides emissions. This is particularly obvious in southern Scandinavia, which receives a heavy burden of air pollution from densely populated, heavily industrialized areas in Europe. The strong geographic dependence of acid precipitation is illustrated, representing the pH of precipitation in the continental U.S. The preponderance of acidic rainfall in the northeastern U.S. is obvious. Acid rain is not a new phenomenon; it has been observed for well over a century, with many of the older observations from Great Britain. The first manifestations of this phenomenon were elevated levels of SO_4^{2-} in precipitation collected in industrialized areas.

More modern evidence was obtained from analyses of precipitation in Sweden in the 1950s and of U.S. precipitation a decade or so later. A vast research effort on acid rain was conducted in North America by the National Acid Precipitation Assessment Program, which resulted from the U.S. Acid Precipitation Act of 1980.

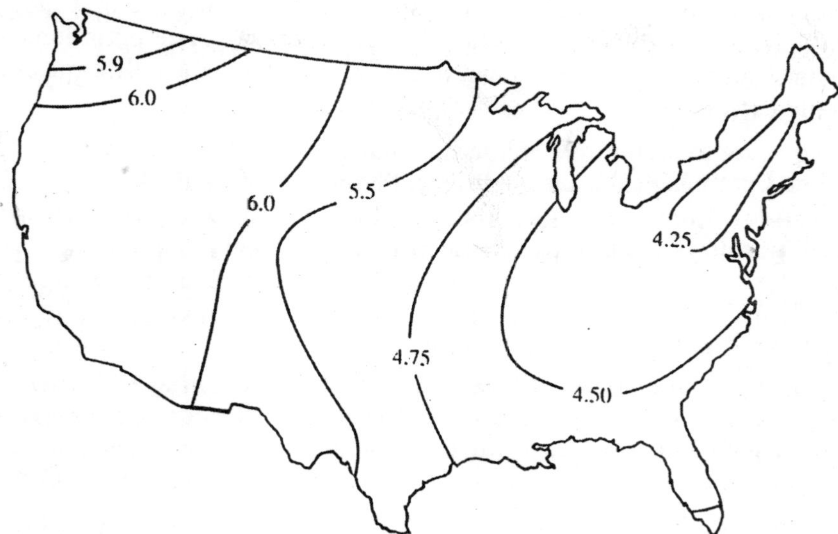

Fig. 6.7. Hypothetical precipitation-pH pattern in the continental United States. Actual values found may vary with the time of year and climatic conditions.

Ample evidence exists of the damaging effects of acid rain. The major such effects are the following:

- Direct phytotoxicity to plants from excessive acid concentrations. (Evidence of direct or indirect phytoxicity of acid rain is provided by the declining health of Eastern U.S. and Scandinavian forests and especially by damage to Germany's Black Forest.)
- Phytotoxicity from acid-forming gases, particularly SO_2 and NO_2, that accompany acid rain
- Indirect phytotoxicity, such as from $A1^{3+}$ liberated from acidified soil
- Destruction of sensitive forests
- Respiratory effects on humans and other animals
- Acidification of lake water with toxic effects to lake flora and fauna, especially fish fingerlings
- Corrosion to exposed structures, electrical relays, equipment, and ornamental materials. Because of the effect of hydrogen ion,

  $$2H^+ + CaCO_3(s) \rightarrow Ca^{2+} + CO_2(g) + H_2O$$

  limestone, $CaCO_3$, is especially susceptible to damage from acid rain
- Associated effects, such as reduction of visibility by sulfate aerosols and the influence of sulfate aerosols on physical and optical properties of clouds.

FLUORINE, CHLORINE, AND THEIR GASEOUS INORGANIC COMPOUNDS

Fluorine, hydrogen fluoride, and other volatile fluorides are produced in the manufacture of aluminum, and hydrogen fluoride is a by-product in the conversion of fluorapatite (rock phosphate) to phosphoric acid, superphosphate fertilizers, and other phosphorus products. Hydrogen fluoride gas is a dangerous substance that is so corrosive that it even reacts with glass. It is irritating to body tissues, and the respiratory tract is very sensitive to it. Brief exposure to HF vapors at the part-per-thousand level may be fatal. The acute toxicity of F_2 is even higher than that of HF.

Chronic exposure to high levels of fluorides causes fluorosis, the symptoms of which include mottled teeth and pathological bone conditions. Plants are particularly susceptible to the effects of gaseous fluorides. Fluorides from the atmosphere appear to enter the leaf tissue through the stomata. Fluoride is a cumulative poison in plants, and exposure of sensitive plants to even very low levels of fluorides for prolonged periods results in damage. Characteristic symptoms of fluoride poisoning are chlorosis (fading of green color due to conditions other than the absence of light), edge burn, and tip burn.

Conifers (such as pine trees) afflicted with fluoride poisoning may have reddish-brown necrotic needle tips as far as 10 miles from the plant. Silicon tetrafluoride gas, SiF_4, is a gaseous fluoride pollutant produced during some steel and metal smelting operations that employ CaF_2, fluorspar. Fluorspar reacts with silicon dioxide (sand), releasing SiF_4 gas:

$$2CaF_2 + 3SiO_2 \rightarrow 2CaSiO_3 + SiF_4 \qquad ....(6.28)$$

Another gaseous fluorine compound, sulfur hexafluoride, SiF_6, occurs in the atmosphere at levels of about 0.3 parts per trillion. It is extremely unreactive and is used as an atmospheric tracer. It does not absorb ultraviolet light in either the troposphere or stratosphere and is probably destroyed above 50 km by reactions beginning with its capture of free electrons.

Chlorine and Hydrogen Chloride

Chlorine gas, Cl_2, does not occur as an air pollutant on a large scale but can be quite damaging on a local scale. Chlorine was the first poisonous gas deployed in World War I. It is widely used as a manufacturing chemical in the plastics industry, for example, as well as for water treatment and as a bleach. Therefore, possibilities for its release exist in a number of locations. Chlorine is quite toxic and is a mucous-membrane irritant. It is very reactive and a powerful oxidizing agent. Chlorine dissolves in atmospheric water droplets, yielding hydrochloric acid and hypochlorous acid, an oxidizing agent:

$$H_2O + Cl_2 \rightarrow H^+ + Cl^- + HOCl \qquad ...(6.29)$$

Spills of chlorine gas have caused fatalities among exposed persons.

Hydrogen chloride, HCl, is emitted from a number of sources. Incineration of chlorinated plastics, such as polyvinylchloride, releases HC1 as a combustion product.

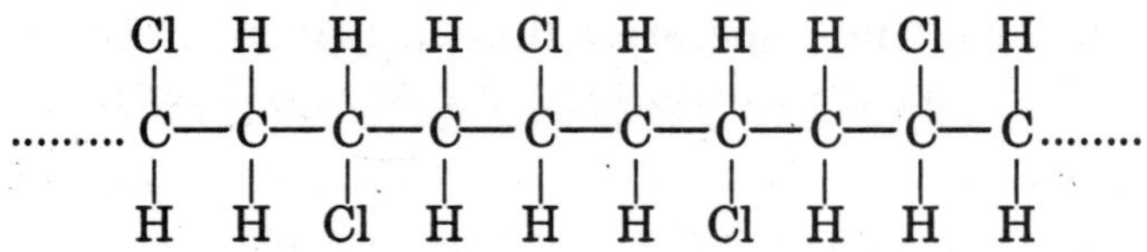

```
          Cl  H   H   H   Cl  H   H   H   Cl  H
          |   |   |   |   |   |   |   |   |   |
 .........C—C—C—C—C—C—C—C—C—C........ Polyvinylchloride
          |   |   |   |   |   |   |   |   |   |
          H   H   Cl  H   H   H   Cl  H   H   H
```

Some compounds released to the atmosphere as air pollutants hydrolyze to form HCl. One such incident occurred on April 26, 1974, when a storage tank containing 750,000 gallons of liquid silicon tetrachloride, $SiCl_4$, began to leak in South Chicago, Illinois. This compound reacted with water in the atmosphere to form a choking fog of hydrochloric acid droplets:

$$SiCl_4 + 2H_2O \rightarrow SiO_2 + 4HCl \qquad ...(6.30)$$

Many people became ill from inhaling the vapor.

In February, 1981, in Stroudsburg, Pennsylvania, a wrecked truck dumped 12 tons of powdered aluminum chloride during a rainstorm. This compound produces HC1 gas when wet,

$$AlCl_3 + 3H_2O \rightarrow Al(OH)_3 + 3HCl \quad ...(6.31)$$

and more than 1,200 residents had to be evacuated from their homes because of the fumes generated.

HYDROGEN SULFIDE, CARBONYL SULFIDE, AND CARBON DISULFIDE

Hydrogen sulfide is produced by microbial decay of sulfur compounds and microbial reduction of sulfate, from geothermal steam, from wood pulping, and from a number of miscellaneous natural and anthropogenic sources. Most atmospheric hydrogen sulfide is rapidly converted to SO_2 and to sulfates by oxidizing processes in the atmosphere. The organic homologs of hydrogen sulfide, the mercaptans or thiols, enter the atmosphere from decaying organic matter and have particularly objectionable odors. Hydrogen sulfide pollution from artificial sources is not as much of an overall air pollution problem as sulfur dioxide pollution.

However, there have been several acute incidents of hydrogen sulfide emissions resulting in damage to human health and even fatalities. The most notorious such incident occurred in Poza Rica, Mexico, in 1950. Accidental release of hydrogen sulfide from a plant used for the removal of sulfur from natural gas caused the deaths of 22 people and the hospitalization of over 300. Hydrogen sulfide at levels well above ambient concentrations destroys immature plant tissue. This type of plant injury is readily distinguished from that due to other phytotoxins.

More sensitive species are killed by continuous exposure to around 3000 ppb H_2S, whereas other species exhibit reduced growth, leaf lesions, and defoliation. Damage to certain kinds of materials is a very expensive effect of hydrogen sulfide pollution. Paints containing lead pigments, $2PbCO_3 \cdot Pb(OH)_2$ (no longer usdd), were particularly susceptible to darkening by H_2S. A black layer of copper sulfide forms on copper metal exposed to H_2S. Eventually, this layer is replaced by a green coating of basic copper sulfate such as $CuSO_4 \cdot 3Cu(OH)_2$. The green "patina," as it is called, is very resistant to further corrosion. Such layers of corrosion can seriously impair the function of copper contacts on electrical equipment.

Hydrogen sulfide also forms a black sulfide coating on silver. Carbonyl sulfide, COS, is now recognized as a component of the atmosphere at a tropospheric concentration of approximately 500 parts per trillion by volume, corresponding to a global burden of about 2.4 teragrams. It is, therefore, a significant sulfur species in the atmosphere. Both COS and CS_2 are oxidized in the atmosphere by reactions initiated by the hydroxyl radical. The initial reactions are

$$HO\cdot + COS \rightarrow CO_2 + HS\cdot \quad ...(6.32)$$

$$HO\cdot + CS_2 \rightarrow COS + HS\cdot \quad ...(6.33)$$

The sulfur-containing products undergo further reactions to sulfur dioxide and, eventually, to sulfate species.